It's another great book from CGP...

GCSE Maths isn't just about learning facts and formulas — you'll also need to know how to use them to answer questions.

But don't panic. This book explains everything you need to know, with plenty of worked examples and practice questions. It even includes a **free** Online Edition you can read on your computer or tablet.

How to get your free Online Edition

Just go to **cgpbooks.co.uk/extras** and enter this code...

2049 8137 6165 5369

By the way, this code only works for one person. If somebody else has used this book before you, they might have already claimed the Online Edition.

CGP — still the best! ☺

Our sole aim here at CGP is to produce the highest quality books — carefully written, immaculately presented and dangerously close to being funny.

Then we work our socks off to get them out to you — at the cheapest possible prices.

Contents

Section One — Numbers

Topic	Page	Unit 1	Unit 2	Unit 3
Calculating Tips	2	✓	✓	✓
Types of Number	4		✓	
Multiples, Factors and Prime Factors	5		✓	
LCM and HCF	6		✓	
Fractions	7	✓	✓	✓
Fractions, Decimals and Percentages	9	✓	✓	✓
Fractions and Recurring Decimals	10		✓	
Percentages	12	✓	✓	✓
Compound Growth and Decay	14	✓		
Ratios	15	✓	✓	✓
Rounding Numbers	17	✓	✓	✓
Bounds	19	✓		
Standard Form	20	✓	✓	
Revision Questions for Section One	22			

Section Two — Algebra

Topic	Page	Unit 1	Unit 2	Unit 3
Sequences	23		✓	
Powers and Roots	24		✓	
Algebra Basics	25	✓	✓	✓
Multiplying Out Brackets	26		✓	✓
Factorising	27		✓	✓
Manipulating Surds	28		✓	
Solving Equations	29		✓	✓
Rearranging Formulas	31		✓	
Factorising Quadratics	33		✓	✓
The Quadratic Formula	35			✓
Completing the Square	36		✓	
Quadratic Equations — Tricky Ones	37		✓	✓
Algebraic Fractions	38		✓	
Inequalities	39		✓	
Graphical Inequalities	40		✓	
Trial and Improvement	41			✓
Simultaneous Equations and Graphs	42			✓
Simultaneous Equations	43		✓	✓
Direct and Inverse Proportion	45			✓
Proof	46		✓	
Revision Questions for Section Two	47			

Section Three — Graphs

Topic	Page	Unit 1	Unit 2	Unit 3
X, Y and Z Coordinates	48		✓	✓
Straight-Line Graphs	49		✓	✓
Plotting Straight-Line Graphs	50		✓	✓
Finding the Gradient	51		✓	
"y = mx + c"	52		✓	✓
Parallel and Perpendicular Lines	53		✓	
Quadratic Graphs	54			✓
Harder Graphs	55			✓
Graph Transformations	57			✓
Real-Life Graphs	59		✓	✓
Revision Questions for Section Three	60			

Section Four — Geometry and Measures

Topic	Page	Unit 1	Unit 2	Unit 3
Geometry	61			✓
Parallel Lines	62			✓
Geometry Problems	63			✓
Polygons	64			✓
Symmetry	65			✓
Circle Geometry	66			✓
The Four Transformations	68			✓
More Transformation Stuff	70			✓
Congruent Shapes	71			✓
Similar Shapes	72			✓
Projections	73			✓
Areas	74			✓
Surface Area	76			✓
Volume	77			✓
Density and Speed	79			✓
Distance-Time Graphs	80		✓	✓
Unit Conversions	81			✓
Triangle Construction	83			✓
Loci and Constructions	84			✓
Loci and Constructions — Worked Examples	86			✓
Bearings	87			✓
Revision Questions for Section Four	88			

The columns for Units 1-3 above might be useful if you're studying AQA's 'Mathematics A' Specification (the one divided into units — ask your teacher if you're not sure which one you do). They show which units the pages cover.

Throughout this book you'll see grade stamps like these: A* C B A
You can use these to focus your revision on easier or harder work.
But remember — to get a top grade you have to know **everything**, not just the hardest topics.

Section Five — Pythagoras and Trigonometry

Topic	Page	Unit 1	Unit 2	Unit 3
Pythagoras' Theorem	90			✓
Trigonometry — Sin, Cos, Tan	91			✓
The Sine and Cosine Rules	93			✓
3D Pythagoras	95			✓
3D Trigonometry	96			✓
Vectors	97			✓
Revision Questions for Section Five	99			

Section Six — Statistics and Probability

Topic	Page	Unit 1	Unit 2	Unit 3
Sampling and Bias	100	✓		
Sampling Methods	101	✓		
Collecting Data	102	✓		
Mean, Median, Mode and Range	103	✓		
Averages and Spread	104	✓		
Frequency Tables — Finding Averages	106	✓		
Grouped Frequency Tables	107	✓		
Cumulative Frequency	108	✓		
Histograms and Frequency Density	109	✓		
Other Graphs and Charts	110	✓		
Scatter Graphs	111	✓		
Probability Basics	112	✓		
Listing Outcomes and Expected Frequency	113	✓		
The AND / OR Rules	114	✓		
Tree Diagrams	115	✓		
Relative Frequency	117	✓		
Revision Questions for Section Six	118			

Answers 119
Index 126

Published by CGP

Written by Richard Parsons

Updated by: Paul Jordin, Sharon Keeley-Holden, Simon Little, Alison Palin, Andy Park, Caley Simpson, Ruth Wilbourne

With thanks to Glenn Rogers and Janet Dickinson for the proofreading

ISBN: 978 1 84146 540 1

Groovy website: www.cgpbooks.co.uk
Printed by Elanders Ltd, Newcastle upon Tyne.
Jolly bits of clipart from CorelDRAW®

Calculating Tips

Ah, the glorious world of GCSE Maths. OK maybe it's more like whiffy socks at times, but learn it you must. And there's plenty of it. Here are some nifty exam tricks that could net you quite a few lovely marks.

BODMAS

Brackets, Other, Division, Multiplication, Addition, Subtraction

(D)

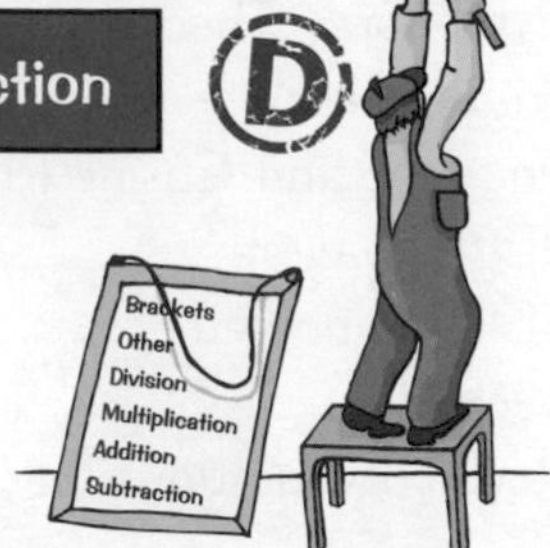

BODMAS tells you the ORDER in which these operations should be done:
Work out Brackets first, then Other things like squaring, then Divide / Multiply groups of numbers before Adding or Subtracting them.

This set of rules works really well, so remember the word BODMAS.

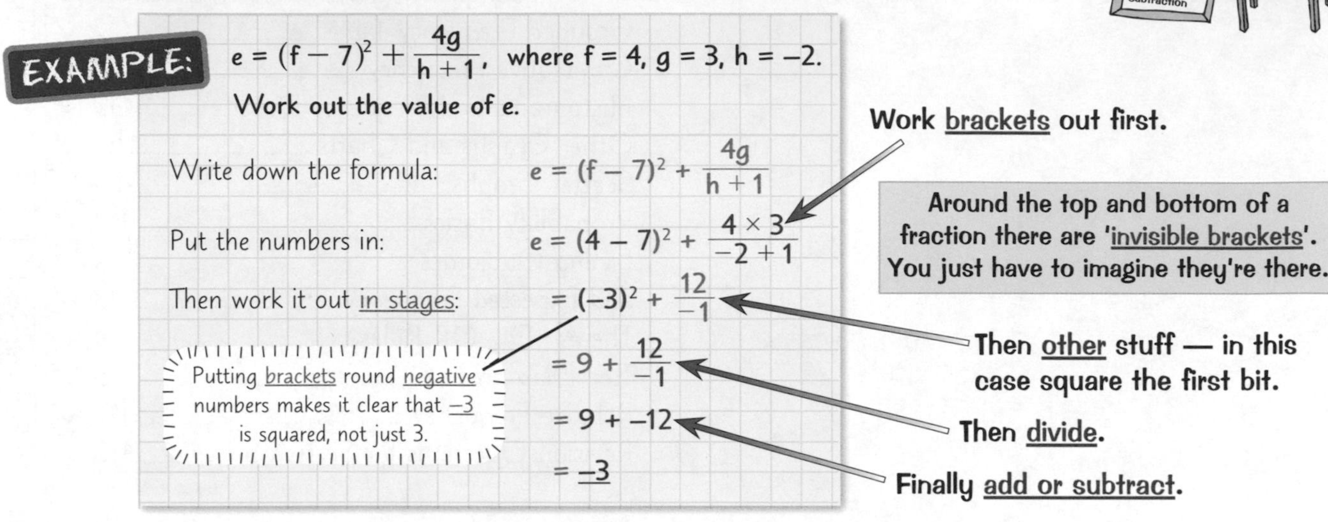

EXAMPLE: $e = (f - 7)^2 + \frac{4g}{h+1}$, where f = 4, g = 3, h = –2. Work out the value of e.

Write down the formula: $e = (f - 7)^2 + \frac{4g}{h+1}$

Put the numbers in: $e = (4 - 7)^2 + \frac{4 \times 3}{-2+1}$ — Work brackets out first.

Around the top and bottom of a fraction there are 'invisible brackets'. You just have to imagine they're there.

Then work it out in stages: $= (-3)^2 + \frac{12}{-1}$ — Then other stuff — in this case square the first bit.

$= 9 + \frac{12}{-1}$ — Then divide.

$= 9 + -12$ — Finally add or subtract.

$= \underline{-3}$

Putting brackets round negative numbers makes it clear that –3 is squared, not just 3.

Don't Be Scared of Wordy Questions

A lot of the marks in your exam are for answering wordy, real-life questions. For these you don't just have to do the maths — you've got to work out what the question's asking you to do. Relax and work through them step by step.

1) READ the question carefully. Work out what bit of maths you need to answer it.
2) Underline the INFORMATION YOU NEED to answer the question — you might not have to use all the numbers they give you.
3) Write out the question IN MATHS and answer it, showing all your working clearly.

EXAMPLE: The table shows rates of depreciation over a three year period for three different motorbikes. Helen bought a B260 for £6300 three years ago. How much is the motorbike worth now?.

Model	Depreciation over 3 years
A125	37%
B260	45%
F400	42%

1) The word "depreciation" tells you this is a percentages question.
2) You need the initial value of £6300 and the B260 depreciation of 45%.
3) "Depreciation" is a percentage decrease, so in maths: 0.45 × £6300 = £2835

£6300 – £2835 = £3465

What's your BODMAS? About 50 kg, dude...

It's really important to check your working on BODMAS questions. You might be certain you did it right, but it's surprisingly easy to make a slip. Try your hand at this Exam Practice Question and see how you do.

Q1 $d = \frac{3a^2 + 2b}{4(c+3)}$, where a = –2, b = 2 and c = –4. Work out the value of d. [2 marks] (D)

Calculating Tips

This page covers some mega-important stuff about using calculators.

Know Your Buttons

Look for these buttons on your calculator — they might be a bit different on yours.

 The reciprocal button. The reciprocal of a number is 1 divided by it. So the reciprocal of 2 is ½.

 The cube root button. You might have to press shift first.

 This uses your last answer in your current calculation. Super useful.

 Flips your answer from a fraction or surd to a decimal and vice versa.

BODMAS on Your Calculator

BODMAS questions on the calculator paper will be packed with tricky decimals and maybe a square root and sin/cos/tan. You could do it on your calculator in one go, but that runs the risk of losing precious marks.

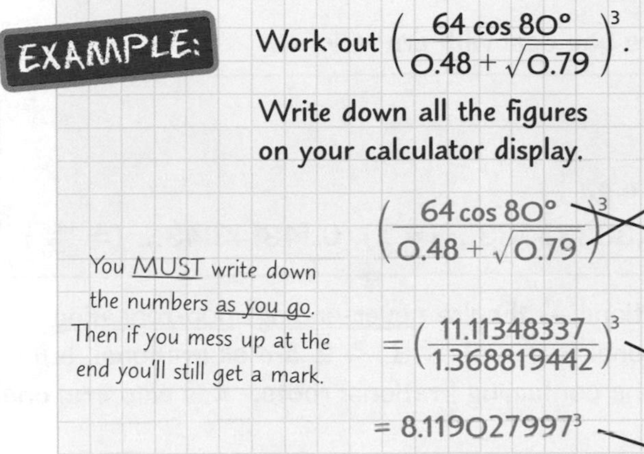

There are lots of slightly different ways of working out this type of calculation. Here's one:

1) Work out the bottom of the fraction: 0.48 + √■ 0.79 =

Write the answer down and store it in the memory by pressing: STO M+

2) Now work out the top of the fraction: 64 cos 80 =

3) Do the division:  Ans ÷ RCL M+ =

This gets the value of the bottom of the fraction out of the memory.

4) And cube: Ans $x^{■}$ 3 =

NOTE:

1) On some calculators, a bracket opens when you use a trig function or the square/cube root function. So to enter something like tan 40° + 1, you have to close the bracket: tan 40) + 1

2) On some calculators, the cursor stays under the square root bar until you nudge it out by pressing the right arrow.

Check Your Answer using Brackets  (and)

Check your answer to a question like the one above by plugging it into your calculator in fewer steps.

1) To work out $\frac{64\cos 80°}{0.48+\sqrt{0.79}}$ you CAN'T just press 64 cos 80 ÷ 0.48 + √■ 0.79 =

2) The calculator follows BODMAS, so it'll think you mean $\frac{64\cos 80°}{0.48}+\sqrt{0.79}$.

3) The secret is to OVERRIDE the automatic BODMAS using the BRACKETS BUTTONS.

4) The calculator will do the bits in brackets first. So you'd press:

(64 cos 80) ÷ (0.48 + √■ 0.79) =   (Cube this to check the question above.)

Your calculator might need you to add an extra ")" here. See the note above. And maybe an extra ")" or a right arrow nudge here.

Calculators — only as clever as the button presser...

Different calculators behave differently, so get to know your own. Try everything above on your calculator.

Q1 Work out $\frac{\sqrt{8.67-4.94}}{4\tan 87°}$. Write down all the figures on your calculator display. [2 marks] 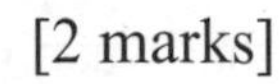

Types of Number

This section is about numbers, so before we get stuck into the maths, there are a few definitions of different types of number that you need to know.

Integers:

You need to make sure you know the meaning of this word — it'll come up all the time in GCSE Maths. An integer is another name for a whole number — either a positive or negative number, or zero.

Examples

Integers: –365, 0, 1, 17, 989, 1 234 567 890

Not integers: 0.5, $\frac{2}{3}$, $\sqrt{7}$, $13\frac{3}{4}$, –1000.1, 66.66, π

Rational and Irrational Numbers:

All numbers fall into one of these two categories.

Rational numbers can be written as fractions. Most numbers you deal with are rational.

Rational numbers come in 3 different forms:

1) Integers e.g. $4\ (=\frac{4}{1})$, $-5\ (=\frac{-5}{1})$, $-12\ (=\frac{-12}{1})$
2) Fractions p/q, where p and q are (non-zero) integers, e.g. $\frac{1}{4}$, $-\frac{1}{2}$, $\frac{3}{4}$
3) Terminating or recurring decimals e.g. $0.125\ (=\frac{1}{8})$, $0.33333333...\ (=\frac{1}{3})$, $0.143143143...\ (=\frac{143}{999})$

Irrational numbers are messy. They can't be written as fractions — they're never-ending, non-repeating decimals. Roots of +ve integers are either integers or irrational (e.g. $\sqrt{2}$, $\sqrt{3}$, $\sqrt[3]{2}$ are all irrational, but $\sqrt{4} = 2$ isn't). Surds (see p.28) are numbers or expressions containing irrational roots. π is also irrational.

Squares and Cubes:

Make sure you know these squares and cubes by heart — they could come up on a non-calculator paper.

THE SPECIAL SQUARES:

1^2	2^2	3^2	4^2	5^2	6^2	7^2	8^2	9^2	10^2	11^2	12^2	13^2	14^2	15^2
1	4	9	16	25	36	49	64	81	100	121	144	169	196	225
(1×1)	(2×2)	(3×3)	(4×4)	(5×5)	(6×6)	(7×7)	(8×8)	(9×9)	(10×10)	(11×11)	(12×12)	(13×13)	(14×14)	(15×15)

THE CRUCIAL CUBES:

1^3	2^3	3^3	4^3	5^3	10^3
1	8	27	64	125	1000
(1×1×1)	(2×2×2)	(3×3×3)	(4×4×4)	(5×5×5)	(10×10×10)

Admit it, you never knew maths could be this exciting, did you?

Prime Numbers:

2 3 5 7 11 13 17 19 23 29 31 37 41 43...

1) A prime number is a number which doesn't divide by anything — apart from itself and 1. (The only exception is 1, which is NOT a prime number.)
2) Apart from 2 and 5, ALL PRIMES END IN 1, 3, 7, OR 9.

So POSSIBLE primes would be 71, 73, 77, 79, 101, 103, 107, 109, etc. But not all of these are primes — you need to check carefully that they don't divide by anything.

And there's me thinking numbers were a kind of squiggly shape...

There, that wasn't too bad — a nice, gentle introduction. Now it's time to get stuck into the real maths...

Multiples, Factors and Prime Factors

If you think 'factor' is short for 'fat actor', I suggest you give this page a read. Stop thinking about fat actors now. Stop it...

Multiples and Factors 

The MULTIPLES of a number are just its times table.

EXAMPLE: Find the first 8 multiples of 13.

You just need to find the first 8 numbers in the 13 times table:
13 26 39 52 65 78 91 104

The FACTORS of a number are all the numbers that divide into it.

There's a method that guarantees you'll find them all:

1) Start off with 1 × the number itself, then try 2 ×, then 3 × and so on, listing the pairs in rows.
2) Try each one in turn. Cross out the row if it doesn't divide exactly.
3) Eventually, when you get a number repeated, stop.
4) The numbers in the rows you haven't crossed out make up the list of factors.

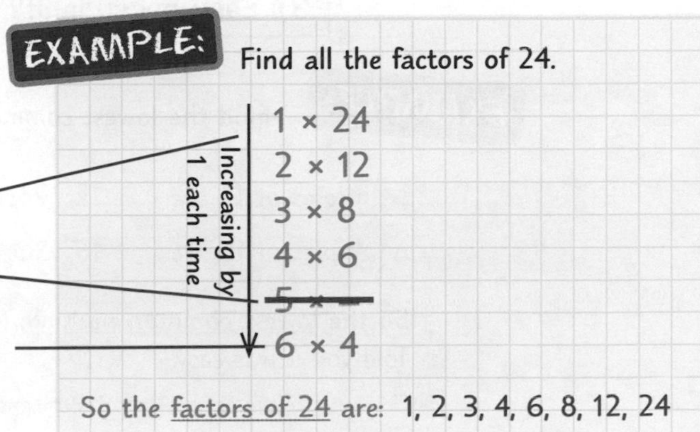

EXAMPLE:
Find all the factors of 24.
Increasing by 1 each time
1 × 24
2 × 12
3 × 8
4 × 6
~~5 × —~~
6 × 4
So the factors of 24 are: 1, 2, 3, 4, 6, 8, 12, 24

Finding Prime Factors — The Factor Tree 

Any number can be broken down into a string of prime numbers all multiplied together — this is called 'expressing it as a product of prime factors'.

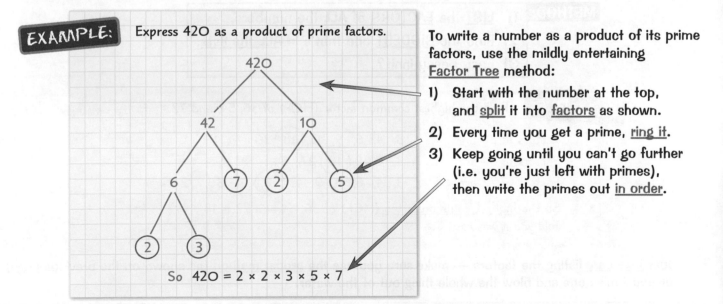

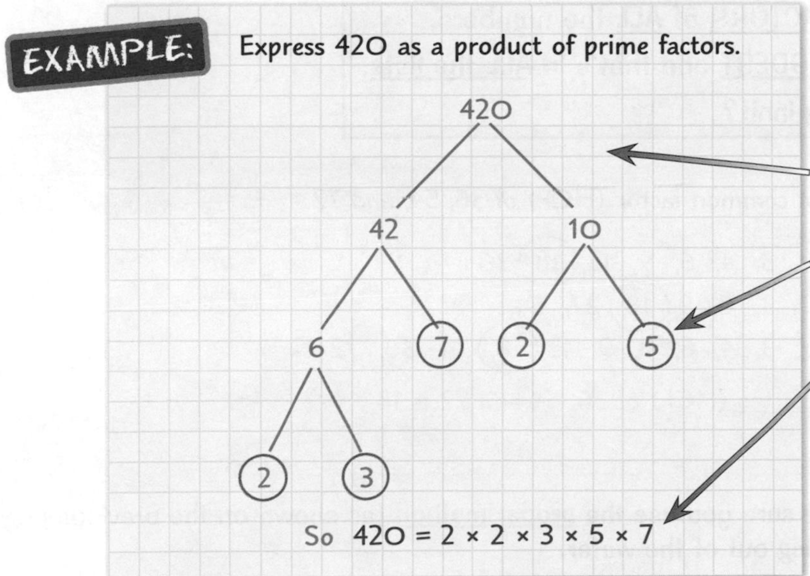

To write a number as a product of its prime factors, use the mildly entertaining Factor Tree method:

1) Start with the number at the top, and split it into factors as shown.
2) Every time you get a prime, ring it.
3) Keep going until you can't go further (i.e. you're just left with primes), then write the primes out in order.

Takes me back, scrumping prime factors from the orchard...

Make sure you know the Factor Tree method inside out, then give these Exam Practice Questions a go...

Q1 Express 990 as a product of its prime factors. [2 marks] 

Q2 Express 160 as a product of its prime factors. [2 marks] (C)

LCM and HCF

Two big fancy names but don't be put off — they're both real easy.

LCM — 'Lowest Common Multiple'

'Lowest Common Multiple' — sure, it sounds kind of complicated but all it means is this:

The SMALLEST number that will DIVIDE BY ALL the numbers in question.

METHOD:
1) LIST the MULTIPLES of ALL the numbers.
2) Find the SMALLEST one that's in ALL the lists.
3) Easy peasy innit?

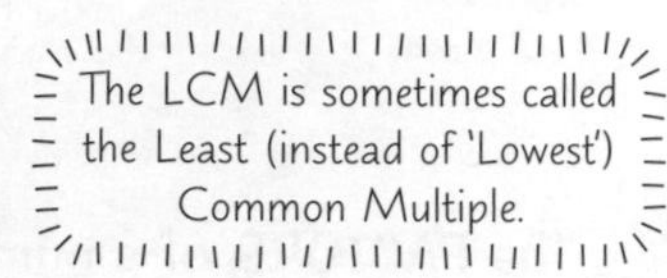

The LCM is sometimes called the Least (instead of 'Lowest') Common Multiple.

EXAMPLE: Find the lowest common multiple (LCM) of 12 and 15.

Multiples of 12 are: 12, 24, 36, 48, (60), 72, 84, 96, ...
Multiples of 15 are: 15, 30, 45, (60), 75, 90, 105, ...

So the lowest common multiple (LCM) of 12 and 15 is 60.
Told you it was easy.

HCF — 'Highest Common Factor'

'Highest Common Factor' — all it means is this:

The BIGGEST number that will DIVIDE INTO ALL the numbers in question.

METHOD:
1) LIST the FACTORS of ALL the numbers.
2) Find the BIGGEST one that's in ALL the lists.
3) Easy peasy innit?

EXAMPLE: Find the highest common factor (HCF) of 36, 54, and 72.

Factors of 36 are: 1, 2, 3, 4, 6, 9, 12, (18), 36
Factors of 54 are: 1, 2, 3, 6, 9, (18), 27, 54
Factors of 72 are: 1, 2, 3, 4, 6, 8, 9, 12, (18), 24, 36, 72

So the highest common factor (HCF) of 36, 54 and 72 is 18.
Told you it was easy.

Just take care listing the factors — make sure you use the proper method (as shown on the previous page) or you'll miss one and blow the whole thing out of the water.

LCM and HCF live together — it's a House of Commons...

You need to learn what LCM and HCF are, and how to find them. Turn over and write it all down. And after that, some lovely Exam Practice Questions — bonus.

Q1 Find the Lowest Common Multiple (LCM) of 9 and 12. [2 marks] 

Q2 Find the Highest Common Factor (HCF) of 36 and 84. [2 marks] 

Fractions

These pages show you how to cope with fraction calculations without your beloved calculator.

1) Cancelling down — easy

To cancel down or simplify a fraction, divide top and bottom by the same number, till they won't go further:

EXAMPLE: Simplify $\frac{18}{24}$.

Cancel down in a series of easy steps — keep going till the top and bottom don't have any common factors.

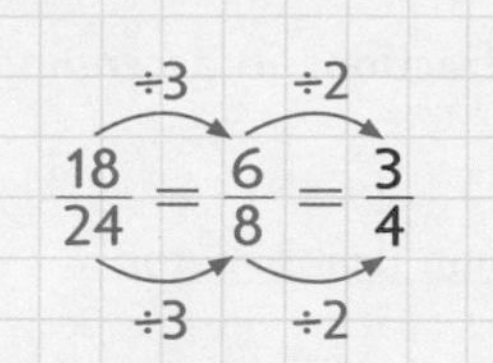

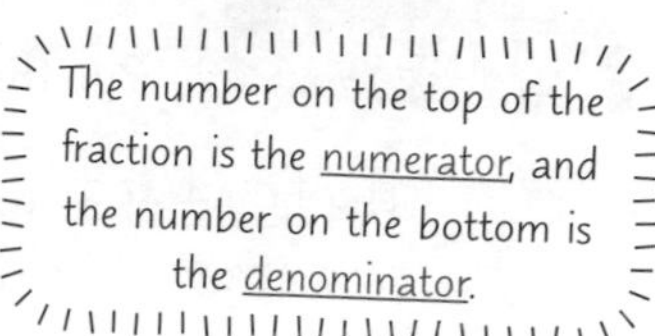

The number on the top of the fraction is the numerator, and the number on the bottom is the denominator.

2) Mixed numbers — quite easy

Mixed numbers are things like $3\frac{1}{3}$, with an integer part and a fraction part. Improper fractions are ones where the top number is larger than the bottom number. You need to be able to convert between the two.

EXAMPLES:

1. Write $4\frac{2}{3}$ as an improper fraction.

1) Think of the mixed number as an addition:

$$4\frac{2}{3} = 4 + \frac{2}{3}$$

2) Turn the integer part into a fraction:

$$4 + \frac{2}{3} = \frac{12}{3} + \frac{2}{3} = \frac{12+2}{3} = \frac{14}{3}$$

2. Write $\frac{31}{4}$ as a mixed number.

Divide the top number by the bottom.

1) The answer gives the whole number part.
2) The remainder goes on top of the fraction.

$31 \div 4 = 7$ remainder 3 so $\frac{31}{4} = 7\frac{3}{4}$

3) Multiplying — easy

Multiply top and bottom separately. It usually helps to cancel down first if you can.

EXAMPLE: Find $\frac{8}{15} \times \frac{5}{12}$.

Cancel down by dividing top and bottom by any common factors you find in either fraction:

8 and 12 both divide by 4

$$\frac{\not{8}^{\,2}}{15} \times \frac{5}{{}_{3}\not{12}} = \frac{2}{\not{15}_{\,3}} \times \frac{{}^{1}\not{5}}{3}$$

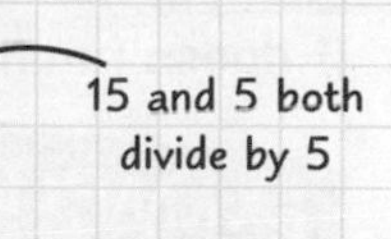

Now multiply the top and bottom numbers separately:

$$= \frac{2}{3} \times \frac{1}{3} = \frac{2 \times 1}{3 \times 3} = \frac{2}{9}$$

4) Dividing — quite easy

C

Turn the 2nd fraction UPSIDE DOWN and then multiply:

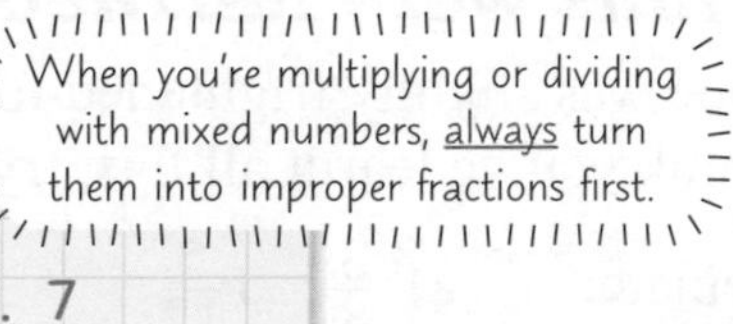

When you're multiplying or dividing with mixed numbers, always turn them into improper fractions first.

EXAMPLE: Find $2\frac{1}{3} \div 3\frac{1}{2}$.

Rewrite the mixed numbers as fractions: $2\frac{1}{3} \div 3\frac{1}{2} = \frac{7}{3} \div \frac{7}{2}$

Turn $\frac{7}{2}$ upside down and multiply: $= \frac{7}{3} \times \frac{2}{7}$

Simplify by cancelling the 7s: $= \frac{1}{3} \times \frac{2}{1} = \frac{2}{3}$

Fractions

5) Common denominators — slightly trickier

This comes in handy for ordering fractions by size, and for adding or subtracting fractions.
You need to find a number that all the denominators divide into — this will be your common denominator.
The simplest way is to find the lowest common multiple of the denominators:

EXAMPLE: Put these fractions in ascending order of size: $\frac{8}{3}, \frac{5}{4}, \frac{12}{5}$

The LCM of 3, 4 and 5 is 60, so make 60 the common denominator:

$$\frac{8}{3} = \frac{160}{60} \;(\times 20) \qquad \frac{5}{4} = \frac{75}{60} \;(\times 15) \qquad \frac{12}{5} = \frac{144}{60} \;(\times 12)$$

So the correct order is $\frac{75}{60}, \frac{144}{60}, \frac{160}{60}$ i.e. $\frac{5}{4}, \frac{12}{5}, \frac{8}{3}$

Don't forget to use the original fractions in the final answer.

6) Adding, subtracting — sort the denominators first

1) Make sure the denominators are the same (see above).
2) Add (or subtract) the top lines (numerators) only.

If you're adding or subtracting mixed numbers, it usually helps to convert them to improper fractions first.

EXAMPLE: Calculate $2\frac{1}{5} - 1\frac{1}{2}$.

Rewrite the mixed numbers as fractions: $2\frac{1}{5} - 1\frac{1}{2} = \frac{11}{5} - \frac{3}{2}$

Find a common denominator: $= \frac{22}{10} - \frac{15}{10}$

Combine the top lines: $= \frac{22-15}{10} = \frac{7}{10}$

People usually find adding and subtracting fractions harder than multiplying and dividing — but it's actually pretty easy as long as you remember to make sure the denominators are the same.

7) Finding a fraction of something — just multiply

Multiply the 'something' by the TOP of the fraction, and divide it by the BOTTOM.
It doesn't matter which order you do those two steps in — just start with whatever's easiest.

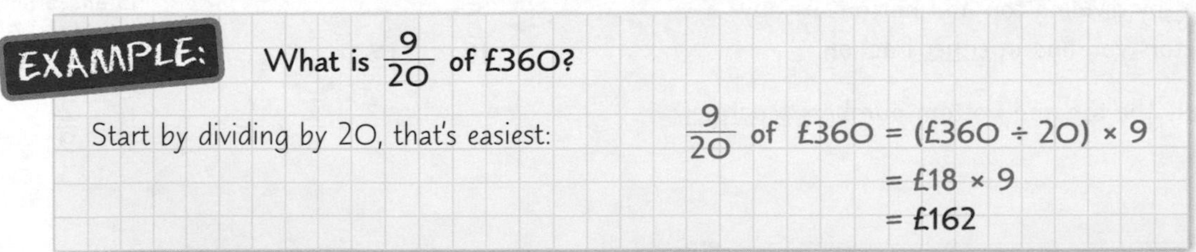

EXAMPLE: What is $\frac{9}{20}$ of £360?

Start by dividing by 20, that's easiest:

$$\frac{9}{20} \text{ of } £360 = (£360 \div 20) \times 9$$
$$= £18 \times 9$$
$$= £162$$

No fractions were harmed in the making of these pages...

...although one was slightly frightened for a while, and several were tickled.
When you think you've learnt all this, try all of these Exam Practice Questions without a calculator.

Q1 Calculate: a) $\frac{3}{8} \times 1\frac{5}{12}$ [3 marks] b) $1\frac{7}{9} \div 2\frac{2}{3}$ [3 marks]

c) $4\frac{1}{9} + 2\frac{2}{27}$ [3 marks] d) $5\frac{2}{3} - 9\frac{1}{4}$ [3 marks]

Q2 Dean has baked 550 muffins. $\frac{2}{5}$ of the muffins are chocolate, $\frac{3}{11}$ are lemon and the rest are beetroot. How many beetroot muffins has Dean baked? [4 marks]

Fractions, Decimals and Percentages

The one word that describes all these three is PROPORTION. Fractions, decimals and percentages are simply three different ways of expressing a proportion of something — and it's pretty important you should see them as closely related and completely interchangeable with each other. This table shows the really common conversions which you should know straight off without having to work them out:

Fraction	Decimal	Percentage
$\frac{1}{2}$	0.5	50%
$\frac{1}{4}$	0.25	25%
$\frac{3}{4}$	0.75	75%
$\frac{1}{3}$	0.333333...	$33\frac{1}{3}\%$
$\frac{2}{3}$	0.666666...	$66\frac{2}{3}\%$
$\frac{1}{10}$	0.1	10%
$\frac{2}{10}$	0.2	20%
$\frac{1}{5}$	0.2	20%
$\frac{2}{5}$	0.4	40%

The more of those conversions you learn, the better — but for those that you don't know, you must also learn how to convert between the three types. These are the methods:

Fraction —Divide→ **Decimal** —× by 100→ **Percentage**

E.g. $\frac{7}{20}$ is 7÷20 = 0.35 e.g. 0.35 × 100 = 35%

Fraction ←The awkward one— **Decimal** ←÷ by 100— **Percentage**

Converting decimals to fractions is awkward, because it's different for different types of decimal. There are two different methods you need to learn:

1) Terminating decimals to fractions — this is fairly easy. The digits after the decimal point go on the top, and a power of 10 on the bottom — with the same number of zeros as there were decimal places.

$0.6 = \frac{6}{10}$ $0.3 = \frac{3}{10}$ $0.7 = \frac{7}{10}$ etc.

$0.12 = \frac{12}{100}$ $0.78 = \frac{78}{100}$ $0.05 = \frac{5}{100}$ etc.

$0.345 = \frac{345}{1000}$ $0.908 = \frac{908}{1000}$ $0.024 = \frac{24}{1000}$ etc.

These can often be cancelled down — see p7.

2) Recurring decimals to fractions — this is trickier. See next page...

Eight out of ten cats prefer the perfume Eighty Purr Scent...

Learn the whole of the top table and the 4 conversion processes. Then it's time to break into a mild sweat...

Q1 Turn the following decimals into fractions and reduce them to their simplest form.
a) 0.4 b) 0.02 c) 0.77 d) 0.555 e) 5.6 [5 marks] 

Q2 Which is greater: a) 57% or $\frac{5}{9}$, b) 0.2 or $\frac{6}{25}$, c) $\frac{7}{8}$ or 90%? [3 marks] 

Fractions and Recurring Decimals

You might think that a decimal is just a decimal. But oh no — things get a lot more juicy than that...

Recurring or Terminating...

1) Recurring decimals have a pattern of numbers which repeats forever, e.g. $\frac{1}{3}$ is the decimal 0.333333... Note, it doesn't have to be a single digit that repeats. You could have, for instance: 0.143143143....
2) The repeating part is usually marked with dots or a bar on top of the number. If there's one dot, then only one digit is repeated. If there are two dots, then everything from the first dot to the second dot is the repeating bit. E.g. $0.2\dot{5} = 0.2555555...$, $0.\dot{2}\dot{5} = 0.25252525...$, $0.\dot{2}5\dot{5} = 0.255255255...$
3) Terminating decimals are finite (they come to an end), e.g $\frac{1}{20}$ is the decimal 0.05.

The denominator (bottom number) of a fraction in its simplest form tells you if it converts to a recurring or terminating decimal. Fractions where the denominator has prime factors of only 2 or 5 will give terminating decimals. All other fractions will give recurring decimals.

	Only prime factors: 2 and 5				Also other prime factors			
Fraction	$\frac{1}{5}$	$\frac{1}{125}$	$\frac{1}{2}$	$\frac{1}{20}$	$\frac{1}{7}$	$\frac{1}{35}$	$\frac{1}{3}$	$\frac{1}{6}$
Equivalent Decimal	0.2	0.008	0.5	0.05	$0.\dot{1}4285\dot{7}$	$0.0\dot{2}8571\dot{4}$	$0.\dot{3}$	$0.1\dot{6}$
	Terminating decimals				Recurring decimals			

For prime factors, see p5.

Converting terminating decimals into fractions was covered on the previous page.
Converting recurring decimals is quite a bit harder — but you'll be OK once you've learnt the method...

Recurring Decimals into Fractions

1) Basic Ones

Turning a recurring decimal into a fraction uses a really clever trick. Just watch this...

EXAMPLE: Write $0.\dot{2}3\dot{4}$ as a fraction.

1) Name your decimal — I've called it r. $\quad$ Let $r = 0.\dot{2}3\dot{4}$
2) Multiply r by a power of ten to move it past the decimal point by one full repeated lump — here that's 1000: $\quad 1000r = 234.\dot{2}3\dot{4}$
3) Now you can subtract to get rid of the decimal part:

$$\begin{aligned} 1000r &= 234.\dot{2}3\dot{4} \\ - \quad r &= 0.\dot{2}3\dot{4} \\ \hline 999r &= 234 \end{aligned}$$

4) Then just divide to leave r, and cancel if possible: $\quad r = \frac{234}{999} = \frac{26}{111}$

The 'Just Learning the Result' Method:

1) For converting recurring decimals to fractions, you could just learn the result that the fraction always has the repeating unit on the top and the same number of nines on the bottom...
2) BUT this only works if the repeating bit starts straight after the decimal point (see the next page for an example where it doesn't).
3) AND some exam questions will ask you to 'show that' or 'prove' that a fraction and a recurring decimal are equivalent — and that means you have to use the proper method.

Fractions and Recurring Decimals

2) The Trickier Type

If the recurring bit doesn't come right after the decimal point, things are slightly trickier — but only slightly.

EXAMPLE: Write $0.1\dot{6}$ as a fraction.

1) Name your decimal. — Let $r = 0.1\dot{6}$
2) Multiply r by a power of ten to move the non-repeating part past the decimal point. — $10r = 1.\dot{6}$
3) Now multiply again to move one full repeated lump past the decimal point. — $100r = 16.\dot{6}$
4) Subtract to get rid of the decimal part:

$$\begin{array}{r} 100r = 16.\dot{6} \\ -\quad 10r = 1.\dot{6} \\ \hline 90r = 15 \end{array}$$

5) Divide to leave r, and cancel if possible: — $r = \frac{15}{90} = \frac{1}{6}$

Fractions into Recurring Decimals

You might find this cropping up in your exam too — and if they're being really unpleasant, they'll stick it in a non-calculator paper.

EXAMPLE: Write $\frac{8}{33}$ as a recurring decimal.

There are two ways you can do this:

1 Find an equivalent fraction with all nines on the bottom. The number on the top will tell you the recurring part.

$$\frac{8}{33} \overset{\times 3}{=} \frac{24}{99}$$

Watch out — the number of nines on the bottom tells you the number of digits in the recurring part. E.g. $\frac{24}{99} = 0.\dot{2}\dot{4}$, but $\frac{24}{999} = 0.\dot{0}2\dot{4}$

$$\frac{24}{99} = 0.\dot{2}\dot{4}$$

2 Remember, $\frac{8}{33}$ means $8 \div 33$, so you could just do the division: (This is OK if you're allowed your calculator, but could be a bit of a nightmare if not... you could use long division if you're feeling bold, but I recommend sticking with method 1 instead.)

$8 \div 33 = 0.24242424...$

$$\frac{8}{33} = 0.\dot{2}\dot{4}$$

Oh, what's recurrin'?...

Learn how to tell whether a fraction will be a terminating or recurring decimal, and all the methods above. Then turn over and write it all down. Now, try to answer these beauties...

Q1 Express $0.\dot{1}2\dot{6}$ as a fraction in its simplest form. [2 marks] 

Q2 Show that $0.\dot{0}\dot{7} = \frac{7}{99}$ [3 marks]

Q3 Without using a calculator, convert $\frac{5}{111}$ to a recurring decimal. [2 marks]

Percentages

You shouldn't have any trouble with the simple types of percentage question.
Watch out for the trickier types and make sure you know the proper method for each of them.

Three Simple Question Types

Type 1 — "Find x% of y"

Turn the percentage into a decimal, then multiply.

EXAMPLE: Find 15% of £46.

1) Write 15% as a decimal: 15% = 15 ÷ 100 = 0.15
2) Multiply £46 by 0.15: 0.15 × £46 = **£6.90**

Type 2 — "Find the new amount after a % increase/decrease"

Turn the percentage into a decimal, then multiply. Add this on (or subtract from) the original value.

EXAMPLE: A toaster is reduced in price by 40% in the sales. It originally cost £68. What is the new price of the toaster?

1) Write 40% as a decimal: 40% = 40 ÷ 100 = 0.4
2) Multiply to find 40% of £68: 0.4 × £68 = £27.20
3) It's a decrease, so subtract from the original: £68 – £27.20 = **£40.80**

You should also know how to use the multiplier method:
multiplier = 1 – 0.4
= 0.6
68 × 0.6 = £40.80

Type 3 — "Express x as a percentage of y"

Divide x by y, then multiply by 100.

EXAMPLE: Give 40p as a percentage of £3.34.

1) Make sure both amounts are in the same units — convert £3.34 to pence: £3.34 = 334p
2) Divide 40p by 334p, then multiply by 100: (40 ÷ 334) × 100 = **12.0% (1 d.p.)**

Three Trickier Question Types

Type 1 — Finding the percentage change

1) This is the formula for giving a change in value as a percentage — LEARN IT, AND USE IT:

$$\text{PERCENTAGE 'CHANGE'} = \frac{\text{'CHANGE'}}{\text{ORIGINAL}} \times 100$$

2) This is similar to Type 3 above, because you end up with a percentage rather than an amount.
3) Typical questions will ask 'Find the percentage increase/profit/error' or 'Calculate the percentage decrease/loss/discount', etc.

EXAMPLE: A trader buys watches for £5 and sells them for £7. Find his profit as a percentage.

1) Here the 'change' is profit, so the formula looks like this: $\text{percentage profit} = \frac{\text{profit}}{\text{original}} \times 100$
2) Work out the actual value of the profit: profit = £7 – £5 = £2
3) Calculate the percentage profit: $\text{percentage profit} = \frac{2}{5} \times 100 = 40\%$

Percentages

Type 2 — Finding the original value

This is the type that most people get wrong — but only because they don't recognise it as this type and don't apply this simple method:

1) Write the amount in the question as a percentage of the original value.
2) Divide to find 1% of the original value.
3) Multiply by 100 to give the original value (= 100%).

EXAMPLE: A house increases in value by 20% to £72 000. Find what it was worth before the rise.

Note: The new, not the original value is given.

1) An increase of 20% means £72 000 represents 120% of the original value.
2) Divide by 120 to find 1% of the original value.
3) Then multiply by 100.

÷120: £72 000 = 120%
£600 = 1%
×100: £60 000 = 100%

If it was a decrease of 20%, then you'd put '£72 000 = 80%' and divide by 80 instead of 120.

So the original value was **£60 000**

Always set them out exactly like this example. The trickiest bit is deciding the top % figure on the right-hand side — the 2nd and 3rd rows are always 1% and 100%.

Type 3 — Simple Interest vs Compound Interest

Compound interest is covered on the next page.

1) There are two types of interest you could get asked about — simple and compound. Funnily enough, simple interest is the simpler of the two.
2) Simple interest means a certain percentage of the original amount only is paid at regular intervals (usually once a year). So the amount of interest is the same every time it's paid.

EXAMPLE: Regina invests £380 in an account which pays 3% simple interest per annum. How much interest will she earn in 4 years?

'Per annum' just means 'each year'.

1) Work out the amount of interest earned in one year: 3% = 3 ÷ 100 = 0.03
3% of £380 = 0.03 × £380 = **£11.40**
2) Multiply by 4 to get the total interest for 4 years: 4 × £11.40 = **£45.60**

Fact: 70% of people understand percentages, the other 40% don't...

Learn the details for each type of percentage question, then turn over and write it all down.
Then try these Exam Practice Questions:

Q1 A normal bottle of Kenny's Kiwi Juice contains 450 ml of juice. A special offer bottle contains 22% extra. How much juice is in the special offer bottle? [2 marks] 

Q2 Jenny bought a llama for £4500. She later sold it for £3285. Calculate Jenny's percentage loss. [3 marks] 

Q3 A car is reduced in price by 30% to £14 350. What did it cost before? [3 marks] 

Q4 Benny invests £1900 for 5 years in an account which pays simple interest at a rate of 2.2% per annum. How much interest will Benny earn in total? [3 marks] D

Compound Growth and Decay

One more sneaky % type for you... Compound growth shows how a thing increases over time, e.g. money in a savings account. Compound decay shows the opposite, e.g. a shiny new car losing its value with age.

The Formula

This topic is simple if you LEARN THIS FORMULA. If you don't, it's pretty well impossible:

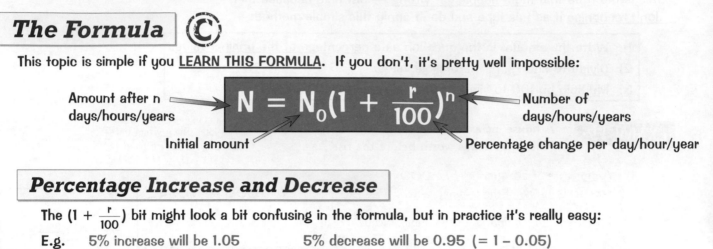

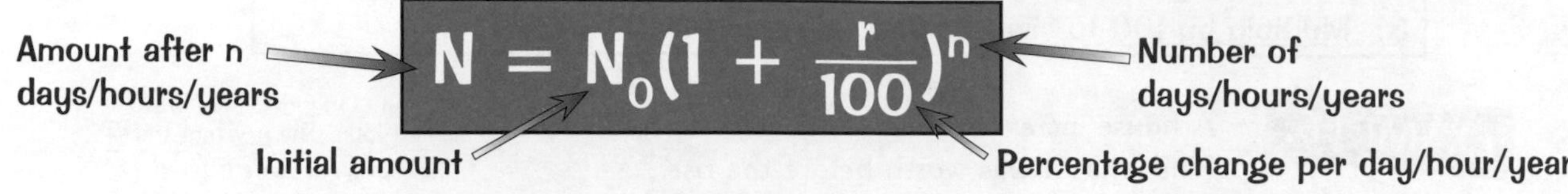

Percentage Increase and Decrease

The $(1 + \frac{r}{100})$ bit might look a bit confusing in the formula, but in practice it's really easy:

E.g. 5% increase will be 1.05 — 5% decrease will be 0.95 (= 1 – 0.05)

26% increase will be 1.26 — 26% decrease will be 0.74 (= 1 – 0.26)

3 Examples to show you how EASY it is:

The most popular context for these is compound interest. Compound interest means the interest is added on each time, and the next lot of interest is calculated using the new total rather than the original amount.

EXAMPLE: A man invests £1000 in a savings account which pays 8% compound interest per annum. How much will there be after 6 years?

Use the formula: Amount = $1000(1.08)^6$ = £1586.87

(initial amount; 8% increase; 6 years)

'Per annum' just means 'each year'.

Depreciation questions are about things (e.g. cars) which decrease in value over time.

EXAMPLE: Susan has just bought a car for £6500. If the car depreciates by 9% each year, how much will it be worth in 3 years' time?

Just use the formula again: Value = $6500(0.91)^3$ = £4898.21

(initial value; 9% decrease; 3 years)

The compound growth and decay formula isn't just used for money questions.

EXAMPLE: In a sample of bacteria, there are initially 500 cells and they increase in number by 15% each day. Find the formula relating the number of cells, c, and the number of days, d.

Well stone me, it's the same old easy-peasy compound growth formula again:

$c = N_0(1 + 0.15)^d \Rightarrow c = 500 \times (1.15)^d$

Oh man, that last joke has still got me increases...

Bleurgh. What a horrible looking formula. Make sure you learn it... learn it real good. Oh, and try these:

Q1 Josie's savings account pays 5% compound interest. If she invests £400, how much will she have in the account after 4 years? [3 marks] 

Q2 The value of Naveen's favourite painting has been depreciating by 11% per year. 6 years ago, the painting was worth £200 000. What is it worth now? [3 marks] 

Ratios

Ratios can be a grisly subject, no doubt about it — but work your way through the examples on the next two pages, and the whole murky business should become crystal clear...

Reducing Ratios to their Simplest Form 

To reduce a ratio to a simpler form, divide all the numbers in the ratio by the same thing (a bit like simplifying a fraction). It's in its simplest form when there's nothing left you can divide by.

EXAMPLE: Write the ratio 15:18 in its simplest form.

For the ratio 15:18, both numbers have a factor of 3, so divide them by 3.

We can't reduce this any further. So the simplest form of 15:18 is **5:6**.

$$15:18 \xrightarrow{\div 3} = 5:6$$

A handy trick for the calculator paper — use the fraction button

If you enter a fraction with the ⊟ or $a\frac{b}{c}$ button, the calculator automatically cancels it down when you press .

So for the ratio 8:12, just enter $\frac{8}{12}$ as a fraction, and you'll get the reduced fraction $\frac{2}{3}$.

Now you just change it back to ratio form, i.e. 2 : 3. Ace.

The More Awkward Cases: 

1) If the ratio contains decimals or fractions — multiply

EXAMPLES:

1. Simplify the ratio 2.4:3.6 as far as possible.

1) Multiply both sides by 10 to get rid of the decimal parts.
2) Now divide to reduce the ratio to its simplest form.

$$2.4:3.6 \xrightarrow{\times 10} = 24:36 \xrightarrow{\div 12} = 2:3$$

2. Give the ratio $\frac{5}{4}:\frac{7}{2}$ in its simplest form.

1) Put the fractions over a common denominator (see p8).
2) Multiply both sides by 4 to get rid of the fractions.
3) This ratio won't cancel further, so we're done.

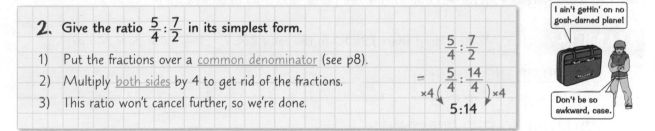

$$\frac{5}{4}:\frac{7}{2} = \frac{5}{4}:\frac{14}{4} \xrightarrow{\times 4} 5:14$$

I ain't gettin' on no gosh-darned plane!
Don't be so awkward, case.

2) If the ratio has mixed units — convert to the smaller unit

EXAMPLE: Reduce the ratio 24 mm : 7.2 cm to its simplest form.

1) Convert 7.2 cm to millimetres.
2) Simplify the resulting ratio. Once the units on both sides are the same, get rid of them for the final answer.

$$24\text{ mm}:7.2\text{ cm} = 24\text{ mm}:72\text{ mm} \xrightarrow{\div 24} = 1:3$$

3) To get the form 1 : n or n : 1 — just divide

EXAMPLE: Reduce 3:56 to the form 1:n.

Divide both sides by 3:

$$3:56 \xrightarrow{\div 3} = 1:\frac{56}{3} = 1:18\frac{2}{3} \text{ (or } 1:18.\dot{6}\text{)}$$

This form is often the most useful, since it shows the ratio very clearly.

Ratios

There's just so much great stuff to say about ratios. I couldn't possibly fit it onto only one page...

Scaling Up Ratios

If you know the ratio between parts and the actual size of one part, you can scale the ratio up to find the other parts.

EXAMPLE: Mortar is made from sand and cement in the ratio 7:2. If 21 buckets of sand are used, how much cement is needed?

You need to multiply by 3 to go from 7 to 21 on the left-hand side (LHS) — so do that to both sides:

sand : cement
= 7:2
= 21:6
(×3 on each side)

So **6 buckets** of cement are needed.

EXAMPLE: Mrs Miggins owns tabby cats and ginger cats in the ratio 3:5. All her cats are either tabby or ginger, and she has 12 tabby cats. How many cats does Mrs Miggins have in total?

Multiply both sides by 4 to go from 3 to 12 on the LHS:

tabby : ginger
= 3:5
= 12:20
(×4 on each side)

So Mrs Miggins has 12 tabby cats and 20 ginger cats.
So in total she has 12 + 20 = **32 cats**

Proportional Division

In a proportional division question a TOTAL AMOUNT is split into parts in a certain ratio.
The key word here is PARTS — concentrate on 'parts' and it all becomes quite painless:

EXAMPLE: Jess, Mo and Greg share £9100 in the ratio 2:4:7. How much does Mo get?

1) **ADD UP THE PARTS:**
   The ratio 2:4:7 means there will be a total of 13 parts: 2 + 4 + 7 = 13 parts

2) **DIVIDE TO FIND ONE "PART":**
   Just divide the total amount by the number of parts: £9100 ÷ 13 = £700 (= 1 part)

3) **MULTIPLY TO FIND THE AMOUNTS:**
   We want to know Mo's share, which is 4 parts: 4 parts = 4 × £700 = £2800

Ratio Nelson — he proportionally divided the French at Trafalgar...

Learn the rules for simplifying, how to scale ratios up and the three steps for proportional division.
Now turn over and write down what you've learned. Then try these:

Q1 Simplify: a) 25 : 35 b) 3.4 : 5.1 c) $\frac{9}{4} : \frac{15}{2}$ [4 marks]

Q2 Porridge and ice cream are mixed in the ratio 7 : 4.
How much porridge should go with 12 bowls of ice cream? [1 mark]

Q3 Divide £8400 in the ratio 5 : 3 : 4. [3 marks]

Rounding Numbers

There are two different ways of specifying where a number should be rounded.
They are: 'Decimal Places' and 'Significant Figures'.
We'll do decimal places first, but the basic method is the same for both...

Decimal Places (d.p.)

To round to a given number of decimal places:

1) Identify the position of the 'last digit' from the number of decimal places.

'Last digit' = last one in the rounded version, not the original number.

2) Then look at the next digit to the right — called the decider.

3) If the decider is 5 or more, then round up the last digit.
If the decider is 4 or less, then leave the last digit as it is.

4) There must be no more digits after the last digit (not even zeros).

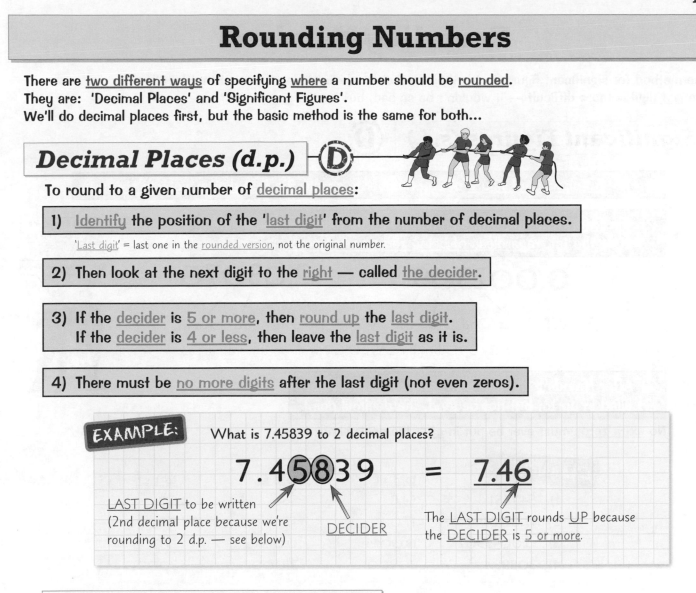

EXAMPLE: What is 7.45839 to 2 decimal places?

7.45839 = 7.46

LAST DIGIT to be written (2nd decimal place because we're rounding to 2 d.p. — see below)

DECIDER

The LAST DIGIT rounds UP because the DECIDER is 5 or more.

Trickier cases with pesky nines

1) If you have to round up a 9 (to 10), replace the 9 with 0, and carry 1 to the left.
2) Remember to keep enough zeros to fill the right number of decimal places.

EXAMPLES:

1. Round 45.699 to 2 d.p.

45.699 (decider; last digit — round up) ⟶ 45.699 (6→7, 9→0) ⟶ 45.70 to 2 d.p.

45.7 has the same value as 45.70, but 45.7 isn't rounded to 2 d.p. so it would be marked wrong.

2. Round 64.996 to 2 d.p.

64.996 (decider; last digit) ⟶ 64.996 (4→5, 9→0, 9→0) ⟶ 65.00 to 2 d.p.

When you carry the 1, this 9 rounds up to 10 too, so carry 1 to the left again.

It's official — this is the most exciting page of revision ever...

OK, maybe not, but it is important stuff, so learn the steps of the basic method, and make sure you know what to do with those cheeky nines. Then have a crack at these:

Q1 a) Round 3.5743 to 2 decimal places. b) Give 0.0481 to 2 decimal places.
c) Express 12.9096 to 3 d.p. d) Express 3546.054 to 1 d.p. [4 marks]

Rounding Numbers

The method for significant figures is identical to that for decimal places except that locating the last digit is more difficult — it wouldn't be so bad, but for the zeros...

Significant Figures (s.f.)

1) The 1st significant figure of any number is simply the first digit which isn't a zero.

2) The 2nd, 3rd, 4th, etc. significant figures follow on immediately after the 1st, regardless of being zeros or not zeros.

I asked Stacey for her significant figures.
Data?
Nah, we're just friends.

	0.002309	2.03070
SIG. FIGS:	1st 2nd 3rd 4th	1st 2nd 3rd 4th

(If we're rounding to say, 3 s.f., then the LAST DIGIT is simply the 3rd sig. fig.)

3) After rounding the last digit, end zeros must be filled in up to, but not beyond, the decimal point.

No extra zeros must ever be put in after the decimal point.

EXAMPLES:

	to 3 s.f.	to 2 s.f.	to 1 s.f.
1) 54.7651	54.8	55	50
2) 17.0067	17.0	17	20
3) 0.0045902	0.00459	0.0046	0.005
4) 30895.4	30900	31000	30000

Estimating

This is very easy, so long as you don't over-complicate it.

1) Round everything off to nice, easy, convenient numbers.
2) Then work out the answer using these nice easy numbers — that's it!

In the exam you'll need to show all the steps, to prove you didn't just use a calculator.

EXAMPLE: **Estimate the value of $\frac{127.8 + 41.9}{56.5 \times 3.2}$, showing all your working.**

1) Round all the numbers to easier ones — 1 or 2 s.f. usually does the trick.
2) You can round again to make later steps easier if you need to.

$$\frac{127.8 + 41.9}{56.5 \times 3.2} \approx \frac{130 + 40}{60 \times 3} = \frac{170}{180} \approx 1$$

Julius Caesar, Henry VIII, Einstein — all significant figures...

If a question says 'give your answer to an appropriate degree of accuracy', work out how many significant figures the numbers in the question are rounded to, and use the same number of s.f. in your answer.
Now, learn the whole of this page, turn over and write down everything you've learned. And for pudding...

Q1 Round these to 3 s.f. : a) 567.78 b) 23445 c) 0.04563 d) 0.90876 [4 marks] 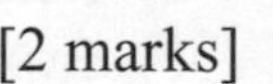 

Q2 Estimate the value of $\frac{4.23 \times 11.8}{7.7}$ [2 marks]

Bounds

Rounding and bounds go hand in hand, and not just because they sort-of rhyme...

Upper and Lower Bounds

Whenever a measurement is rounded to a given UNIT, the actual measurement can be anything up to HALF A UNIT bigger or smaller.

EXAMPLE: A room is 9 m long to the nearest metre. Find upper and lower bounds for its length.

The actual length could be half a metre either side of 9 m.

lower bound = 8.5 m
upper bound = 9.5 m

Note that the actual value is greater than or equal to the lower bound but less than the upper bound. In the example above, the actual length could be exactly 8.5 m, but if it was exactly 9.5 m it would round up to 10 m instead. Or, written as an inequality (see p39), 8.5 m $\leq$ actual length $<$ 9.5 m.

EXAMPLE: The mass of a cake is given as 2.4 kg to the nearest 0.1 kg.
What are the upper and lower bounds for the actual mass of the cake?

The rounding unit here is 0.1 kg, so the actual value could be anything between 2.4 kg ± 0.05 kg.

lower bound = 2.4 – 0.05 = 2.35 kg
upper bound = 2.4 + 0.05 = 2.45 kg

Maximum and Minimum Values for Calculations

When a calculation is done using rounded values there will be a DISCREPANCY between the CALCULATED VALUE and the ACTUAL VALUE:

EXAMPLES:

1. A floor is measured as being 5.3 m by 4.2 m, to the nearest 10 cm.
Calculate minimum and maximum possible values for the area of the floor.

The actual dimensions of the floor could be anything from 5.25 m to 5.35 m and 4.15 m to 4.25 m.

Find the minimum area by multiplying the lower bounds, and the maximum by multiplying the upper bounds.

minimum possible floor area = 5.25 × 4.15
= 21.7875 m^2

maximum possible floor area = 5.35 × 4.25
= 22.7375 m^2

2. a = 5.3 and b = 4.2, both given to 1 d.p. What are the maximum and minimum values of a ÷ b?

First find the bounds for a and b. → $5.25 \leq a < 5.35$, $4.15 \leq b < 4.25$

Now the tricky bit... The bigger the number you divide by, the smaller the answer, so:

max(a ÷ b) = max(a) ÷ min(b) → max. value of a ÷ b = 5.35 ÷ 4.15 = 1.289 (to 3 d.p.)

and min(a ÷ b) = min(a) ÷ max(b) → min. value of a ÷ b = 5.25 ÷ 4.25 = 1.235 (to 3 d.p.)

Bound, bound, get a bound, I get a bound...

This is bound to come up in the exam — or at least, it's not beyond the bounds of possibility that it could. When you think you know this page, try an Exam Practice Question:

Q1 x and y are measured as 2.32 m and 0.45 m, both to the nearest 0.01 m.

a) Find the upper and lower bounds of x and y. [2 marks]

b) If $z = x + 1/y$, find the maximum and minimum possible values of z. [2 marks]

Standard Form

Standard form (or 'standard index form') is useful for writing VERY BIG or VERY SMALL numbers in a more convenient way, e.g.

56 000 000 000 would be 5.6×10^{10} in standard form.

0.000 000 003 45 would be 3.45×10^{-9} in standard form.

But ANY NUMBER can be written in standard form and you need to know how to do it:

What it Actually is:

A number written in standard form must always be in exactly this form:

This number must always be between 1 and 10. → $A \times 10^n$ ← This number is just the number of places the decimal point moves.

(The fancy way of saying this is $1 \leq A < 10$)

Learn the Three Rules:

1) The front number must always be between 1 and 10.
2) The power of 10, n, is how far the decimal point moves.
3) n is positive for BIG numbers, n is negative for SMALL numbers.
   (This is much better than rules based on which way the decimal point moves.)

Four Important Examples:

1 Express 35 600 in standard form.

1) Move the decimal point until 35 600 becomes 3.56 ($1 \leq A < 10$)
2) The decimal point has moved 4 places so n = 4, giving: 10^4
3) 35 600 is a big number so n is +4, not −4

3.5 6 0 0

$= 3.56 \times 10^4$

2 Express 0.0000623 in standard form.

1) The decimal point must move 5 places to give 6.23 ($1 \leq A < 10$).
   So the power of 10 is 5.
2) Since 0.0000623 is a small number it must be 10^{-5} not 10^{+5}

0.0 0 0 0 6 2 3

$= 6.23 \times 10^{-5}$

3 Express 4.95×10^{-3} as an ordinary number.

1) The power of 10 is negative, so it's a small number — the answer will be less than 1.
2) The power is −3, so the decimal point moves 3 places.

0 0 0 4.9 5 $\times 10^{-3}$

= 0.00495

4 What is 146.3 million in standard form?

Too many people get this type of question wrong.
Just take your time and do it in two stages:

146.3 million

= 146 300 000

$= 1.463 \times 10^8$

The two favourite wrong answers for this are:

146.3×10^6 — which is kind of right but it's not in standard form because 146.3 is not between 1 and 10

1.463×10^6 — this one is in standard form but it's not big enough

Standard Form

Calculations with Standard Form

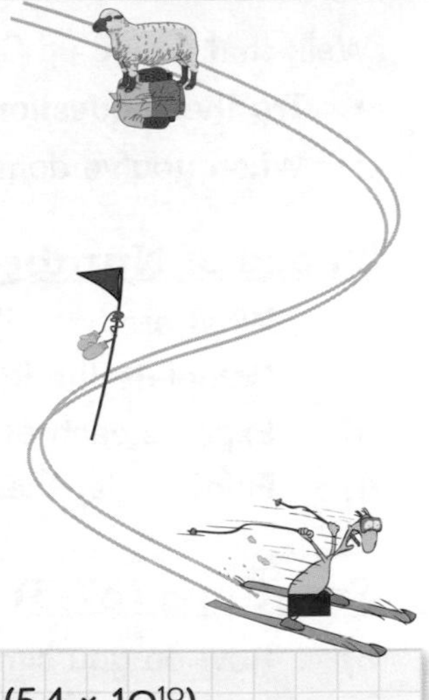

These are really popular exam questions — you might be asked to add, subtract, multiply or divide using numbers written in standard form.

Multiplying and Dividing — not too bad

1) Rearrange to put the front numbers and the powers of 10 together.
2) Multiply or divide the front numbers, and use the power rules (see p24) to multiply or divide the powers of 10.
3) Make sure your answer is still in standard form.

EXAMPLES:

1. Find $(2.24 \times 10^3) \times (6.75 \times 10^5)$. Give your answer in standard form.

$(2.24 \times 10^3) \times (6.75 \times 10^5)$
$= (2.24 \times 6.75) \times (10^3 \times 10^5)$ — Multiply front numbers and powers separately
$= 15.12 \times 10^{3+5}$ — Add the powers (see p24)
$= 15.12 \times 10^8$ — Not in standard form — convert it
$= 1.512 \times 10 \times 10^8$
$= 1.512 \times 10^9$

2. Calculate 189 000 ÷ (5.4×10^{10}). Give your answer in standard form.

$189\,000 \div (5.4 \times 10^{10})$ — Convert 189 000 to standard form
$= \frac{1.89 \times 10^5}{5.4 \times 10^{10}} = \frac{1.89}{5.4} \times \frac{10^5}{10^{10}}$ — Divide front numbers and powers separately
$= 0.35 \times 10^{5-10}$ — Subtract the powers (see p24)
$= 0.35 \times 10^{-5}$ — Not in standard form — convert it
$= 3.5 \times 10^{-1} \times 10^{-5}$
$= 3.5 \times 10^{-6}$

Adding and Subtracting — a bit trickier

1) Make sure the powers of 10 are the same — you'll probably need to rewrite one of them.
2) Add or subtract the front numbers.
3) Convert the answer to standard form if necessary.

EXAMPLE: Calculate $(9.8 \times 10^4) + (6.6 \times 10^3)$. Give your answer in standard form.

$(9.8 \times 10^4) + (6.6 \times 10^3)$

1) Rewrite one number so both powers of 10 are equal: $= (9.8 \times 10^4) + (0.66 \times 10^4)$
2) Now add the front numbers: $= (9.8 + 0.66) \times 10^4$
3) 10.46×10^4 isn't in standard form, so convert it: $= 10.46 \times 10^4 = 1.046 \times 10^5$

To put standard form numbers into your calculator, use the EXP or the $\times 10^x$ button.
E.g. enter 2.67×10^{15} by pressing 2.67 EXP 15 = or 2.67 $\times 10^x$ 15 = .

Your calculator might display an answer such as 7.986×10^{15} as 7.986 [15]. If so, don't forget to add in the "×10" bit when you write it down. Some calculators do display a little "×10" so check what yours does.

Or for just £25, you can upgrade to luxury form...

Make sure you understand all the examples on these pages. Then try these Exam Practice Questions:

Q1 Express 0.854 million and 0.00018 in standard form. [2 marks] (B)

Q2 Express 4.56×10^{-3} and 2.7×10^5 as ordinary numbers. [2 marks] (B)

Q3 Work out the following. Give your answers in standard form.
a) $(3.2 \times 10^7) \div (1.6 \times 10^{-4})$ [2 marks] b) $(6.7 \times 10^{10}) + (5.8 \times 10^{11})$ [2 marks] (B)

Revision Questions for Section One

Well, that wraps up Section One — time to put yourself to the test and find out how much you really know.

- Try these questions and tick off each one when you get it right.
- When you've done all the questions for a topic and are completely happy with it, tick off the topic.

Types of Number, Factors and Multiples (p4-6)

1) What are: a) integers b) rational numbers c) prime numbers?
2) Complete the following: a) 13^2 = __ b) $\_\_^2 = 49$ c) $\_\_^3 = 27$ d) 5^3 = __
3) Express each of these as a product of prime factors: a) 210 b) 1050
4) Find: a) the HCF of 42 and 28 b) the LCM of 8 and 10

Fractions (p7-8)

5) How do you simplify a fraction?
6) a) Write $\frac{74}{9}$ as a mixed number b) Write $4\frac{5}{7}$ as an improper fraction
7) What are the rules for multiplying, dividing and adding/subtracting fractions?
8) Calculate: a) $\frac{2}{11} \times \frac{7}{9}$ b) $5\frac{1}{2} \div 1\frac{3}{4}$ c) $\frac{5}{8} - \frac{1}{6}$ d) $3\frac{3}{10} + 4\frac{1}{4}$

Fractions, Decimals and Percentages (p9-11)

9) How do you convert: a) a fraction to a decimal? b) a terminating decimal to a fraction?
10) Write: a) 0.04 as: (i) a fraction (ii) a percentage b) 65% as: (i) a fraction (ii) a decimal
11) How can you tell if a fraction will convert to a terminating or recurring decimal?
12) Show that $0.5\dot{1} = \frac{17}{33}$

Percentages (p12-14)

13) What's the method for finding one amount as a percentage of another?
14) What's the formula for finding a change in value as a percentage?
15) A tree's height has increased by 15% in the last year to 20.24 m. What was its height a year ago?
16) I have £850 to invest for 4 years. Which will pay more interest, and how much more: an account paying 6% simple interest, or an account paying 4% compound interest?

Ratios (p15-16)

17) Sarah is in charge of ordering stock for a clothes shop. The shop usually sells red scarves and blue scarves in the ratio 5:8. Sarah orders 150 red scarves. How many blue scarves should she order?
18) What are the three steps of the method of proportional division?
19) Divide 3000 in the ratio 5 : 8 : 12.

Rounding and Bounds (p17-19)

20) Round 427.963 to: a) 2 d.p. b) 1 d.p. c) 2 s.f. d) 4 s.f.
21) Estimate the value of (124.6 + 87.1) ÷ 9.4
22) How do you determine the upper and lower bounds of a rounded measurement?
23) A rectangle measures 15.6 m by 8.4 m, to the nearest 0.1 m. Find its maximum possible area.

Standard Form (p20-21)

24) What are the three rules for writing numbers in standard form?
25) Write these numbers in standard form: a) 970 000 b) 3 560 000 000 c) 0.00000275
26) Calculate: a) $(2.54 \times 10^6) \div (1.6 \times 10^3)$ b) $(1.75 \times 10^{12}) + (9.89 \times 10^{11})$
Give your answers in standard form.

Sequences

You'll often be asked to "find an expression for the nth term of a sequence" — this is just a formula with n in, like 5n – 3. It gives you every term in a sequence when you put in different values for n.

Finding the nth Term of a Sequence 

The two methods below work for sequences with a common difference — where the sequence increases or decreases by the same number each time (i.e. the difference between each pair of terms is the same).

EXAMPLE:

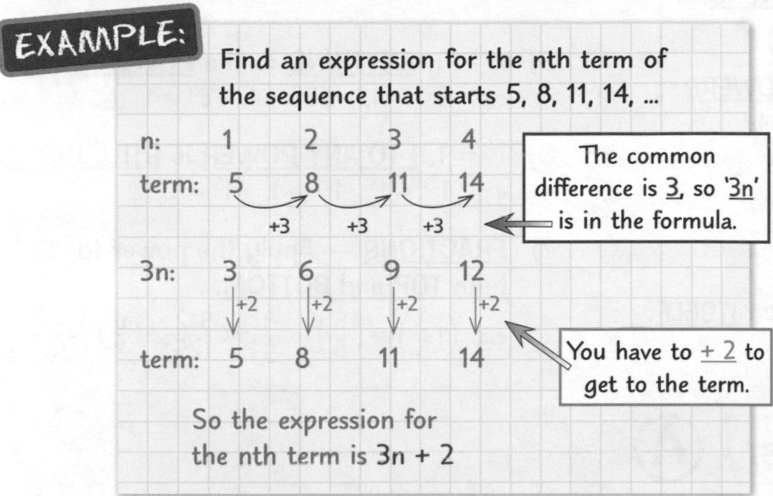

Find an expression for the nth term of the sequence that starts 5, 8, 11, 14, ...
n: 1 2 3 4
term: 5 8 11 14
+3 +3 +3
The common difference is 3, so '3n' is in the formula.
3n: 3 6 9 12
+2 +2 +2 +2
term: 5 8 11 14
You have to + 2 to get to the term.
So the expression for the nth term is 3n + 2

Method 1 — Work it out

1) Find the common difference — this tells you what to multiply n by. So here, 3 gives '3n'.
2) Work out what to add or subtract. So for n = 1, '3n' is 3 so add 2 to get to the term (5).
3) Put both bits together. So you get 3n + 2.

Always check your formula by putting the first few values of n back in, e.g. putting n = 1 into 3n + 2 gives 5, n = 2 gives 8, etc. which is the original sequence you were given — hooray!

Method 2 — Learn the formula

The other approach is to simply learn this formula and stick in the values of a and d (you don't need to replace the n though):

$$\text{nth term} = dn + (a - d)$$

d is the common difference and a is the first term.

So for the example above, d = 3 and a = 5. Putting these in the formula gives:
nth term = 3n + (5 – 3) = 3n + 2. Again, check it by putting in values for n.

Deciding if a Term is in a Sequence 

You might be given the nth term and asked if a certain value is in the sequence. The trick here is to set the expression equal to that value and solve to find n. If n is a whole number, the value is in the sequence.

EXAMPLE: The nth term of a sequence is given by $n^2 - 2$.

a) Find the 6th term in the sequence.

This is dead easy — just put n = 6 into the expression:

$6^2 - 2 = 36 - 2$
$= 34$

b) Is 45 a term in this sequence?

Set it equal to 45... $n^2 - 2 = 45$
$n^2 = 47$ ...and solve for n.
$n = \sqrt{47} = 6.8556...$

n is not a whole number, so 47 is not in the sequence $n^2 - 2$.

Have a look at p29 for more on solving equations.

If I've told you n times, I've told you n + 1 times — learn this page...

Right, I've given you two methods for working out the rule, so pick your favourite out of Method 1 and Method 2 and make sure you learn it. Then have a go at this Exam Practice Question.

Q1 A sequence starts 2, 9, 16, 23, ...
a) Find an expression for the *n*th term of the sequence. [2 marks]
b) Use your expression to find the 8th term in the sequence. [1 mark] (D)
c) Is 63 a term in the sequence? Explain your answer. [2 marks]

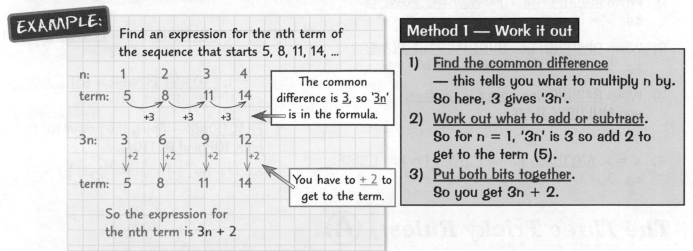

Powers and Roots

Powers are a very useful shorthand: $2\times2\times2\times2\times2\times2\times2 = 2^7$ ('two to the power 7')

That bit is easy to remember. Unfortunately, there are also ten special rules for powers that you need to learn.

The Seven Easy Rules:

Warning: Rules 1 & 2 don't work for things like $2^3 \times 3^7$, only for powers of the same number.

1) **When MULTIPLYING, you ADD THE POWERS.**
   e.g. $3^4 \times 3^6 = 3^{6+4} = 3^{10}$, $a^2 \times a^7 = a^{2+7} = a^9$

2) **When DIVIDING, you SUBTRACT THE POWERS.**
   e.g. $5^4 \div 5^2 = 5^{4-2} = 5^2$, $b^8 \div b^5 = b^{8-5} = b^3$

3) **When RAISING one power to another, you MULTIPLY THEM.**
   e.g. $(3^2)^4 = 3^{2\times4} = 3^8$, $(c^3)^6 = c^{3\times6} = c^{18}$

4) **$x^1 = x$, ANYTHING to the POWER 1 is just ITSELF.**
   e.g. $3^1 = 3$, $d \times d^3 = d^1 \times d^3 = d^{1+3} = d^4$

5) **$x^0 = 1$, ANYTHING to the POWER 0 is just 1.**
   e.g. $5^0 = 1$, $67^0 = 1$, $e^0 = 1$

6) **$1^x = 1$, 1 TO ANY POWER is STILL JUST 1.**
   e.g. $1^{23} = 1$, $1^{89} = 1$, $1^2 = 1$

7) **FRACTIONS — Apply the power to both TOP and BOTTOM.**
   e.g. $\left(1\frac{3}{5}\right)^3 = \left(\frac{8}{5}\right)^3 = \frac{8^3}{5^3} = \frac{512}{125}$, $\left(\frac{u}{v}\right)^5 = \frac{u^5}{v^5}$

The Three Tricky Rules:

8) **NEGATIVE Powers — Turn it Upside-Down**

   People have real difficulty remembering this — whenever you see a negative power you need to immediately think: "Aha, that means turn it the other way up and make the power positive".

   e.g. $7^{-2} = \frac{1}{7^2} = \frac{1}{49}$, $a^{-4} = \frac{1}{a^4}$, $\left(\frac{3}{5}\right)^{-2} = \left(\frac{5}{3}\right)^{+2} = \frac{5^2}{3^2} = \frac{25}{9}$

9) **FRACTIONAL POWERS**

   The power $\frac{1}{2}$ means Square Root,
   The power $\frac{1}{3}$ means Cube Root,
   The power $\frac{1}{4}$ means Fourth Root etc.

   e.g. $25^{\frac{1}{2}} = \sqrt{25} = 5$
   $64^{\frac{1}{3}} = \sqrt[3]{64} = 4$
   $81^{\frac{1}{4}} = \sqrt[4]{81} = 3$
   $z^{\frac{1}{5}} = \sqrt[5]{z}$

   The one to really watch is when you get a negative fraction like $49^{-1/2}$ — people get mixed up and think that the minus is the square root, and forget to turn it upside down as well.

10) **TWO-STAGE FRACTIONAL POWERS**

   With fractional powers like $64^{\frac{5}{6}}$ always split the fraction into a root and a power, and do them in that order: root first, then power: $(64)^{\frac{1}{6}\times5} = (64^{\frac{1}{6}})^5 = (2)^5 = 32$.

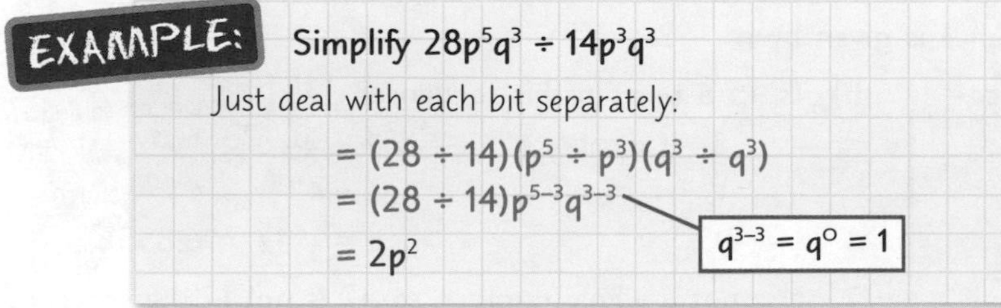

EXAMPLE: Simplify $28p^5q^3 \div 14p^3q^3$

Just deal with each bit separately:

$= (28 \div 14)(p^5 \div p^3)(q^3 \div q^3)$
$= (28 \div 14)p^{5-3}q^{3-3}$
$= 2p^2$

$q^{3-3} = q^0 = 1$

You simplify algebraic fractions using the power rules (though you might not realise it). So if you had to simplify e.g. $\frac{p^3q^6}{p^2q^3}$, you'd just cancel using the power rules to get $p^{3-2}q^{6-3} = pq^3$.

Don't let the power go to your head...

Learn all ten exciting rules on this page, then have a go at these Exam Practice Questions.

Q1 Simplify: a) $e^4 \times e^7$ [1 mark] b) $f^9 \div f^5$ [1 mark]
c) $(g^6)^{\frac{1}{2}}$ [1 mark] d) $2h^5j^{-2} \times 3h^2j^4$ [2 marks]

Q2 Evaluate:
a) $625^{\frac{3}{4}}$ [2 marks] b) $25^{-\frac{1}{2}}$ [2 marks] c) $\left(\frac{27}{216}\right)^{-\frac{1}{3}}$ [2 marks]

Algebra Basics

Before you can really get your teeth into algebra, there are some basics you need to get your head around.

Negative Numbers

Negative numbers crop up everywhere so you need to learn these rules for dealing with them:

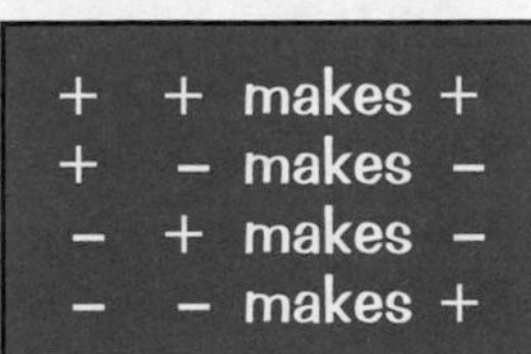

Use these rules when:

1) Multiplying or dividing.
   e.g. $-2 \times 3 = -6$, $-8 \div -2 = +4$, $-4p \times -2 = +8p$
2) Two signs are together.
   e.g. $5 - -4 = 5 + 4 = 9$, $x + -y - -z = x - y + z$

Letters Multiplied Together

Watch out for these combinations of letters in algebra that regularly catch people out:

1) abc means $a \times b \times c$. The ×'s are often left out to make it clearer.
2) gn^2 means $g \times n \times n$. Note that only the n is squared, not the g as well — e.g. πr^2 means $\pi \times r \times r$.
3) $(gn)^2$ means $g \times g \times n \times n$. The brackets mean that BOTH letters are squared.
4) $p(q - r)^3$ means $p \times (q - r) \times (q - r) \times (q - r)$. Only the brackets get cubed.
5) -3^2 is a bit ambiguous. It should either be written $(-3)^2 = 9$, or $-(3^2) = -9$ (you'd usually take -3^2 to be -9).

Terms

Before you can do anything else with algebra, you must understand what a term is:

A TERM IS A COLLECTION OF NUMBERS, LETTERS AND BRACKETS, ALL MULTIPLIED/DIVIDED TOGETHER

Terms are separated by + and – signs. Every term has a + or – attached to the front of it.

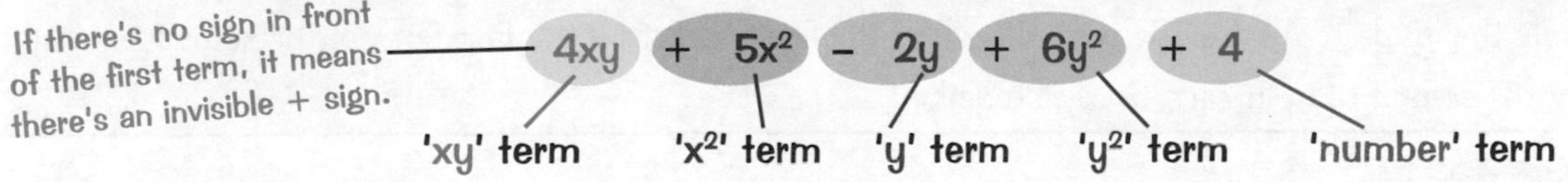

Simplifying or 'Collecting Like Terms'

To simplify an algebraic expression, you combine 'like terms' — terms that have the same combination of letters (e.g. all the x terms, all the y terms, all the number terms etc.).

EXAMPLE: Simplify $2x - 4 + 5x + 6$

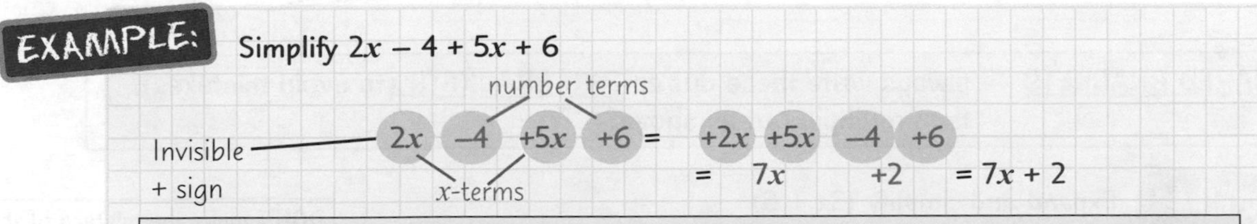

1) Put bubbles round each term — be sure you capture the +/– sign in front of each.
2) Then you can move the bubbles into the best order so that like terms are together.
3) Combine like terms.

Ahhh algebra, it's as easy as abc, or 2(ab) + c, or something like that...

Nothing too tricky on this page — but simplifying questions do come up in the exam so here's some practice:

Q1 Simplify $5x + y - 2x + 7y$ 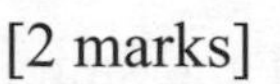 [2 marks] 

Multiplying Out Brackets

I usually use brackets to make witty comments (I'm very witty), but in algebra they're useful for simplifying things. First of all, you need to know how to expand brackets (multiply them out).

Single Brackets 

There are a few key things to remember before you start multiplying out brackets:

1) The thing outside the brackets multiplies each separate term inside the brackets.
2) When letters are multiplied together, they are just written next to each other, e.g. pq.
3) Remember, $r \times r = r^2$, and xy^2 means $x \times y \times y$, but $(xy)^2$ means $x \times x \times y \times y$.
4) Remember, a minus outside the bracket REVERSES ALL THE SIGNS when you multiply.

EXAMPLE: Expand the following:

a) $3(2x + 5)$
$= (3 \times 2x) + (3 \times 5)$
$= 6x + 15$

b) $4a(3b - 2c)$
$= (4a \times 3b) + (4a \times -2c)$
$= 12ab - 8ac$

c) $-4(3p^2 - 7q^3)$
$= (-4 \times 3p^2) + (-4 \times -7q^3)$
$= -12p^2 + 28q^3$

Note: both signs have been reversed — see point 4.

Double Brackets 

Double brackets are a bit more tricky than single brackets — this time, you have to multiply everything in the first bracket by everything in the second bracket.

5) DOUBLE BRACKETS — you get 4 terms, and usually 2 of them combine to leave 3 terms.

There's a handy way to multiply out double brackets — it's called the FOIL method and works like this:

First — multiply the first term in each bracket together
Outside — multiply the outside terms (i.e. the first term in the first bracket by the second term in the second bracket)
Inside — multiply the inside terms (i.e. the second term in the first bracket by the first term in the second bracket)
Last — multiply the second term in each bracket together

EXAMPLE: Expand and simplify $(2p - 4)(3p + 1)$

$(2p - 4)(3p + 1) = (2p \times 3p) + (2p \times 1) + (-4 \times 3p) + (-4 \times 1)$
$= 6p^2 + 2p - 12p - 4$
$= 6p^2 - 10p - 4$

The two p terms combine together.

6) SQUARED BRACKETS — always write these out as TWO BRACKETS (to avoid mistakes), then multiply out as above.

EXAMPLE: Expand and simplify $(3x + 5)^2$

$(3x + 5)^2 = (3x + 5)(3x + 5)$
Using the FOIL method — $= 9x^2 + 15x + 15x + 25 = 9x^2 + 30x + 25$

DON'T make the mistake of thinking that $(3x + 5)^2 = 9x^2 + 25$ (this is wrong wrong wrong).

Go forth and multiply out brackets...

The FOIL method is a foolproof way of multiplying out a pair of brackets so learn it. Here's some practice:

Q1 Expand: a) $-5(3x - 2y)$ [2 marks] b) $2x(x - 5)$ [2 marks] 

Q2 Expand and simplify: a) $(y + 4)(y - 5)$ [2 marks] b) $(2p - 3)^2$ [2 marks] (C)

Factorising

Right, now you know how to expand brackets, it's time to put them back in. This is known as <u>factorising</u>.

Factorising — Putting Brackets In

This is the <u>exact reverse</u> of multiplying out brackets. Here's the method to follow:

1) Take out the <u>biggest number</u> that goes into all the terms.
2) <u>For each letter in turn</u>, take out the <u>highest power</u> (e.g. x, x^2 etc.) that will go into EVERY term.
3) Open the brackets and fill in all the bits needed to <u>reproduce each term</u>.
4) <u>Check</u> your answer by <u>multiplying out</u> the brackets and making sure it matches the original expression.

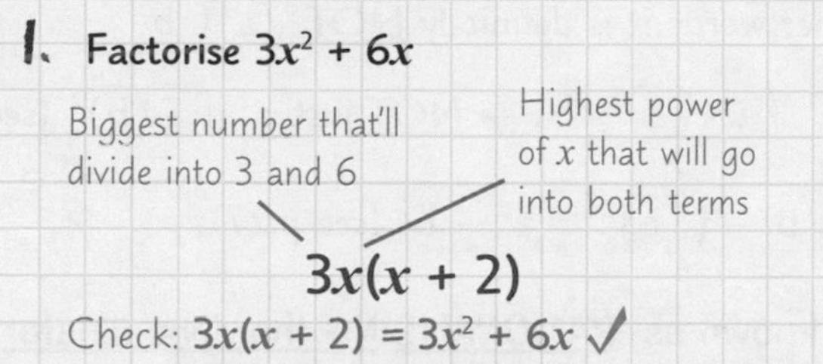

1. Factorise $3x^2 + 6x$
Biggest number that'll divide into 3 and 6
Highest power of x that will go into both terms
$3x(x + 2)$
Check: $3x(x + 2) = 3x^2 + 6x$ ✓

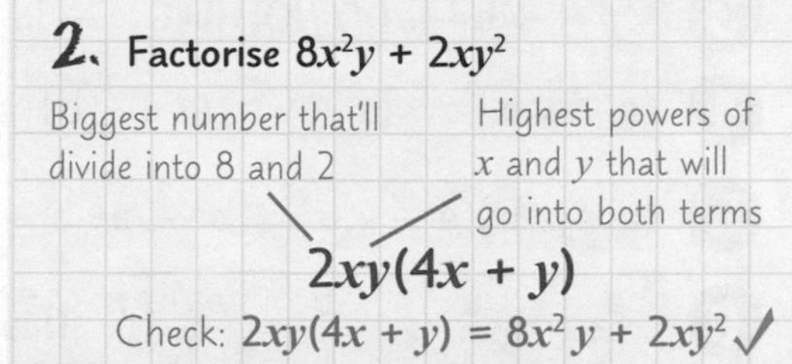

2. Factorise $8x^2y + 2xy^2$
Biggest number that'll divide into 8 and 2
Highest powers of x and y that will go into both terms
$2xy(4x + y)$
Check: $2xy(4x + y) = 8x^2y + 2xy^2$ ✓

<u>REMEMBER</u>: The bits <u>taken out</u> and put at the front are the <u>common factors</u>. The bits <u>inside the brackets</u> are what's needed to get back to the <u>original terms</u> if you multiply the brackets out again.

D.O.T.S. — The Difference Of Two Squares

The 'difference of two squares' (D.O.T.S. for short) is where you have 'one thing squared' <u>take away</u> 'another thing squared'. There's a quick and easy way to factorise it — just use the rule below:

$$a^2 - b^2 = (a + b)(a - b)$$

EXAMPLE:

Factorise:

a) $x^2 - 1$ Answer: $x^2 - 1 = (x + 1)(x - 1)$
Don't forget that 1 is a square number (it's 1^2).

b) $9p^2 - 16q^2$ Answer: $9p^2 - 16q^2 = (3p + 4q)(3p - 4q)$
Here you had to spot that 9 and 16 are square numbers.

c) $3x^2 - 75y^2$ Answer: $3x^2 - 75y^2 = 3(x^2 - 25y^2) = 3(x + 5y)(x - 5y)$
This time, you had to take out a factor of 3 first.

Watch out — the difference of two squares can creep into other algebra questions. A popular <u>exam question</u> is to put a difference of two squares on the top or bottom of a <u>fraction</u> and ask you to simplify it. There's more on algebraic fractions on p.38.

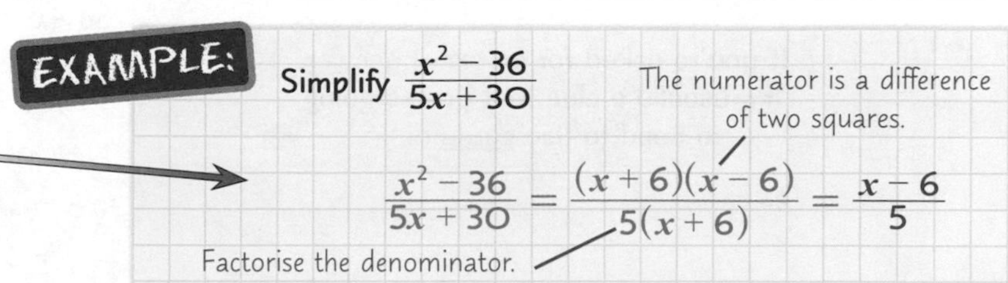

EXAMPLE:
Simplify $\frac{x^2 - 36}{5x + 30}$
The numerator is a difference of two squares.
$\frac{x^2 - 36}{5x + 30} = \frac{(x+6)(x-6)}{5(x+6)} = \frac{x-6}{5}$
Factorise the denominator.

Well, one's green and one's yellow...

As factorising is the reverse process of expanding brackets, you <u>must check</u> your answer by multiplying out the brackets. Make sure you can spot differences of two squares as well — they can be a bit sneaky.

Q1 Factorise: a) $7x - 14$ [2 marks] b) $6xy + 15y^2$ [2 marks] (C)

Q2 Factorise $8x^2 - 2y^2$ [2 marks] (B) Q3 Simplify $\frac{6x - 42}{x^2 - 49}$ [3 marks] (A)

Manipulating Surds

__Surds__ are expressions with __irrational square roots__ in them (remember from p4 that irrational numbers are ones which __can't__ be written as __fractions__, such as most square roots, cube roots and π).

Manipulating Surds — 6 Rules to Learn

There are 6 rules you need to learn for dealing with surds...

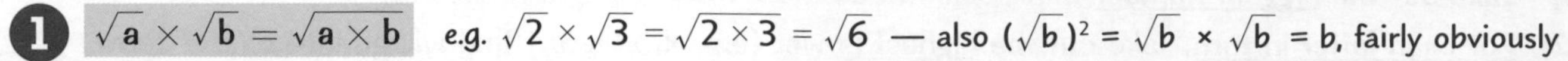

1. $\sqrt{a} \times \sqrt{b} = \sqrt{a \times b}$ e.g. $\sqrt{2} \times \sqrt{3} = \sqrt{2 \times 3} = \sqrt{6}$ — also $(\sqrt{b})^2 = \sqrt{b} \times \sqrt{b} = b$, fairly obviously
2. $\dfrac{\sqrt{a}}{\sqrt{b}} = \sqrt{\dfrac{a}{b}}$ e.g. $\dfrac{\sqrt{8}}{\sqrt{2}} = \sqrt{\dfrac{8}{2}} = \sqrt{4} = 2$

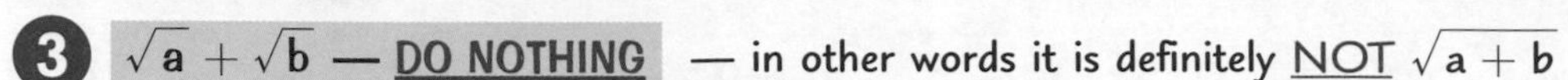

3. $\sqrt{a} + \sqrt{b}$ — __DO NOTHING__ — in other words it is definitely __NOT__ $\sqrt{a + b}$

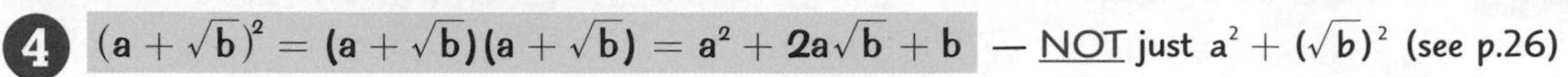

4. $(a + \sqrt{b})^2 = (a + \sqrt{b})(a + \sqrt{b}) = a^2 + 2a\sqrt{b} + b$ — __NOT__ just $a^2 + (\sqrt{b})^2$ (see p.26)

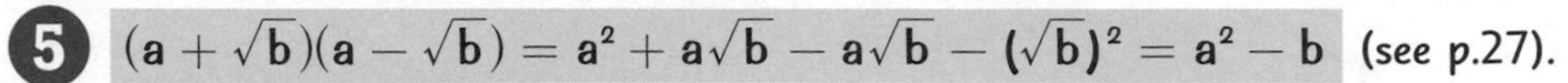

5. $(a + \sqrt{b})(a - \sqrt{b}) = a^2 + a\sqrt{b} - a\sqrt{b} - (\sqrt{b})^2 = a^2 - b$ (see p.27).

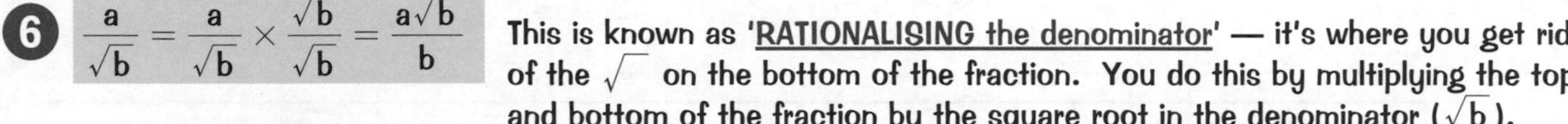

6. $\dfrac{a}{\sqrt{b}} = \dfrac{a}{\sqrt{b}} \times \dfrac{\sqrt{b}}{\sqrt{b}} = \dfrac{a\sqrt{b}}{b}$ This is known as '__RATIONALISING the denominator__' — it's where you get rid of the $\sqrt{\ }$ on the bottom of the fraction. You do this by multiplying the top and bottom of the fraction by the square root in the denominator ($\sqrt{b}$).

EXAMPLE: Write $\dfrac{3}{\sqrt{5}}$ in the form $\dfrac{a\sqrt{5}}{b}$, where a and b are whole numbers.

You have to rationalise the denominator — so multiply top and bottom by $\sqrt{5}$:

$$\frac{3\sqrt{5}}{\sqrt{5}\sqrt{5}} = \frac{3\sqrt{5}}{5}$$ — so a = 3 and b = 5

Leave Surds and π in Exact Answers

π is an __irrational__ number that comes up in calculations like finding the area of a circle. Most of the time you can use the π button on your calculator, but if you're asked to give an __exact__ answer, just __leave__ the π symbol in your answer. The same goes for __surds__ — if you're asked for an exact answer, __leave the surds in__.

EXAMPLE: Find the exact area of a circle with radius 4 cm.

Area = $\pi r^2 = \pi \times 4^2 = 16\pi$ cm^2

If you're asked for an __exact answer__, it's usually a clue that you're going to need to use __surds__ or __π__.

EXAMPLE: A rectangle has area 32 cm^2. It has length x cm and width $4x$ cm. Find the exact value of x, giving your answer in its simplest form.

Area of rectangle = length × width = $x \times 4x = 4x^2$

So $4x^2 = 32$

$x^2 = 8$

$x = \pm\sqrt{8}$ — You can ignore the negative square root (see p.30) as length must be positive.

Now get $\sqrt{8}$ into its simplest form:

$\sqrt{8} = \sqrt{4 \times 2} = \sqrt{4}\sqrt{2}$ (using rule 1)

$= 2\sqrt{2}$ So $x = 2\sqrt{2}$

Rationalise the denominator? How absurd...

Learn the 6 rules for manipulating surds, then give these Exam Practice Questions a go...

Q1 Simplify: a) $(1 + \sqrt{2})(4 - \sqrt{2})$ [2 marks] b) $(2 - \sqrt{5})^2$ [2 marks] (A)

Q2 Write $\dfrac{2}{7\sqrt{3}}$ in the form $\dfrac{a\sqrt{3}}{b}$, where a and b are integers. [3 marks] (A*)

Solving Equations

The basic idea of solving equations is very simple — keep rearranging until you end up with x = number. The two most common methods for rearranging equations are: 1) 'same to both sides' and 2) do the opposite when you cross the '='. We'll use the 'same to both sides' method on these pages.

Rearrange Until You Have x = Number

The easiest ones to solve are where you just have a mixture of x's and numbers.

1) First, rearrange the equation so that all the x's are on one side and the numbers are on the other. Combine terms where you can.
2) Then divide both sides by the number multiplying x to find the value of x.

EXAMPLE: Solve $5x + 4 = 8x - 5$

(+5) $5x + 4 + 5 = 8x - 5 + 5$ — This means 'add 5 to both sides'.

$5x + 9 = 8x$

(−5x) $5x + 9 - 5x = 8x - 5x$

$9 = 3x$ — Numbers on left, x's on right.

(÷3) $9 \div 3 = 3x \div 3$ — Divide by number multiplying x.

$3 = x$

Once you're happy with the method, you don't have to write everything out in full — your working might be:

$5x + 9 = 8x$

$9 = 3x$

$3 = x$

Multiply Out Brackets First

If your equation has brackets in it...

1) Multiply them out before rearranging.
2) Solve it in the same way as above.

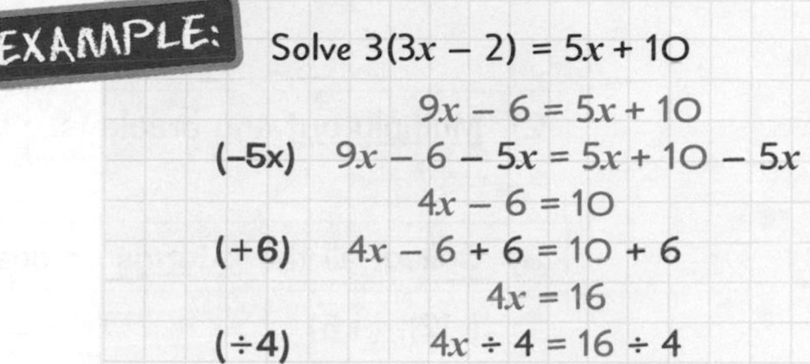

EXAMPLE: Solve $3(3x - 2) = 5x + 10$

$9x - 6 = 5x + 10$

(−5x) $9x - 6 - 5x = 5x + 10 - 5x$

$4x - 6 = 10$

(+6) $4x - 6 + 6 = 10 + 6$

$4x = 16$

(÷4) $4x \div 4 = 16 \div 4$

$x = 4$

Get Rid of Fractions (before they take over the world)

1) Fractions make everything more complicated — so you need to get rid of them before doing anything else (yep, even before multiplying out brackets).
2) To get rid of fractions, multiply every term of the equation by whatever's on the bottom of the fraction. If there are two fractions, you'll need to multiply by both denominators.

EXAMPLES:

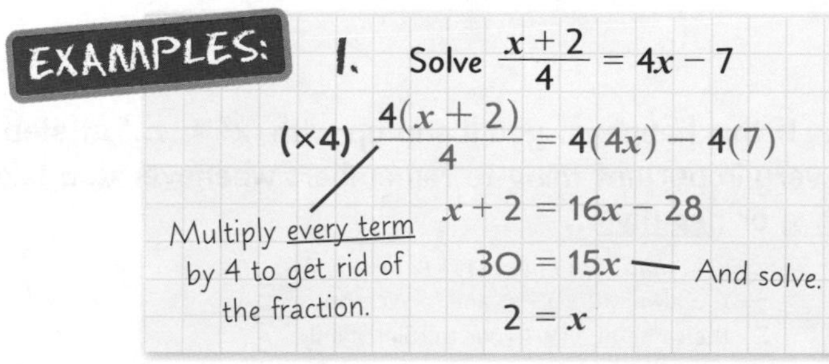

1. Solve $\frac{x+2}{4} = 4x - 7$

(×4) $\frac{4(x+2)}{4} = 4(4x) - 4(7)$ — Multiply every term by 4 to get rid of the fraction.

$x + 2 = 16x - 28$

$30 = 15x$ — And solve.

$2 = x$

2. Solve $\frac{3x+5}{2} = \frac{4x+10}{3}$ — Multiply everything by 2 then by 3.

(×2), (×3) $\frac{2 \times 3 \times (3x+5)}{2} = \frac{2 \times 3 \times (4x+10)}{3}$

$3(3x + 5) = 2(4x + 10)$ — And solve.

$9x + 15 = 8x + 20$

$x = 5$

Solving equations — more fun than greasing a seal...

Here's a handy final tip — you can always check your answer by sticking it in both sides of the original equation. They should both give the same number. Now practise what you've learned on these beauts:

Q1 Solve $2x + 5 = 17 - 4x$ [2 marks] (D)

Q2 Solve $4(y + 3) = 3y + 16$ [3 marks] (C)

Q3 Solve $\frac{3x+2}{5} = \frac{5x+6}{9}$ [3 marks]

Solving Equations

Now you know the basics of solving equations, it's time to put it all together into a handy step-by-step method.

Solving Equations Using the 6-Step Method

Here's the method to follow (just ignore any steps that don't apply to your equation):

1) Get rid of any fractions.
2) Multiply out any brackets.
3) Collect all the x-terms on one side and all number terms on the other.
4) Reduce it to the form '$Ax = B$' (by combining like terms).
5) Finally divide both sides by A to give '$x =$ ', and that's your answer.
6) If you had '$x^2 =$ ' instead, square root both sides to end up with '$x = \pm$ '.

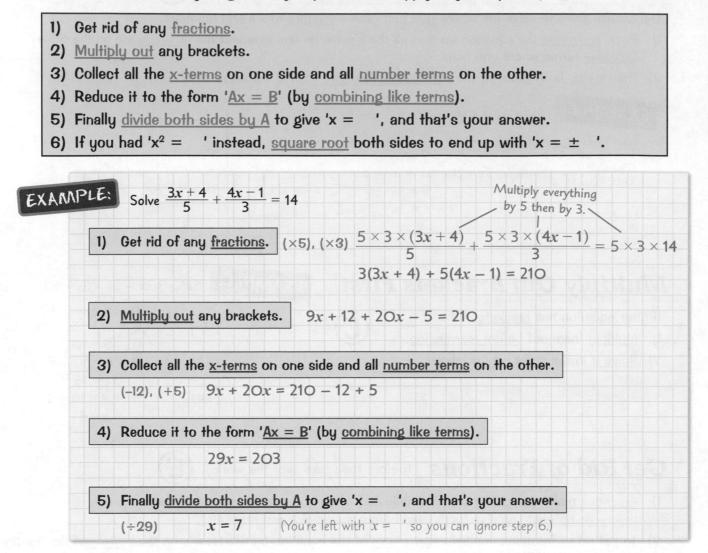

EXAMPLE: Solve $\frac{3x+4}{5} + \frac{4x-1}{3} = 14$

1) Get rid of any fractions. (×5), (×3) $\frac{5 \times 3 \times (3x+4)}{5} + \frac{5 \times 3 \times (4x-1)}{3} = 5 \times 3 \times 14$ (Multiply everything by 5 then by 3.)

$3(3x + 4) + 5(4x - 1) = 210$

2) Multiply out any brackets. $9x + 12 + 20x - 5 = 210$

3) Collect all the x-terms on one side and all number terms on the other.

(−12), (+5) $9x + 20x = 210 - 12 + 5$

4) Reduce it to the form '$Ax = B$' (by combining like terms).

$29x = 203$

5) Finally divide both sides by A to give '$x =$ ', and that's your answer.

(÷29) $x = 7$ (You're left with '$x =$ ' so you can ignore step 6.)

Dealing With Squares

If you're unlucky, you might get an x^2 in an equation. If this happens, you'll end up with '$x^2 = ...$' at step 5, and then step 6 is to take square roots. There's one very important thing to remember: whenever you take the square root of a number, the answer can be positive or negative...

EXAMPLE: Solve $3x^2 = 75$.

(÷3) $x^2 = 25$

($\sqrt{\ }$) $x = \pm 5$

You always get a +ve and -ve version of the same number (your calculator only gives the +ve answer). This shows why: $5^2 = 5 \times 5 = 25$ but also $(-5)^2 = (-5) \times (-5) = 25$.

Square Roots? Must be a geomer-tree...*

Learn the 6-step method, then try these questions.

Q1 Solve $2x^2 + 8 = 80$ [2 marks] (B)

Q2 Solve $\frac{3x-2}{2} - \frac{4x-5}{3} = 2$ [3 marks] 

*winner of Best Maths Gag in a Supporting Role, International Algebra Awards 2013

Rearranging Formulas

Rearranging formulas means making one letter the subject, e.g. getting 'y = ' from '2x + z = 3(y + 2p)' — you have to get the subject on its own.

Use the Solving Equations Method to Rearrange Formulas

Rearranging formulas is remarkably similar to solving equations. The method below is identical to the method for solving equations, except that I've added an extra step at the start.

1) Get rid of any square root signs by squaring both sides.
2) Get rid of any fractions.
3) Multiply out any brackets.
4) Collect all the subject terms on one side and all non-subject terms on the other.
5) Reduce it to the form 'Ax = B' (by combining like terms). You might have to do some factorising here too.
6) Divide both sides by A to give 'x = '.
7) If you're left with 'x^2 = ', square root both sides to get 'x = ± ' (don't forget the ±).

x is the subject term here. A and B could be numbers or letters (or a mix of both).

What To Do If...

...the Subject Appears in a Fraction

You won't always need to use all 7 steps in the method above — just ignore the ones that don't apply.

EXAMPLE: Make b the subject of the formula $a = \frac{5b + 3}{4}$.

There aren't any square roots, so ignore step 1.

2) Get rid of any fractions. (by multiplying every term by 4, the denominator)

(×4) $4a = \frac{4(5b + 3)}{4}$

$4a = 5b + 3$

There aren't any brackets so ignore step 3.

4) Collect all the subject terms on one side and all non-subject terms on the other.

(remember that you're trying to make b the subject)

(–3) $5b = 4a - 3$

5) It's now in the form Ax = B. (where A = 5 and B = 4a – 3)

6) Divide both sides by 5 to give 'b = '.

(÷5) $b = \frac{4a - 3}{5}$

b isn't squared, so you don't need step 7.

If I could rearrange my subjects, I'd have maths all day every day...

Learn the 7 steps for rearranging formulas. Then get rearrangin' with these snazzy Exam Practice Questions:

Q1 Make q the subject of the formula $p = \frac{q}{7} + 2r$ [2 marks] 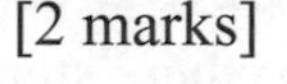

Q2 Make z the subject of the formula $x = \frac{y + 2z}{3}$ [3 marks]

Rearranging Formulas

Carrying straight on from the previous page, now it's time for what to do if...

...there's a Square or Square Root Involved

If the subject appears as a square or in a square root, you'll have to use steps 1 and 7 (not necessarily both).

EXAMPLE: Make v the subject of the formula $u = 4v^2 + 5w$.

There aren't any square roots, fractions or brackets so ignore steps 1-3 (this is pretty easy so far).

4) Collect all the subject terms on one side and all non-subject terms on the other.

$(-5w) \quad 4v^2 = u - 5w$

5) It's now in the form $Ax^2 = B$ (where A = 4 and B = u – 5w)

6) Divide both sides by 4 to give '$v^2 =$ '. $\quad (\div 4) \quad v^2 = \frac{u - 5w}{4}$

7) Square root both sides to get '$v = \pm$ '. $\quad (\sqrt{\ }) \quad v = \pm\sqrt{\frac{u - 5w}{4}}$

Don't forget the ±!

EXAMPLE: Make n the subject of the formula $m = \sqrt{n + 5}$.

1) Get rid of any square roots by squaring both sides. $\quad m^2 = n + 5$

$\sqrt{a}$ means the positive square root, so you don't need a ±.

There aren't any fractions so ignore step 2.

There aren't any brackets so ignore step 3.

4) Collect all the subject terms on one side and all non-subject terms on the other.

$(-5) \quad n = m^2 - 5$ This is in the form 'n = ' so you don't need to do steps 5-7.

...the Subject Appears Twice

Go home and cry. No, not really — you'll just have to do some factorising, usually in step 5.

EXAMPLE: Make p the subject of the formula $q = \frac{p + 1}{p - 1}$.

There aren't any square roots so ignore step 1.

2) Get rid of any fractions. $\quad q(p - 1) = p + 1$

3) Multiply out any brackets. $\quad pq - q = p + 1$

4) Collect all the subject terms on one side and all non-subject terms on the other.

$pq - p = q + 1$

5) Combine like terms on each side of the equation. $\quad p(q - 1) = q + 1$

This is where you factorise — p was in both terms on the LHS so it comes out as a common factor.

6) Divide both sides by (q – 1) to give 'p = '. $\quad p = \frac{q + 1}{q - 1}$ (p isn't squared, so you don't need step 7.)

...there's a pirate invasion — hide in a cupboard...

This is where steps 1 and 7 really come in handy — have a go at these Exam Practice Questions to see how...

Q1 Make y the subject of: a) $x = \frac{y^2}{4}$ [2 marks] b) $x = \frac{y}{y - z}$ [4 marks]

Factorising Quadratics

There are several ways of solving a quadratic equation as detailed on the following pages. You need to know all the methods as they sometimes ask for specific ones in the exam.

Factorising a Quadratic

1) 'Factorising a quadratic' means 'putting it into 2 brackets'.
2) The standard format for quadratic equations is: $ax^2 + bx + c = 0$.
3) Most exam questions have $a = 1$, making them much easier. E.g. $x^2 + 3x + 2 = 0$
4) As well as factorising a quadratic, you might be asked to solve it. This just means finding the values of x that make each bracket 0 (see example below).

See next page for when 'a' is not 1.

Factorising Method when a = 1

1) ALWAYS rearrange into the STANDARD FORMAT: $ax^2 + bx + c = 0$.
2) Write down the TWO BRACKETS with the x's in: $(x \quad)(x \quad) = 0$.
3) Then find 2 numbers that MULTIPLY to give 'c' (the end number) but also ADD/SUBTRACT to give 'b' (the coefficient of x).
4) Fill in the +/– signs and make sure they work out properly.
5) As an ESSENTIAL CHECK, expand the brackets to make sure they give the original equation.
6) Finally, SOLVE THE EQUATION by setting each bracket equal to 0.

Ignore any minus signs at this stage.

You only need to do step 6) if the question asks you to solve the quadratic — if it just tells you to factorise, you can stop at step 5).

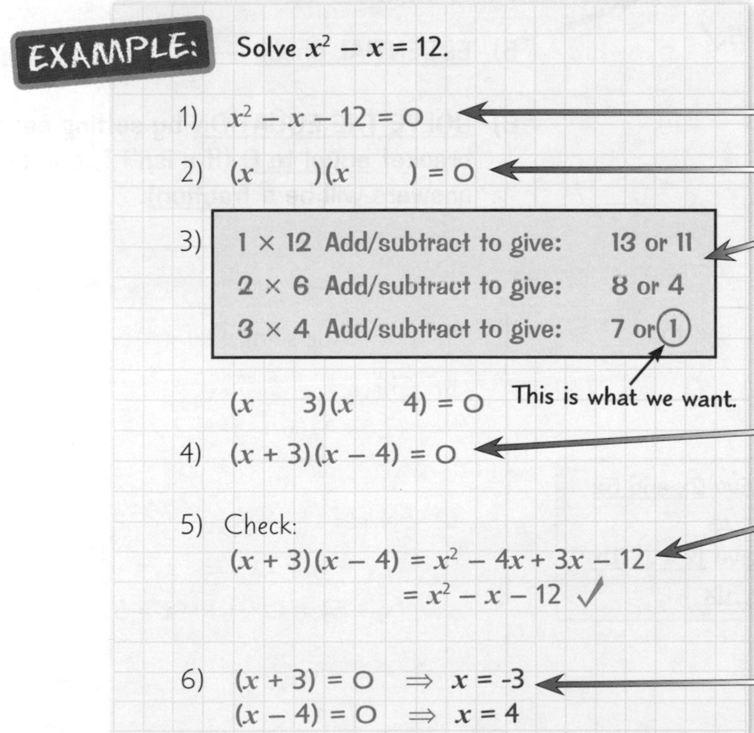

EXAMPLE: Solve $x^2 - x = 12$.

1) $x^2 - x - 12 = 0$
2) $(x \quad)(x \quad) = 0$
3) 1 × 12 Add/subtract to give: 13 or 11
   2 × 6 Add/subtract to give: 8 or 4
   3 × 4 Add/subtract to give: 7 or (1)

   This is what we want.

   $(x \quad 3)(x \quad 4) = 0$
4) $(x + 3)(x - 4) = 0$
5) Check:
   $(x + 3)(x - 4) = x^2 - 4x + 3x - 12$
   $= x^2 - x - 12$ ✓
6) $(x + 3) = 0 \Rightarrow x = -3$
   $(x - 4) = 0 \Rightarrow x = 4$

1) Rearrange into the standard format.
2) Write down the initial brackets.
3) Find the right pairs of numbers that multiply to give c (= 12), and add or subtract to give b (= 1) (remember, we're ignoring the +/– signs for now).
4) Now fill in the +/– signs so that 3 and 4 add/subtract to give -1 (= b).
5) ESSENTIAL check — EXPAND the brackets to make sure they give the original equation.

But we're not finished yet — we've only factorised it, we still need to...

6) SOLVE THE EQUATION by setting each bracket equal to 0.

Bring me a biscuit or I'll factorise your quadratic...

Handy tip: to help you work out which signs you need, look at c. If c is positive, the signs will be the same (both positive or both negative), but if c is negative the signs will be different (one positive and one negative).

Q1 Factorise $x^2 + 2x - 15$ [2 marks] (B)   Q2 Solve $x^2 - 9x + 20 = 0$ [3 marks] 

Factorising Quadratics

So far so good. It gets a bit more complicated when 'a' isn't 1, but it's all good fun, right? Right? Well, I think it's fun anyway.

When 'a' is Not 1

The basic method is still the same but it's a bit messier — the initial brackets are different as the first terms in each bracket have to multiply to give 'a'. This means finding the other numbers to go in the brackets is harder as there are more combinations to try. The best way to get to grips with it is to have a look at an example.

EXAMPLE: Solve $3x^2 + 7x - 6 = 0$.

1) $3x^2 + 7x - 6 = 0$
2) $(3x \quad)(x \quad) = 0$
3) Number pairs: 1×6 and 2×3

(3x 1)(x 6) multiplies to give 18x and 1x which add/subtract to give 17x or 19x

(3x 6)(x 1) multiplies to give 3x and 6x which add/subtract to give 9x or 3x

(3x 3)(x 2) multiplies to give 6x and 3x which add/subtract to give 9x or 3x

(3x 2)(x 3) multiplies to give 9x and 2x which add/subtract to give 11x or 7x ✓

$(3x \quad 2)(x \quad 3)$

4) $(3x - 2)(x + 3)$
5) $(3x - 2)(x + 3) = 3x^2 + 9x - 2x - 6$
$= 3x^2 + 7x - 6$ ✓
6) $(3x - 2) = 0 \Rightarrow x = \frac{2}{3}$
$(x + 3) = 0 \Rightarrow x = -3$

1) Rearrange into the standard format.

2) Write down the initial brackets — this time, one of the brackets will have a 3x in it.

3) The tricky part: first, find pairs of numbers that multiply to give c (= 6), ignoring the minus sign for now.

Then, try out the number pairs you just found in the brackets until you find one that gives 7x. But remember, each pair of numbers has to be tried in 2 positions (as the brackets are different — one has 3x in it).

4) Now fill in the +/– signs so that 9 and 2 add/subtract to give +7 (= b).

5) ESSENTIAL check — EXPAND the brackets.

6) SOLVE THE EQUATION by setting each bracket equal to 0 (if a isn't 1, one of your answers will be a fraction).

EXAMPLE: Solve $2x^2 - 9x = 5$.

1) Put in standard form: $2x^2 - 9x - 5 = 0$
2) Initial brackets: $(2x \quad)(x \quad) = 0$
3) Number pairs: 1×5

(2x 5)(x 1) multiplies to give 2x and 5x which add/subtract to give 3x or 7x

(2x 1)(x 5) multiplies to give 1x and 10x which add/subtract to give 9x or 11x ✓

$(2x \quad 1)(x \quad 5)$

4) Put in the signs: $(2x + 1)(x - 5)$
5) Check:
$(2x + 1)(x - 5) = 2x^2 - 10x + x - 5$
$= 2x^2 - 9x - 5$ ✓
6) Solve:
$(2x + 1) = 0 \Rightarrow x = -\frac{1}{2}$
$(x - 5) = 0 \Rightarrow x = 5$

It's not scary — just think of it as brackets giving algebra a hug ...

Learn the step-by-step method for solving quadratics, then have a go at these Exam Practice Questions.

Q1 Factorise $2x^2 - 5x - 12$ [2 marks] 

Q2 Solve $3x^2 + 10x - 8 = 0$ [3 marks]

Q3 Factorise $3x^2 + 32x + 20$ [2 marks]

Q4 Solve $5x^2 - 13x = 6$ [3 marks]

The Quadratic Formula

The solutions to ANY quadratic equation $ax^2 + bx + c = 0$ are given by this formula:

$$x = \frac{-b \pm \sqrt{b^2 - 4ac}}{2a}$$

<u>LEARN THIS FORMULA</u> — and <u>how to use it</u>. It's usually given in the exam, but if you don't learn it, you won't know how to use it. Using it isn't that hard, but there are a few pitfalls — so <u>TAKE HEED of these crucial details</u>:

Quadratic Formula — Five Crucial Details

1) Take it nice and slowly — always write it down in stages as you go.
2) WHENEVER YOU GET A MINUS SIGN, <u>THE ALARM BELLS SHOULD ALWAYS RING!</u>
3) Remember it's <u>2a</u> on the bottom line, not just a — and you <u>divide ALL of the top line by 2a</u>.
4) The ± sign means you end up with <u>two solutions</u> (by replacing it in the final step with '+' and '–').
5) If you get a <u>negative</u> number inside your square root, go back and <u>check your working</u>. Some quadratics do have a negative value in the square root, but they won't come up at GCSE.

If either 'a' or 'c' is negative, the -4ac effectively becomes +4ac, so watch out. Also, be careful if b is negative, as -b will be positive.

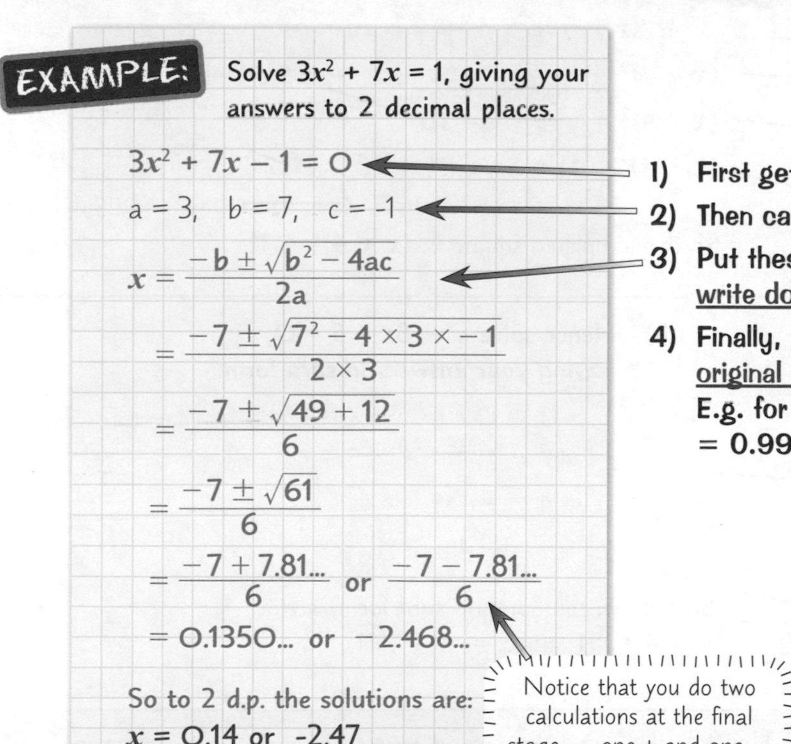

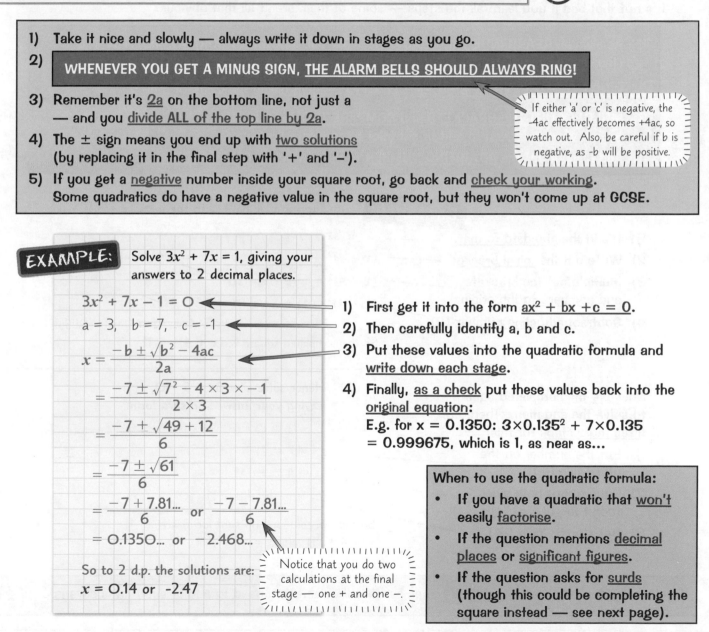

EXAMPLE: Solve $3x^2 + 7x = 1$, giving your answers to 2 decimal places.

$3x^2 + 7x - 1 = 0$

$a = 3, \quad b = 7, \quad c = -1$

$$x = \frac{-b \pm \sqrt{b^2 - 4ac}}{2a}$$

$$= \frac{-7 \pm \sqrt{7^2 - 4 \times 3 \times -1}}{2 \times 3}$$

$$= \frac{-7 \pm \sqrt{49 + 12}}{6}$$

$$= \frac{-7 \pm \sqrt{61}}{6}$$

$$= \frac{-7 + 7.81...}{6} \text{ or } \frac{-7 - 7.81...}{6}$$

$= 0.1350...$ or $-2.468...$

So to 2 d.p. the solutions are: $x = 0.14$ or -2.47

Notice that you do two calculations at the final stage — one + and one –.

1) First get it into the form <u>$ax^2 + bx + c = 0$</u>.
2) Then carefully identify a, b and c.
3) Put these values into the quadratic formula and <u>write down each stage</u>.
4) Finally, <u>as a check</u> put these values back into the <u>original equation</u>:
E.g. for x = 0.1350: $3 \times 0.135^2 + 7 \times 0.135 = 0.999675$, which is 1, as near as...

When to use the quadratic formula:
- If you have a quadratic that <u>won't</u> easily <u>factorise</u>.
- If the question mentions <u>decimal places</u> or <u>significant figures</u>.
- If the question asks for <u>surds</u> (though this could be completing the square instead — see next page).

Enough number crunches? Now it's time to work on your quads...

Learn the crucial details and how to use the Quadratic Formula. Done it? Now it's time to practise your mad new skillz with these handy Exam Practice Questions...

Q1 Solve $x^2 + 10x - 4 = 0$, giving your answers to 2 decimal places. [3 marks]

Q2 Solve $3x^2 - 3x = 2$, giving your answers to 2 decimal places. [3 marks]

Completing the Square

There's just one more method to learn for solving quadratics — and it's a bit of a nasty one. It's called 'completing the square', and takes a bit to get your head round it.

Solving Quadratics by 'Completing the Square'

To 'complete the square' you have to:

1) Write down a SQUARED bracket, and then
2) Stick a number on the end to 'COMPLETE' it.

$$x^2 + 12x - 5 = (x + 6)^2 - 41$$

The SQUARE... ...COMPLETED

It's not that bad if you learn all the steps — some of them aren't all that obvious.

1) As always, REARRANGE THE QUADRATIC INTO THE STANDARD FORMAT: $ax^2 + bx + c$ (the rest of this method is for $a = 1$, as that's what you'll almost certainly be given).
2) WRITE OUT THE INITIAL BRACKET: $(x + \frac{b}{2})^2$ — just divide the value of b by 2.
3) MULTIPLY OUT THE BRACKETS and COMPARE TO THE ORIGINAL to find what you need to add or subtract to complete the square.
4) Add or subtract the ADJUSTING NUMBER to make it MATCH THE ORIGINAL.

If a isn't 1, you have to divide through by 'a' or take out a factor of 'a' before you start.

EXAMPLE: a) Express $x^2 + 8x + 5$ in the form $(x + m)^2 + n$.

1) It's in the standard format. — $x^2 + 8x + 5$
2) Write out the initial bracket — $(x + 4)^2$
3) Multiply out the brackets and compare to the original. — $(x + 4)^2 = x^2 + 8x + 16$ — Original equation had +5 here...
4) Subtract adjusting number (11). — $(x + 4)^2 - 11 = x^2 + 8x + 16 - 11$ — ...so you need -11

$= x^2 + 8x + 5$ ✓ — matches original now!

So the completed square is: $(x + 4)^2 - 11$.

b) Hence solve $x^2 + 8x + 5 = 0$, leaving your answers in surd form.

Now use the completed square to solve the equation. There are three more steps for this:

$(x + 4)^2 - 11 = 0$

1) Put the number on the other side (+11). — $(x + 4)^2 = 11$
2) Square root both sides (don't forget the ±!) ($\sqrt{\ }$). — $x + 4 = \pm\sqrt{11}$
3) Get x on its own (-4). — $x = -4 \pm \sqrt{11}$

So the two solutions (in surd form) are: $x = -4 + \sqrt{11}$ and $x = -4 - \sqrt{11}$

If you really don't like steps 3-4, just remember that the value you need to add or subtract is always $c - \left(\frac{b}{2}\right)^2$.

But if a square's not complete, is it really a square...?

Go over this carefully, 'cos it's pretty gosh darn confusing at first. Learn the method for completing the square, and brush up on your equation-solving skills too. Then try these Exam Practice Questions...

Q1 Write $x^2 - 12x + 18$ in the form $(x + p)^2 + q$. [3 marks] 

Q2 Solve $x^2 + 10x + 7 = 0$, by first writing it in the form $(x + m)^2 + n = 0$. Give your answers as simplified surds. [5 marks] 

Quadratic Equations — Tricky Ones

Now it's time to have a go at some tricky 'hidden quadratic' questions that sometimes pop up in the exam.

Shape Questions Involving Quadratics

Sometimes examiners like to disguise quadratic equations by pretending the question is about a shape — it might look like an area or volume question where you have to find the length of a side. Don't be fooled though.

EXAMPLE: The rectangle on the right has sides of length x cm and $(2x + 1)$ cm. The area of the rectangle is 15 cm². Find the value of x.

x cm

$(2x + 1)$ cm

You're told the side lengths and the area, and you know the formula for the area of a rectangle (A = l × w), so this gives you:

$$x \times (2x + 1) = 15$$
$$2x^2 + x = 15$$

This is a quadratic, so just rearrange into the standard format and solve:

See p33-34 if you need some help factorising.

$$2x^2 + x - 15 = 0$$
$$(2x - 5)(x + 3) = 0 \qquad \text{so } x = \frac{5}{2} \text{ or } x = -3.$$

You couldn't have a shape with sides of length -3 cm and -5 cm.

However, you're looking for a length, which means x can't be negative — so $x = \frac{5}{2}$

Quadratics Hidden as Fractions

This is a nasty exam question — you're given an equation to solve that looks like it's an algebraic fractions question (more about these on the next page), but after some rearranging, it turns out you've got a quadratic.

EXAMPLE: Solve $x - \frac{5}{x-1} = 2$, giving your answers to 3 significant figures.

At first glance, this doesn't look like a quadratic, but wait and see...

The first thing to do is to get rid of the fraction (by multiplying every term by $(x - 1)$):

This is using the method for solving equations from p30.

$$x(x-1) - \frac{5(x-1)}{x-1} = 2(x-1)$$
$$\Rightarrow x(x-1) - 5 = 2(x-1)$$

Next, multiply out the brackets:

$$x^2 - x - 5 = 2x - 2$$

It's starting to look like a quadratic now, so write it out in the standard format:

$$x^2 - 3x - 3 = 0$$

Solve it — you're going to need the quadratic formula (see p35):

The mention of significant figures in the question is a hint that you're going to need to use the quadratic formula.

$a = 1, b = -3, c = -3$

$$x = \frac{-b \pm \sqrt{b^2 - 4ac}}{2a} = \frac{-(-3) \pm \sqrt{(-3)^2 - (4 \times 1 \times -3)}}{2 \times 1}$$
$$= \frac{3 \pm \sqrt{9 - (-12)}}{2} = \frac{3 \pm \sqrt{21}}{2}$$
$$x = \frac{3 + \sqrt{21}}{2} = 3.7912... = 3.79 \text{ (3 s.f.)} \quad \text{or} \quad x = \frac{3 - \sqrt{21}}{2} = -0.7912... = -0.791 \text{ (3 s.f.)}$$

I'd like to hide FROM quadratics...

There's no telling what might come up in an exam — be prepared to spot hidden quadratics and solve them.

Q1 Find the exact solutions of $2x + \frac{3}{x-2} = -2$. Give your answers in surd form. [4 marks]

Algebraic Fractions

Unfortunately, fractions aren't limited to numbers — you can get algebraic fractions too. Fortunately, everything you learnt about fractions on p7-8 can be applied to algebraic fractions as well.

Simplifying Algebraic Fractions

You can simplify algebraic fractions by cancelling terms on the top and bottom — just deal with each letter individually and cancel as much as you can. You might have to factorise first (see pages 27 and 33-34).

EXAMPLES:

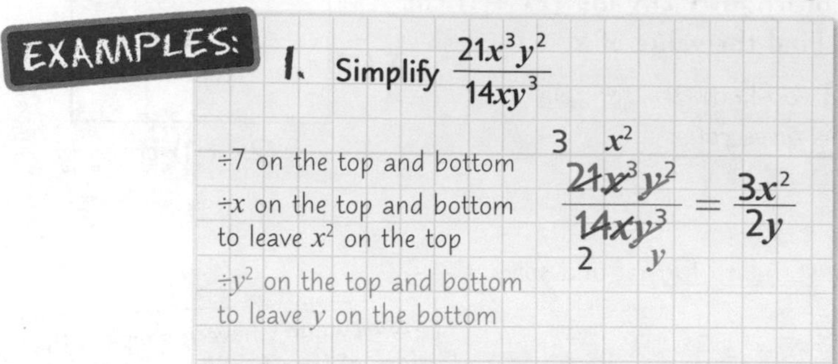

1. Simplify $\frac{21x^3y^2}{14xy^3}$
÷7 on the top and bottom
÷x on the top and bottom to leave x^2 on the top
÷y^2 on the top and bottom to leave y on the bottom
$\frac{21x^3y^2}{14xy^3} = \frac{3x^2}{2y}$

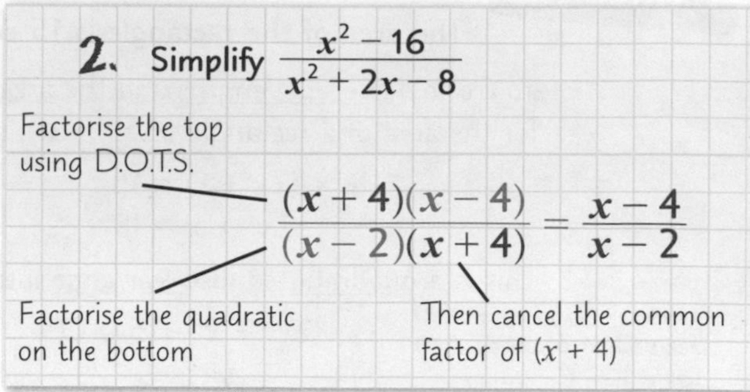

2. Simplify $\frac{x^2-16}{x^2+2x-8}$
Factorise the top using D.O.T.S.
$\frac{(x+4)(x-4)}{(x-2)(x+4)} = \frac{x-4}{x-2}$
Factorise the quadratic on the bottom
Then cancel the common factor of (x + 4)

Multiplying/Dividing Algebraic Fractions

1) **To multiply two fractions, just multiply tops and bottoms separately.**

EXAMPLE:

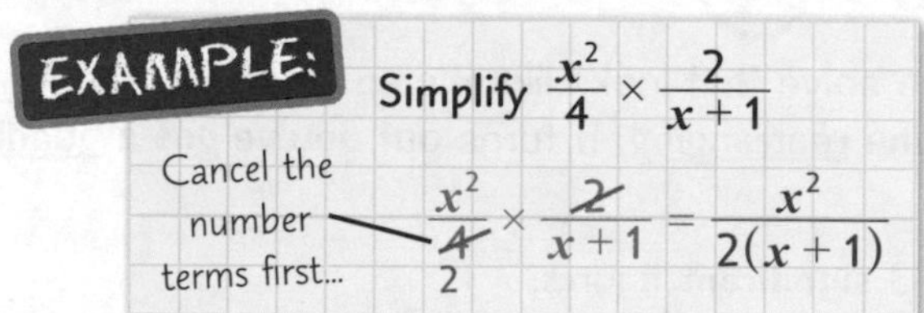

Simplify $\frac{x^2}{4} \times \frac{2}{x+1}$
Cancel the number terms first...
$\frac{x^2}{4} \times \frac{2}{x+1} = \frac{x^2}{2(x+1)}$

2) **To divide, turn the second fraction upside down then multiply.**

EXAMPLE:

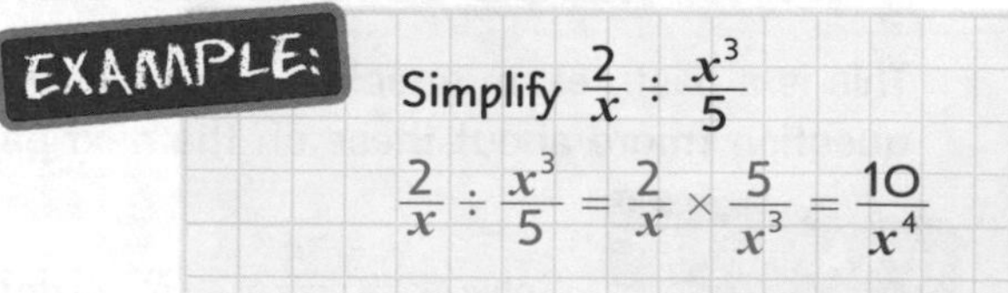

Simplify $\frac{2}{x} \div \frac{x^3}{5}$
$\frac{2}{x} \div \frac{x^3}{5} = \frac{2}{x} \times \frac{5}{x^3} = \frac{10}{x^4}$

Adding/Subtracting Algebraic Fractions

Adding or subtracting is a bit more difficult:

1) **Work out the common denominator (see p8).**
2) **Multiply top and bottom of each fraction by whatever gives you the common denominator.**
3) **Add or subtract the numerators only.**

For the common denominator, find something both denominators divide into.

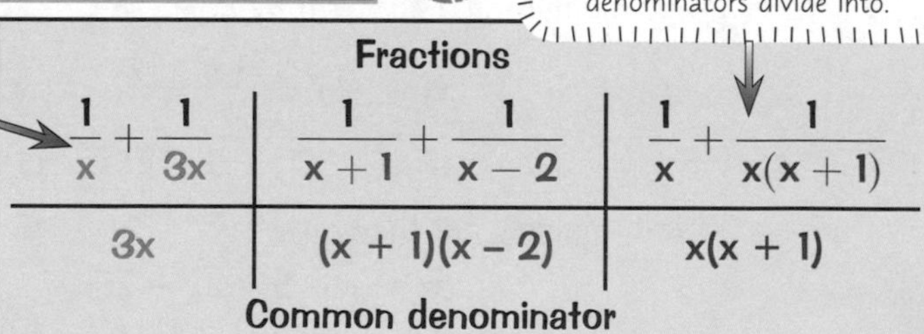

Fractions		
$\frac{1}{x} + \frac{1}{3x}$	$\frac{1}{x+1} + \frac{1}{x-2}$	$\frac{1}{x} + \frac{1}{x(x+1)}$
3x	(x + 1)(x – 2)	x(x + 1)
Common denominator		

EXAMPLE:

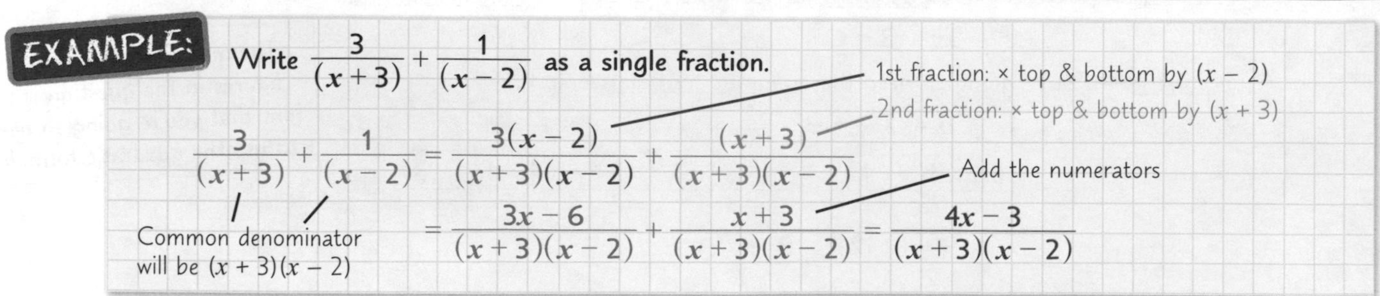

Write $\frac{3}{(x+3)} + \frac{1}{(x-2)}$ as a single fraction.
1st fraction: × top & bottom by (x – 2)
2nd fraction: × top & bottom by (x + 3)
$\frac{3}{(x+3)} + \frac{1}{(x-2)} = \frac{3(x-2)}{(x+3)(x-2)} + \frac{(x+3)}{(x+3)(x-2)}$
Add the numerators
$= \frac{3x-6}{(x+3)(x-2)} + \frac{x+3}{(x+3)(x-2)} = \frac{4x-3}{(x+3)(x-2)}$
Common denominator will be (x + 3)(x – 2)

I'd like to cancel the Summer Term...

One more thing... never do this: $\frac{x}{x+y} = \frac{1}{y}$ ✗ It's wrong wrong WRONG! Got that? Good, now try these:

Q1 Simplify $\frac{18ab^3}{6a^2b}$ [2 marks] (B)

Q2 Simplify $\frac{x+3}{2} \div \frac{2}{x}$ [2 marks] 

Q3 Write $\frac{3}{x+4} - \frac{2}{x-1}$ as a single fraction in its simplest form. [3 marks] (A*)

Inequalities

Inequalities aren't half as difficult as they look. Once you've learned the tricks involved, most of the algebra for them is identical to ordinary equations (have a look back at p29-30 if you need a reminder).

The Inequality Symbols

$>$ means 'Greater than' $\geq$ means 'Greater than or equal to'
$<$ means 'Less than' $\leq$ means 'Less than or equal to'

REMEMBER — the one at the BIG end is BIGGEST so $x > 4$ and $4 < x$ both say: 'x is greater than 4'.

Algebra With Inequalities

The key thing about inequalities — solve them just like regular equations but WITH ONE BIG EXCEPTION:

Whenever you MULTIPLY OR DIVIDE by a NEGATIVE NUMBER, you must FLIP THE INEQUALITY SIGN.

EXAMPLES:

1. x is an integer such that $-4 < x \leq 3$. Write down all the possible values of x.

Work out what each bit of the inequality is telling you:
$-4 < x$ means 'x is greater than -4',
$x \leq 3$ means 'x is less than or equal to 3'.
Now just write down all the values that x can take.
(Remember, integers are just +ve or –ve whole numbers)

–4 isn't included because of the <
but 3 is included because of the ≤.

-3, -2, -1, 0, 1, 2, 3

2. Solve $2x + 7 > x + 11$.

Just solve it like an equation:
(–7) $2x + 7 - 7 > x + 11 - 7$
$2x > x + 4$
(–x) $2x - x > x + 4 - x$
$x > 4$

3. Solve $-2 \leq \frac{x}{4} \leq 5$.

Don't be put off because there are two inequality signs — just do the same thing to each bit of the inequality:
(×4) $4 \times -2 \leq \frac{4 \times x}{4} \leq 4 \times 5$
$-8 \leq x \leq 20$

4. Solve $9 - 2x > 15$.

Again, solve it like an equation:
(–9) $9 - 2x - 9 > 15 - 9$
$-2x > 6$
(÷–2) $-2x \div -2 < 6 \div -2$
$x < -3$

The > has turned into a <, because we divided by a negative number.

You Can Show Inequalities on Number Lines

Drawing inequalities on a number line is dead easy — all you have to remember is that you use an open circle (○) for > or < and a coloured-in circle (●) for ≥ or ≤.

EXAMPLE: Show the inequality $-4 < x \leq 3$ on a number line.

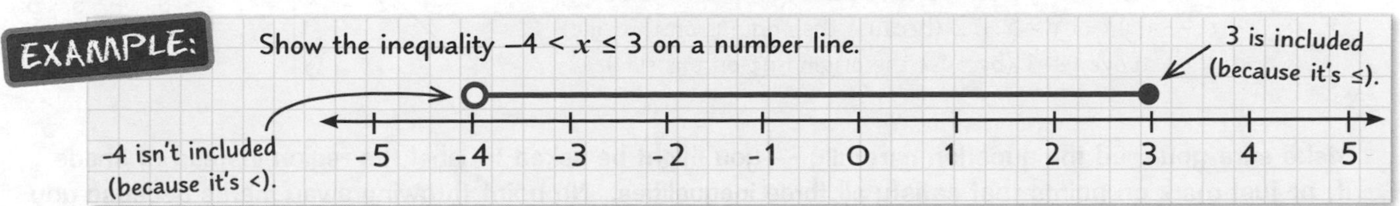

I saw you flip the inequality sign — how rude...

To check you've got the inequality sign right way round, pop in a value for x and check the inequality's true.

Q1 Solve: a) $11x + 3 < 47$ [2 marks] b) $4x \geq 18 - 2x$ [2 marks] 

Q2 Solve the inequality $5x + 3 \leq 2x + 15$ and represent the solution on a number line. [3 marks] (B)

Graphical Inequalities

These questions always involve shading a region on a graph. The method sounds very complicated, but once you've seen it in action with an example, you see that it's OK...

Showing Inequalities on a Graph

Here's the method to follow:

1) CONVERT each INEQUALITY to an EQUATION by simply putting an '=' in place of the inequality sign.
2) DRAW THE GRAPH FOR EACH EQUATION — if the inequality sign is < or > draw a dotted line, but if it's ≥ or ≤ draw a solid line.
3) Work out WHICH SIDE of each line you want — put a point (usually the origin, (0, 0)) into the inequality to see if it's on the correct side of the line.
4) SHADE THE REGION this gives you.

If using the origin doesn't work (e.g. if the origin lies on a line), just pick another point with easy coordinates and use that instead.

EXAMPLE: Shade the region that satisfies all three of the following inequalities:
$x + y < 5$ $\quad y \leq x + 2 \quad$ $y > 1$.

1) **CONVERT EACH INEQUALITY TO AN EQUATION:**
$x + y = 5$, $y = x + 2$ and $y = 1$

2) **DRAW THE GRAPH FOR EACH EQUATION** (see p50)
You'll need a dotted line for $x + y = 5$ and $y = 1$ and a solid line for $y = x + 2$.

3) **WORK OUT WHICH SIDE OF EACH LINE YOU WANT**
This is the fiddly bit. Put $x = 0$ and $y = 0$ (the origin) into each inequality and see if this makes the inequality true or false.

$x + y < 5$:
$x = 0$, $y = 0$ gives $0 < 5$ which is true.
This means the origin is on the correct side of the line.

$y \leq x + 2$:
$x = 0$, $y = 0$ gives $0 \leq 2$ which is true.
So the origin is on the correct side of this line.

$y > 1$:
$x = 0$, $y = 0$ gives $0 > 1$ which is false.
So the origin is on the wrong side of this line.

SHADE THE REGION
You want the region that satisfies all of these:
— below $x + y = 5$ (because the origin is on this side)
— right of $y = x + 2$ (because the origin is on this side)
— above $y = 1$ (because the origin isn't on this side).

Dotted lines mean the region doesn't include the points on the line.

A solid line means the region does include the points on the line

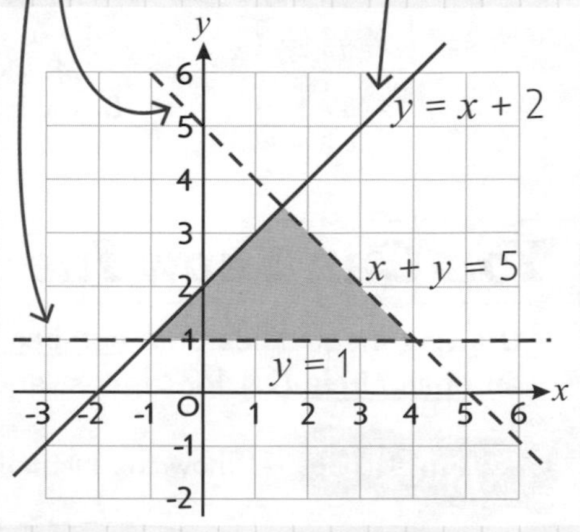

Make sure you read the question carefully — you might be asked to label the region instead of shade it, or just mark on points that satisfy all three inequalities. No point throwing away marks because you didn't read the question properly.

Graphical inequalities — it's a shady business...

Once you've found the region, it's a good idea to pick a point inside it and check that it satisfies ALL the inequalities. Try it out on this Exam Practice Question:

Q1 On a grid, shade the region that satisfies $x \leq 5$, $y > -1$ and $y < x + 1$. [3 marks] 

Trial and Improvement

Trial and improvement is a way of finding an approximate solution to an equation that's too hard to be solved using normal methods. You'll always be told WHEN to use trial and improvement — don't go using it willy-nilly.

Keep Trying Different Values in the Equation

The basic idea of trial and improvement is to keep trying different values of x that are getting closer and closer to the solution. Here's the method to follow:

1) **SUBSTITUTE TWO INITIAL VALUES into the equation that give OPPOSITE CASES.**
   These are usually suggested in the question. 'Opposite cases' means one answer too big, one too small.
2) **Now CHOOSE YOUR NEXT VALUE IN BETWEEN THE PREVIOUS TWO, and SUBSTITUTE it into the equation.**
   Continue this process, always choosing a new value between the two closest opposite cases (and preferably nearer to the one that was closer to the answer).
3) **AFTER ONLY 3 OR 4 STEPS you should have 2 numbers which are to the right degree of accuracy but DIFFER BY 1 IN THE LAST DIGIT.**
   For example, if you had to get your answer to 1 d.p. then you'd eventually end up with say 5.4 and 5.5, with these giving OPPOSITE results of course.
4) **At this point you ALWAYS take the exact middle value to decide which is the answer you want.**
   E.g. for 5.4 and 5.5 you'd try 5.45 to see if the real answer was between 5.4 and 5.45 (so 5.4 to 1 d.p.) or between 5.45 and 5.5 (so 5.5 to 1 d.p.).

You'll be asked for a certain level of accuracy (often 1 d.p.) in the question.

It's a good idea to keep track of your working in a table — see example below.

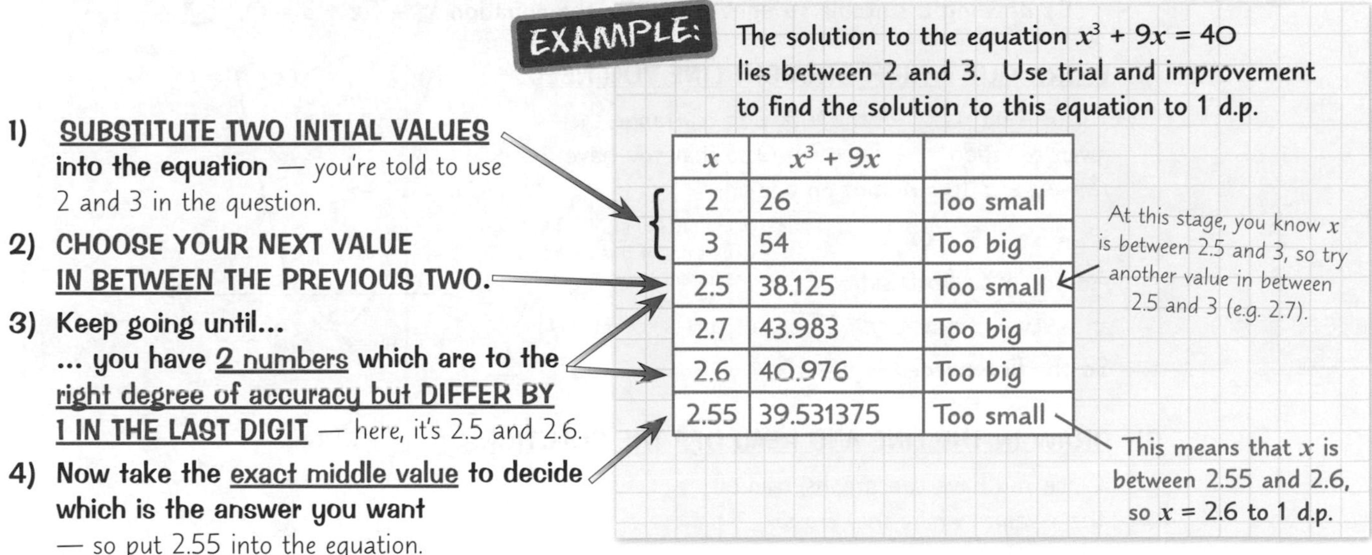

EXAMPLE:
The solution to the equation $x^3 + 9x = 40$ lies between 2 and 3. Use trial and improvement to find the solution to this equation to 1 d.p.
1) SUBSTITUTE TWO INITIAL VALUES into the equation — you're told to use 2 and 3 in the question.
2) CHOOSE YOUR NEXT VALUE IN BETWEEN THE PREVIOUS TWO.
3) Keep going until... ... you have 2 numbers which are to the right degree of accuracy but DIFFER BY 1 IN THE LAST DIGIT — here, it's 2.5 and 2.6.
4) Now take the exact middle value to decide which is the answer you want — so put 2.55 into the equation.
At this stage, you know x is between 2.5 and 3, so try another value in between 2.5 and 3 (e.g. 2.7).
This means that x is between 2.55 and 2.6, so $x = 2.6$ to 1 d.p.

x	$x^3 + 9x$	
2	26	Too small
3	54	Too big
2.5	38.125	Too small
2.7	43.983	Too big
2.6	40.976	Too big
2.55	39.531375	Too small

Make sure you show all your working — otherwise the examiner won't be able to tell what method you've used and you'll lose marks.

Trial and improvement — not a good strategy for lion-taming...

Sorry, it's not the most exciting page in the world — but it is a good way of picking up easy marks in the exam just by putting some numbers into equations. Try it out on these Exam Practice Questions:

Q1 $x^3 + 6x = 69$ has a solution between 3 and 4.
Use trial and improvement to find this solution to 1 d.p. [4 marks] 

Q2 $x^3 - 12x = 100$ has a solution between 5 and 6.
Use trial and improvement to find this solution to 1 d.p. [4 marks] 

Simultaneous Equations and Graphs

You can use graphs to solve simultaneous equations — just plot the graph of each equation, and the solutions are the points where the graphs cross (you can usually just read off the coordinates from the graph).

Plot Both Graphs and See Where They Cross

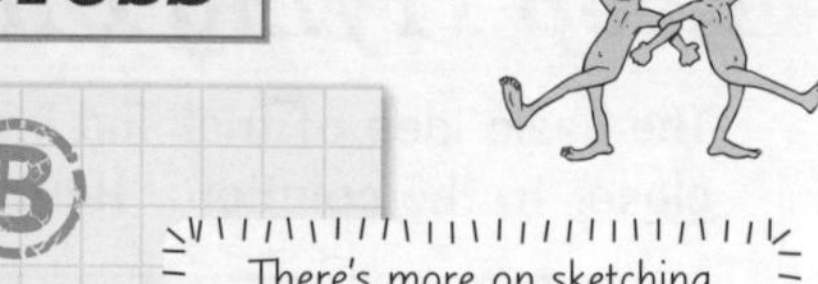

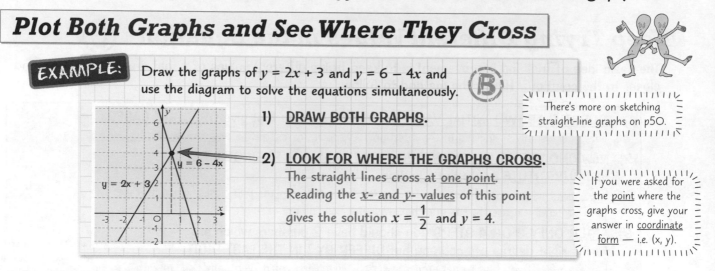

EXAMPLE: Draw the graphs of $y = 2x + 3$ and $y = 6 - 4x$ and use the diagram to solve the equations simultaneously. (B)

There's more on sketching straight-line graphs on p50.

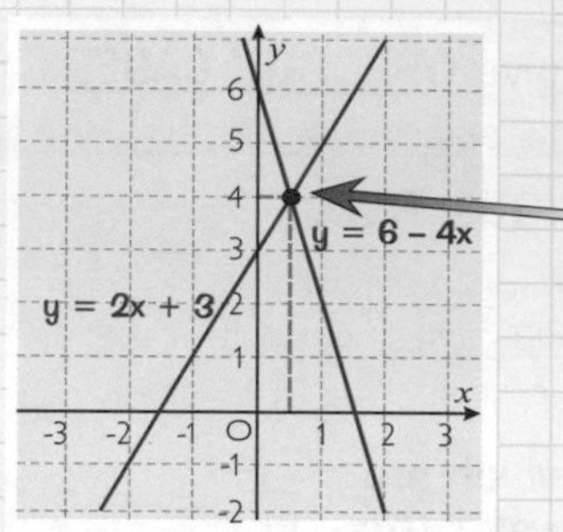

1) DRAW BOTH GRAPHS.

2) LOOK FOR WHERE THE GRAPHS CROSS.

The straight lines cross at one point. Reading the x- and y- values of this point gives the solution $x = \frac{1}{2}$ and $y = 4$.

If you were asked for the point where the graphs cross, give your answer in coordinate form — i.e. (x, y).

The point at which the two graphs cross is actually the solution you'd find if you set the two equations equal to each other (so in the first example, you're actually solving $2x + 3 = 6 - 4x$). This fact comes in handy for the next (trickier) example.

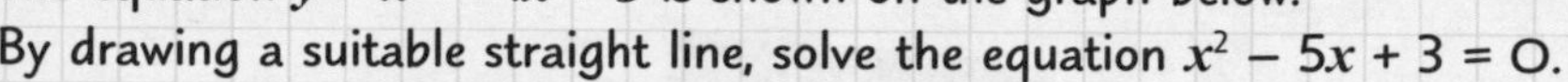

EXAMPLE: The equation $y = x^2 - 4x + 3$ is shown on the graph below. By drawing a suitable straight line, solve the equation $x^2 - 5x + 3 = 0$. (A*)

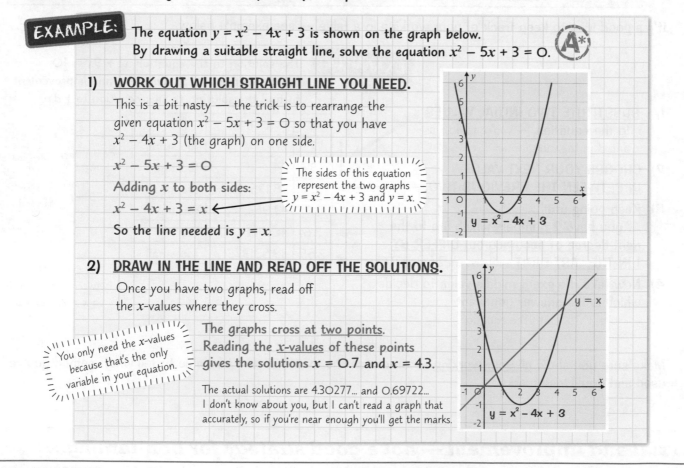

1) WORK OUT WHICH STRAIGHT LINE YOU NEED.

This is a bit nasty — the trick is to rearrange the given equation $x^2 - 5x + 3 = 0$ so that you have $x^2 - 4x + 3$ (the graph) on one side.

$x^2 - 5x + 3 = 0$

Adding x to both sides:

$x^2 - 4x + 3 = x$

So the line needed is $y = x$.

The sides of this equation represent the two graphs $y = x^2 - 4x + 3$ and $y = x$.

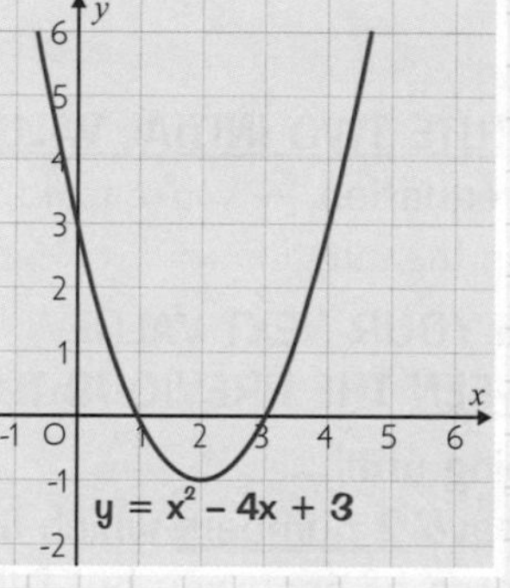

2) DRAW IN THE LINE AND READ OFF THE SOLUTIONS.

Once you have two graphs, read off the x-values where they cross.

The graphs cross at two points. Reading the x-values of these points gives the solutions $x = 0.7$ and $x = 4.3$.

You only need the x-values because that's the only variable in your equation.

The actual solutions are 4.30277... and 0.69722... I don't know about you, but I can't read a graph that accurately, so if you're near enough you'll get the marks.

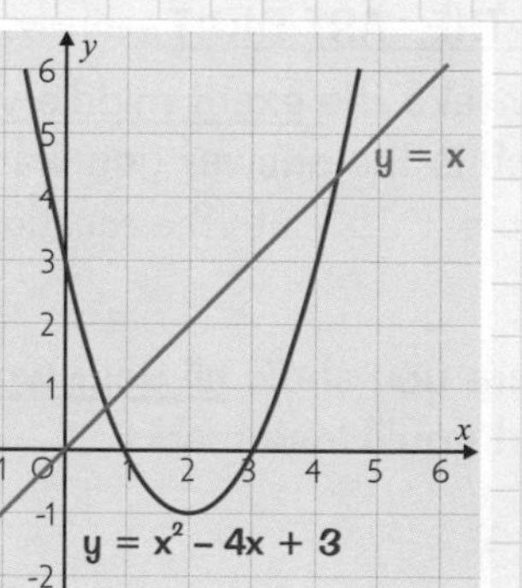

What do you call a giraffe with no eyes? A graph...

Have a go at these Exam Practice Questions to make sure you're a graph expert:

Q1 By sketching the graphs, find the solutions of the simultaneous equations $y = 4x - 4$ and $y = 6 - x$. [3 marks] (B)

Q2 Clare wants to use the graph of $y = x^3 + 4x^2 - 3x + 2$ to solve the equation $x^3 + 4x^2 - 3x - 1 = 0$. Find the equation of the straight line she should draw on the graph. [2 marks] (A*)

Simultaneous Equations

You've seen the easy way to solve simultaneous equations using graphs. Now it's time to learn the less fun algebra methods. The rules are really quite simple, but you must follow ALL the steps, in the right order, and treat them as a strict method.

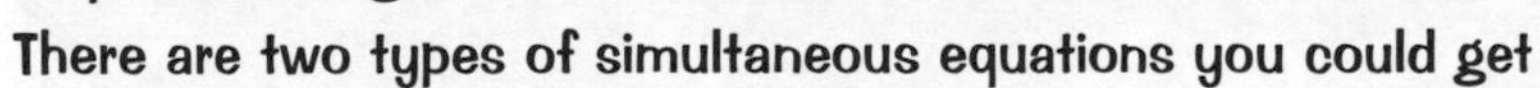

There are two types of simultaneous equations you could get
— EASY ONES (where both equations are linear) and TRICKY ONES (where one's quadratic).

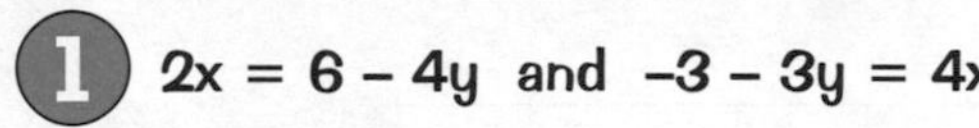

1. $2x = 6 - 4y$ and $-3 - 3y = 4x$

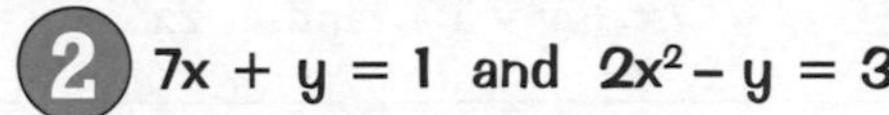

2. $7x + y = 1$ and $2x^2 - y = 3$

1 *Six Steps for Easy Simultaneous Equations* (B)

EXAMPLE: Solve the simultaneous equations $2x = 6 - 4y$ and $-3 - 3y = 4x$

1. Rearrange both equations into the form $ax + by = c$, and label the two equations ① and ②.

a, b and c are numbers (which can be negative)

$2x + 4y = 6$ — ①
$-4x - 3y = 3$ — ②

2. Match up the numbers in front (the 'coefficients') of either the x's or y's in both equations. You may need to multiply one or both equations by a suitable number. Relabel them ③ and ④.

① × 2: $4x + 8y = 12$ — ③
$-4x - 3y = 3$ — ④

3. Add or subtract the two equations to eliminate the terms with the same coefficient.

③ + ④ $0x + 5y = 15$

If the coefficients have the same sign (both +ve or both –ve) then subtract. If the coefficients have opposite signs (one +ve and one –ve) then add.

4. Solve the resulting equation.

$5y = 15 \Rightarrow \underline{y = 3}$

5. Substitute the value you've found back into equation ① and solve it.

Sub $y = 3$ into ①: $2x + (4 \times 3) = 6 \Rightarrow 2x + 12 = 6 \Rightarrow 2x = -6 \Rightarrow \underline{x = -3}$

6. Substitute both these values into equation ② to make sure it works. If it doesn't then you've done something wrong and you'll have to do it all again.

Sub x and y into ②: $(-4 \times -3) - (3 \times 3) = 12 - 9 = 3$, which is right, so it's worked.
So the solutions are: $x = -3$, $y = 3$

Sunday morning, lemon squeezy and simultaneous linear equations...

You need to learn the 6 steps on this page. When you think you've got them, try them out on these Exam Practice Questions.

Q1 Solve the simultaneous equations $3x - 2y = 9$ and $2x + 3y = 19$. [4 marks] (B)

Q2 Find x and y given that $2x - 10 = 4y$ and $3y = 4x - 15$. [4 marks] (B)

Simultaneous Equations

② Seven Steps for TRICKY Simultaneous Equations

EXAMPLE: Solve these two equations simultaneously:

$7x + y = 1$ and $2x^2 - y = 3$

1. Rearrange the quadratic equation so that you have the non-quadratic unknown on its own. Label the two equations ① and ②.

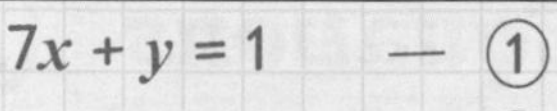

$7x + y = 1$ — ①

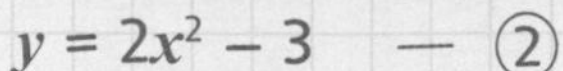

$y = 2x^2 - 3$ — ②

2. Substitute the quadratic expression into the other equation. You'll get another equation — label it ③.

$7x + y = 1$ — ①

$y = 2x^2 - 3$ — ②

$\Rightarrow 7x + (2x^2 - 3) = 1$ — ③

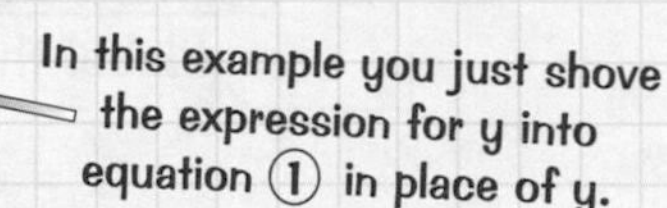

In this example you just shove the expression for y into equation ① in place of y.

3. Rearrange to get a quadratic equation. And guess what... You've got to solve it.

$2x^2 + 7x - 4 = 0$

$(2x - 1)(x + 4) = 0$

So $2x - 1 = 0$ OR $x + 4 = 0$

$x = 0.5$ OR $x = -4$

Remember — if it won't factorise, you can either use the formula or complete the square. Have a look at p35-36 for more details.

4. Stick the first value back in one of the original equations (pick the easy one).

① $7x + y = 1$

Substitute in $x = 0.5$: $3.5 + y = 1$, so $y = 1 - 3.5 = -2.5$

5. Stick the second value back in the same original equation (the easy one again).

① $7x + y = 1$

Substitute in $x = -4$: $-28 + y = 1$, so $y = 1 + 28 = 29$

6. Substitute both pairs of answers back into the other original equation to check they work.

② $y = 2x^2 - 3$

Substitute in $x = 0.5$: $y = (2 \times 0.25) - 3 = -2.5$ — jolly good.

Substitute in $x = -4$: $y = (2 \times 16) - 3 = 29$ — smashing.

7. Write the pairs of answers out again, clearly, at the bottom of your working.

The two pairs of solutions are: $x = 0.5, y = -2.5$ and $x = -4, y = 29$

Simultaneous pain and pleasure — it must be algebra...

Don't make the mistake of thinking that there are 4 separate solutions — you end up with 2 pairs of solutions to the simultaneous equations. Now try these:

Q1 Solve the simultaneous equations $y = x^2 + 4$ and $y - 6x - 4 = 0$ [6 marks] 

Q2 Solve the simultaneous equations $y = 2 - 3x$ and $y = x^2 - 2$ [6 marks]

Direct and Inverse Proportion

Proportion questions involve two variables (often x and y) which are linked in some way. You'll have to figure out the relationship between them, and use this to find values of x or y, given one value.

Simple Proportions

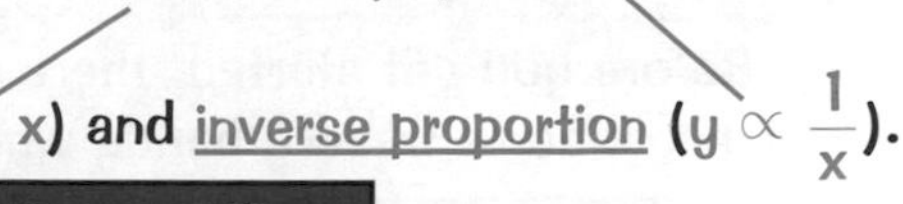

The easiest types of proportions you might get are direct proportion ($y \propto x$) and inverse proportion ($y \propto \frac{1}{x}$).

Direct Proportion BOTH INCREASE TOGETHER

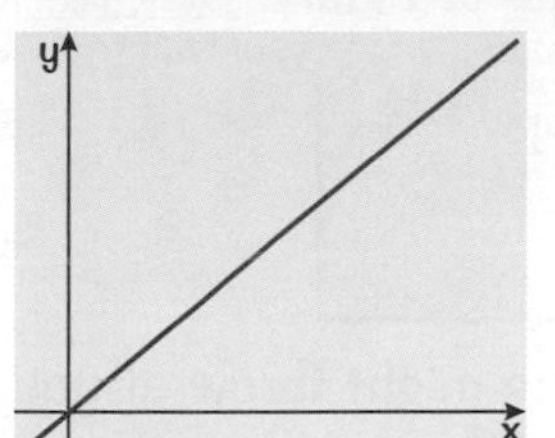

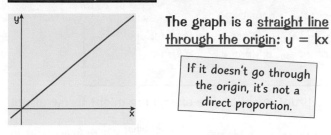

The graph is a straight line through the origin: $y = kx$

If it doesn't go through the origin, it's not a direct proportion.

Inverse Proportion One INCREASES, one DECREASES

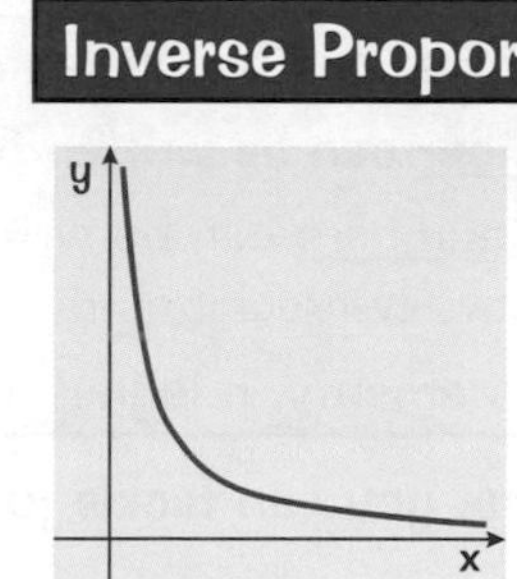

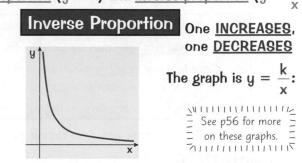

The graph is $y = \frac{k}{x}$:

See p56 for more on these graphs.

Trickier Proportions

More complex proportions involve y varying proportionally or inversely to some function of x, e.g. x^2, x^3, $\sqrt{x}$ etc. You can always turn a proportion statement into an equation by replacing '∝' with '= k' like this:

	Proportionality	Equation
'y is proportional to the square of x'	$y \propto x^2$	$y = kx^2$
't is proportional to the square root of h'	$t \propto \sqrt{h}$	$t = k\sqrt{h}$
'D varies with the cube of t'	$D \propto t^3$	$D = kt^3$
'V is inversely proportional to r cubed'	$V \propto \frac{1}{r^3}$	$V = \frac{k}{r^3}$

k is just some constant (unknown number).

Handling Questions on Proportion

1) Convert the sentence into a proportionality.
2) Replace '∝' with '= k' to make an equation (as above).
3) Find a pair of values of x and y somewhere in the question, and substitute them into the equation with the sole purpose of finding k.
4) Put the value of k into the equation and it's now ready to use, e.g. $y = 3x^2$.
5) Inevitably, they'll ask you to find y, having given you a value for x (or vice versa).

Once you've got it in the form of an equation with k, the rest is easy.

EXAMPLE: G is inversely proportional to the square root of H. When G = 2, H = 16. Find an equation for G in terms of H, and use it to work out the value of G when H = 36.

1) Convert to a proportionality. $G \propto \frac{1}{\sqrt{H}}$

2) Replace ∝ with '= k' to form an equation. $G = \frac{k}{\sqrt{H}}$

3) Use the values of G and H (2 and 16) to find k. $2 = \frac{k}{\sqrt{16}} = \frac{k}{4} \Rightarrow k = 8$

4) Put the value of k back into the equation. $G = \frac{8}{\sqrt{H}}$ ← This is the equation for G in terms of H.

5) Use your equation to find the value of G.

$$G = \frac{8}{\sqrt{H}} = \frac{8}{\sqrt{36}} = \frac{8}{6} = \frac{4}{3}$$

Joy ∝ 1/algebra...

Learn the 5-step method for dealing with proportionality questions, then have a go at this one:

Q1 *t* is proportional to the square of *s*, and when $t = 27$, $s = 3$.
Find the value of *s* when $t = 48$ (given that *s* is positive). [4 marks]

Proof

I'm not going to lie — proof questions can be a bit terrifying. They're usually not as bad as they seem though — you often just have to do a bit of rearranging to show that one thing is equal to another.

Use Algebra to Show That Two Things are Equal

Before you get started, there are a few things you need to know — they'll come in very handy when you're trying to prove things.

- Any even number can be written as $2n$ — i.e. 2 × something.
- Any odd number can be written as $2n + 1$ — i.e. 2 × something + 1.
- Consecutive numbers can be written as n, $n + 1$, $n + 2$ etc. — you can apply this to e.g. consecutive even numbers too (they'd be written as $2n$, $2n + 2$, $2n + 4$).

In all of these statements, n is just any integer.

This can be extended to multiples of other numbers too — e.g. to prove that something is a multiple of 5, show that it can be written as 5 × something.

Armed with these facts, you can tackle just about any proof question the examiners might throw at you.

EXAMPLE: Prove that the sum of any three odd numbers is odd.

So what you're trying to do here is show that the sum of three odd numbers can be written as (2 × something) + 1.

Take three odd numbers:
$2a + 1$, $2b + 1$ and $2c + 1$
(they don't have to be consecutive)

Add them together:

$$2a + 1 + 2b + 1 + 2c + 1 = 2a + 2b + 2c + 2 + 1$$
$$= 2(a + b + c + 1) + 1$$
$$= 2n + 1 \text{ (where } n = a + b + c + 1)$$

You'll see why I've written 3 as 2 + 1 in a second.

So the sum of any three odd numbers is odd.

EXAMPLE: Prove that $(n + 3)^2 - (n - 2)^2 \equiv 5(2n + 1)$.

Take one side of the equation and play about with it until you get the other side:

$$\text{LHS: } (n + 3)^2 - (n - 2)^2 \equiv n^2 + 6n + 9 - (n^2 - 4n + 4)$$
$$\equiv n^2 + 6n + 9 - n^2 + 4n - 4$$
$$\equiv 10n + 5$$
$$\equiv 5(2n + 1) = \text{RHS} ✓$$

$\equiv$ is the identity symbol, and means that two things are identically equal to each other. So $a + b \equiv b + a$ is true for all values of a and b (unlike an equation, which is only true for certain values).

Disprove Things by Finding an Example That Doesn't Work

If you're asked to prove a statement isn't true, all you have to do is find one example that the statement doesn't work for — this is known as disproof by counter example.

EXAMPLE: Ross says "the difference between any two consecutive square numbers is always a prime number". Prove that Ross is wrong.

Just keep trying pairs of consecutive square numbers (i.e. 1^2 and 2^2) until you find one that doesn't work:
1 and 4 — difference = 3 (a prime number)
4 and 9 — difference = 5 (a prime number)
9 and 16 — difference = 7 (a prime number)
16 and 25 — difference = 9 (NOT a prime number) so Ross is wrong.

You don't have to go through loads of examples if you can spot one that's wrong straightaway — you could go straight to 16 and 25.

Prove that maths isn't fun...

The only way to get on top of proof questions is practice — so start with these Exam Practice Questions:

Q1 Prove that the sum of two consecutive even numbers is even. [3 marks] (A*)

Q2 Prove that $(a + b)(a - b) \equiv a^2 - b^2$ [2 marks] (A*)

Revision Questions for Section Two

There's no denying, Section Two is grisly grimsdike algebra — so check now how much you've learned.

- Try these questions and tick off each one when you get it right.
- When you've done all the questions for a topic and are completely happy with it, tick off the topic.

Sequences (p23)

1) Find the expression for the nth term in the following sequences: a) 7, 9, 11, 13 b) 11, 8, 5, 2.
2) The nth term of a sequence is given by $n^2 + 7$. Is 32 a term in this sequence?

Algebra (p24-32)

3) Simplify the following: a) $x^3 \times x^6$ b) $y^7 \div y^5$ c) $(z^3)^4$
4) Simplify by collecting like terms: $3x + 2y - 5 - 6y + 2x$
5) Multiply out these brackets: a) $3(2x + 1)$ b) $(x + 2)(x - 3)$
6) Factorise: a) $x^2 - 16y^2$ b) $49 - 81p^2q^2$ c) $12x^2 - 48y^2$
7) Simplify the following: a) $\sqrt{27}$ b) $\sqrt{125} \div \sqrt{5}$
8) Solve these equations: a) $5(x + 2) = 8 + 4(5 - x)$ b) $x^2 - 21 = 3(5 - x^2)$
9) Make p the subject of these: a) $\frac{p}{p+y} = 4$ b) $\frac{1}{p} = \frac{1}{q} + \frac{1}{r}$

Quadratics (p33-37)

10) Solve the following by factorising them first: a) $x^2 + 9x + 18 = 0$ b) $5x^2 - 17x - 12 = 0$
11) Write down the quadratic formula.
12) Find the solutions of these equations (to 2 d.p.) using the quadratic formula:
    a) $x^2 + x - 4 = 0$ b) $5x^2 + 6x = 2$ c) $(2x + 3)^2 = 15$
13) Find the exact solutions of these equations by completing the square:
    a) $x^2 + 12x + 15 = 0$ b) $x^2 - 6x = 2$

Algebraic Fractions (p38)

14) Write $\frac{2}{x+3} + \frac{1}{x-1}$ as a single fraction.

Inequalities (p39-40)

15) Solve this inequality: $4x + 3 \leq 6x + 7$
16) Show on a graph the region described by these conditions: $x + y \leq 6$, $y > 0.5$, $y \leq 2x - 2$

Trial and Improvement (p41)

17) Given that $x^3 + 8x = 103$ has a solution between 4 and 5, find this solution to 1 d.p.

Simultaneous Equations (p42-44)

18) How can you find the solutions to a pair of simultaneous equations using their graphs?
19) Solve these simultaneous equations: $y = 3x + 4$ and $x^2 + 2y = 0$

Direct and Inverse Proportion (p45)

20) Write the following statement as an equation: "y is proportional to the square of x".
21) p is proportional to the cube of q. When p = 9, q = 3. Find the value of p when q = 6.

Proof (p46)

22) Prove that the product of an odd number and an even number is even.

X, Y and Z Coordinates

What could be more fun than points in one quadrant? Points in four quadrants and 3D space, that's what...

The Four Quadrants

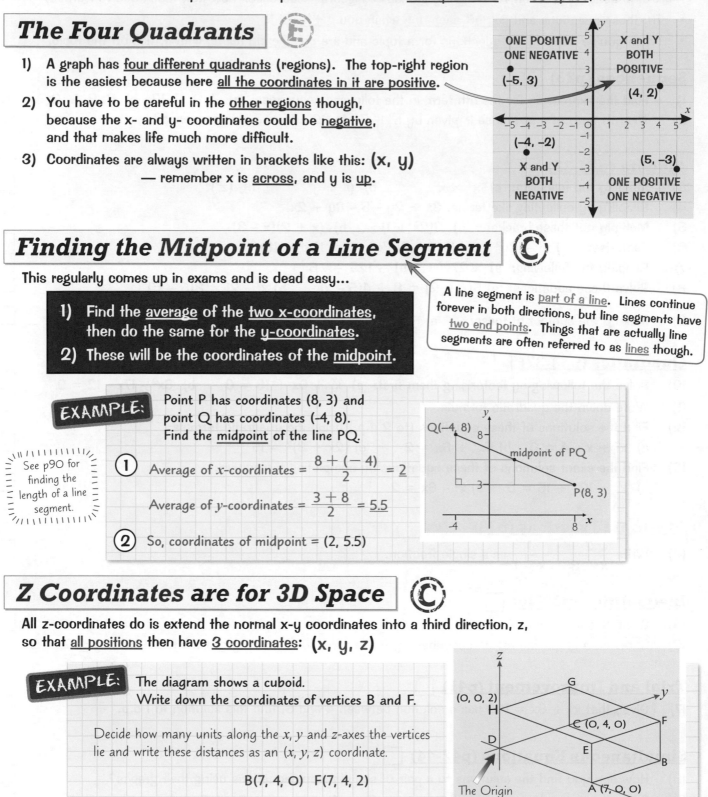

1) A graph has four different quadrants (regions). The top-right region is the easiest because here all the coordinates in it are positive.
2) You have to be careful in the other regions though, because the x- and y- coordinates could be negative, and that makes life much more difficult.
3) Coordinates are always written in brackets like this: (x, y) — remember x is across, and y is up.

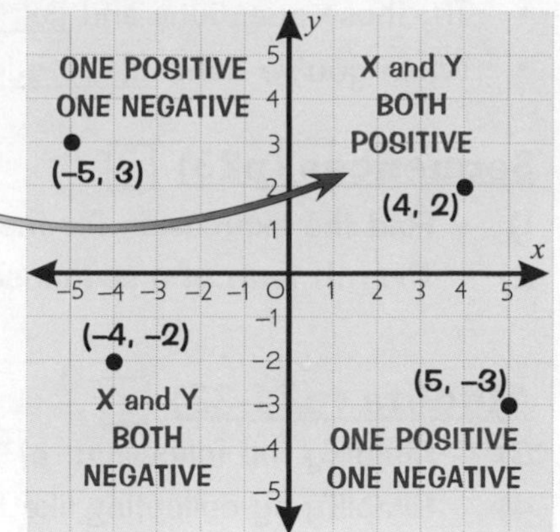

Finding the Midpoint of a Line Segment

This regularly comes up in exams and is dead easy...

1) Find the average of the two x-coordinates, then do the same for the y-coordinates.
2) These will be the coordinates of the midpoint.

A line segment is part of a line. Lines continue forever in both directions, but line segments have two end points. Things that are actually line segments are often referred to as lines though.

EXAMPLE: Point P has coordinates (8, 3) and point Q has coordinates (-4, 8). Find the midpoint of the line PQ.

See p90 for finding the length of a line segment.

① Average of x-coordinates = $\frac{8 + (-4)}{2} = 2$

Average of y-coordinates = $\frac{3 + 8}{2} = 5.5$

② So, coordinates of midpoint = (2, 5.5)

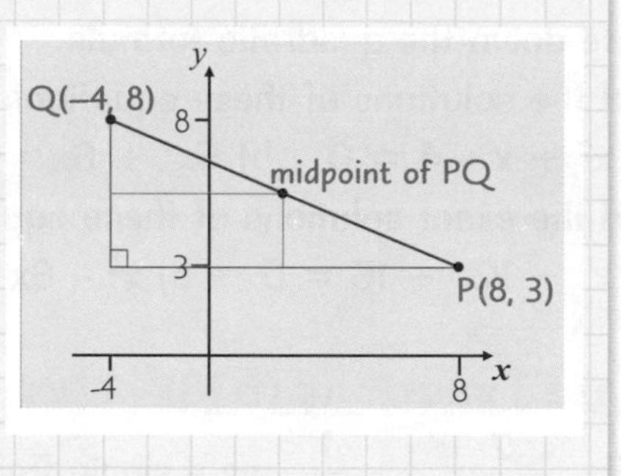

Z Coordinates are for 3D Space

All z-coordinates do is extend the normal x-y coordinates into a third direction, z, so that all positions then have 3 coordinates: (x, y, z)

EXAMPLE: The diagram shows a cuboid. Write down the coordinates of vertices B and F.

Decide how many units along the x, y and z-axes the vertices lie and write these distances as an (x, y, z) coordinate.

B(7, 4, 0) F(7, 4, 2)

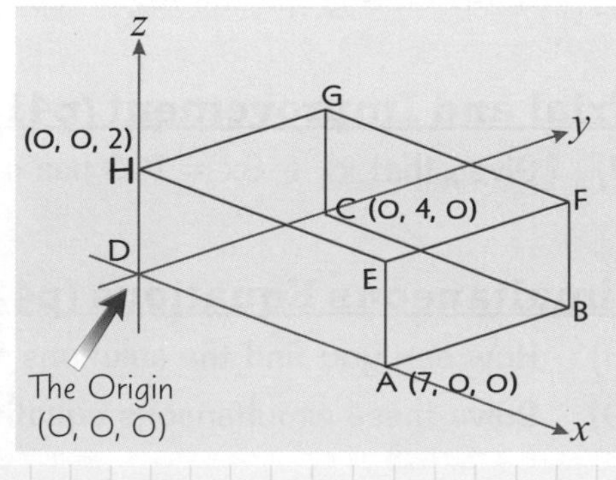

A 3D shape, drawn on 2D paper? What sorcery is this...

Learn how to find the midpoint of a line segment and how to use coordinates in three dimensions.

Q1 Point A has coordinates (–5, –2) and point B has coordinates (6, 0). Find the midpoint of the line AB. [2 marks] 

Q2 Write down the coordinates of point G from the 3D diagram above. [1 mark] 

Straight-Line Graphs

If you thought I-spy was a fun game, wait 'til you play 'recognise the straight-line graph from the equation'.

Horizontal and Vertical lines: 'x = a' and 'y = a'

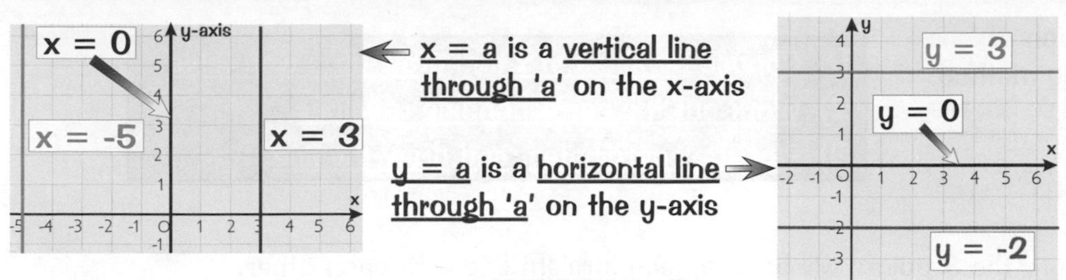

$x = a$ is a vertical line through 'a' on the x-axis

$y = a$ is a horizontal line through 'a' on the y-axis

A common error is to mix up $x = 3$ and $y = 3$, etc. Remember — all the points on $x = 3$ have an x-coordinate of 3, and all the points on $y = 3$ have a y-coordinate of 3.

The Main Diagonals: 'y = x' and 'y = –x'

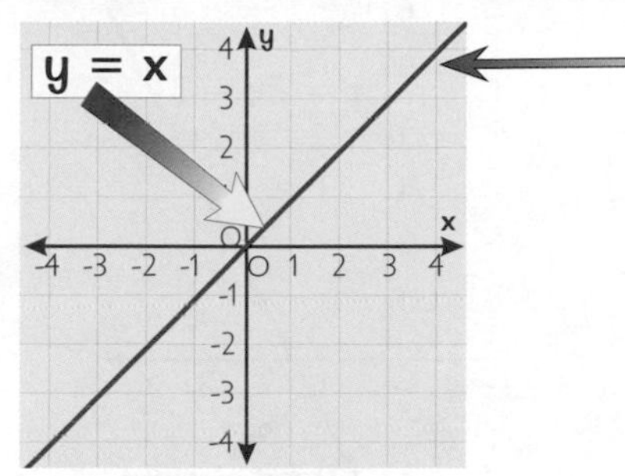

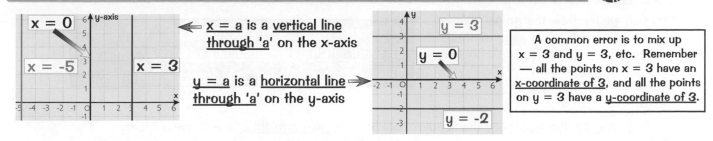

'$y = x$' is the main diagonal that goes UPHILL from left to right.

The x- and y-coordinates of each point are the same.

'$y = -x$' is the main diagonal that goes DOWNHILL from left to right.

The x- and y-coordinates of each point are negatives of each other, e.g. (–4, 4).

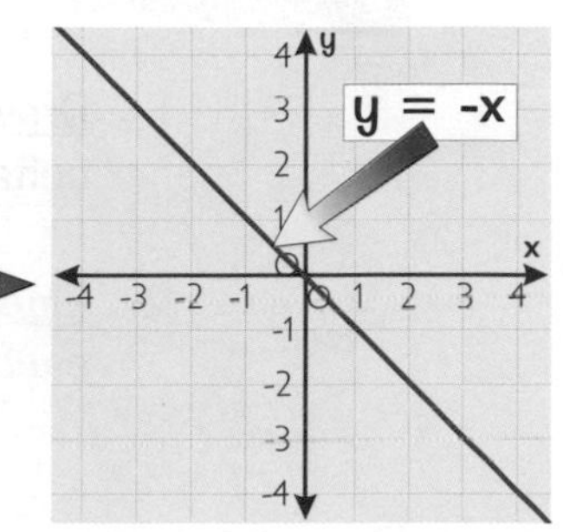

Other Sloping Lines Through the Origin: 'y = ax' and 'y = –ax'

$y = ax$ and $y = -ax$ are the equations for A SLOPING LINE THROUGH THE ORIGIN.

The value of 'a' (known as the gradient) tells you the steepness of the line. The bigger 'a' is, the steeper the slope. A MINUS SIGN tells you it slopes DOWNHILL.

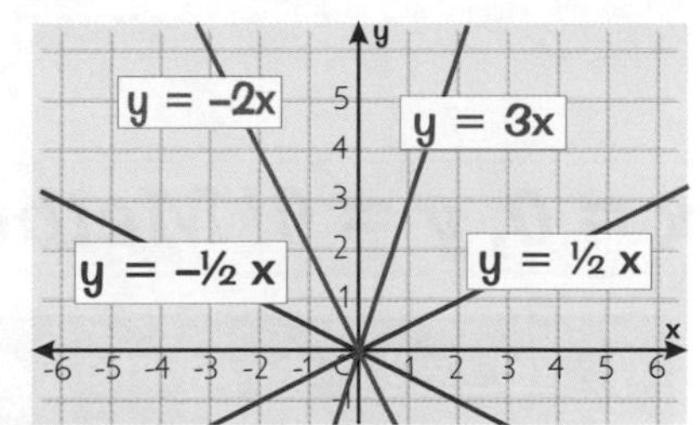

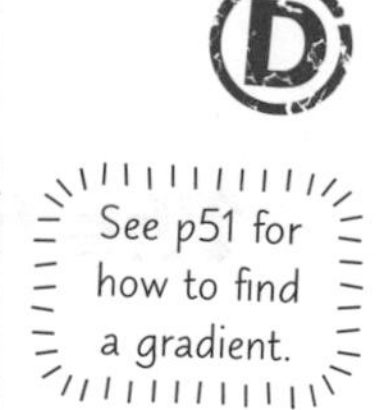
See p51 for how to find a gradient.

Learn to Spot Straight Lines from their Equations

All straight-line equations just contain 'something x, something y and a number'.

EXAMPLE: Decide whether each of the following are equations of straight lines.

$2y - 4x = 7$ $\quad y = x^2 + 3$ $\quad xy + 3 = 0$ $\quad 6y - 8 = x$ $\quad \frac{2}{y} - \frac{1}{x} = 7$

Straight lines: $2y - 4x = 7$, $6y - 8 = x$ ← These equations only have something x, something y and a number. These 'terms' can be added or subtracted in any order.

Not straight lines: $y = x^2 + 3$, $xy + 3 = 0$, $\frac{2}{y} - \frac{1}{x} = 7$ ← 'x^2', 'xy', '2/y' and '1/x' mean that these aren't straight-line equations.

My favourite line's y = 3x — it gets the ladies every time...

It's definitely worth learning all the graphs on this page. Now try this question.

Q1 Write down the equation of the straight line which passes through the points (–1, –2) and (3, –2). [1 mark]

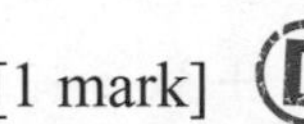

Plotting Straight-Line Graphs

Drawing straight-line graphs is very likely to come up in the exam. We'll cover two methods on this page:

The 'Table of 3 Values' Method

You can easily draw the graph of any equation using this easy method:

1) Choose 3 values of x and draw up a wee table,
2) Work out the corresponding y-values,
3) Plot the coordinates, and draw the line.

Don't forget to use a ruler to draw your line — you can lose exam marks if you don't.

If it's a straight-line equation, the 3 points will be in a dead straight line with each other. If they aren't, you need to go back and CHECK YOUR WORKING.

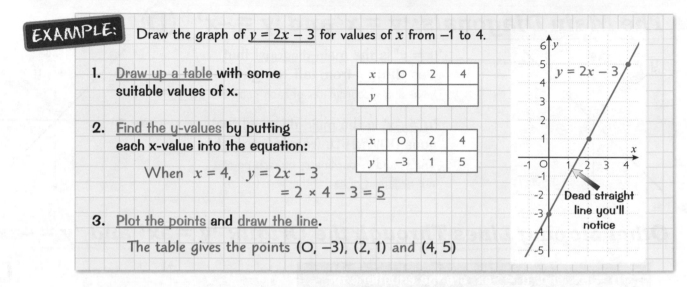

EXAMPLE: Draw the graph of $y = 2x - 3$ for values of x from –1 to 4.

1. Draw up a table with some suitable values of x.

x	0	2	4
y			

2. Find the y-values by putting each x-value into the equation:

x	0	2	4
y	–3	1	5

When $x = 4$, $y = 2x - 3$

$= 2 \times 4 - 3 = 5$

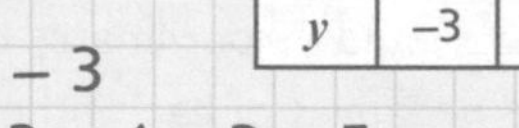

3. Plot the points and draw the line.

The table gives the points (0, –3), (2, 1) and (4, 5)

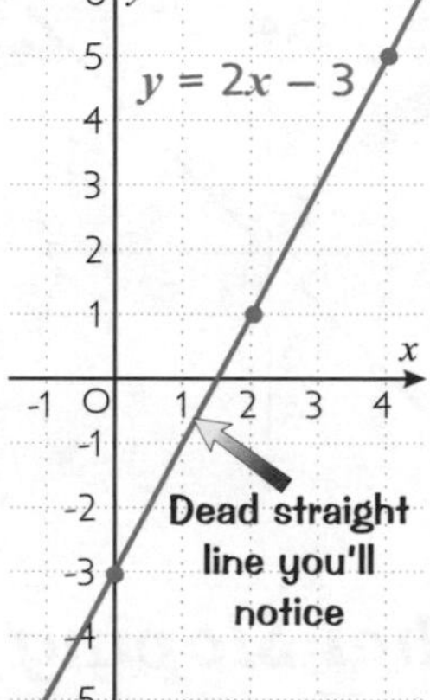

The 'x = 0, y = 0' Method

1) Set x=0 in the equation, and find y — this is where it crosses the y-axis.
2) Set y=0 in the equation and find x — this is where it crosses the x-axis.
3) Plot these two points and join them up with a straight line — and just hope it should be a straight line, since with only 2 points you can't really tell, can you!

EXAMPLE: Draw the graph of $3x + 5y = 15$ between $x = -1$ and $x = 6$.

Only doing 2 points is risky unless you're sure the equation is definitely a straight line — but then that's the big thrill of living life on the edge...

Putting $x = 0$ gives "$5y = 15$" $\Rightarrow$ $y = 3$

Putting $y = 0$ gives "$3x = 15$" $\Rightarrow$ $x = 5$

So plot (0, 3) and (5, 0) on the graph and join them up with a straight line.

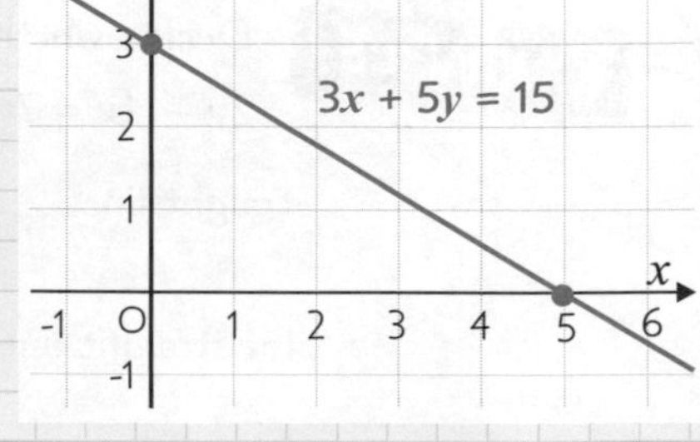

"No!" cried y "You won't cross me again" — extract from a Maths thriller...

Learn the details of these two easy methods. Then you'll be ready for some Exam Practice Questions.

Q1 Draw the graph of $y = 4 + x$ for values of x from –6 to 2. [3 marks] (D)

Q2 Draw the graph of $4y + 3x = 12$ between $x = -4$ and $x = 6$. [3 marks] (C)

Finding the Gradient

Time to hit the slopes. Well, find them anyway...

Finding the Gradient

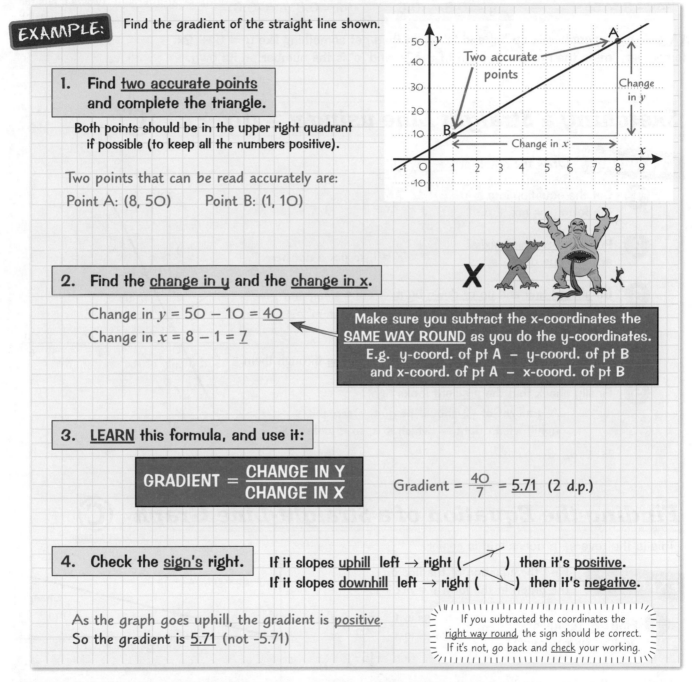

The gradient of a line is a measure of its slope. The bigger the number, the steeper the line.

EXAMPLE: Find the gradient of the straight line shown.

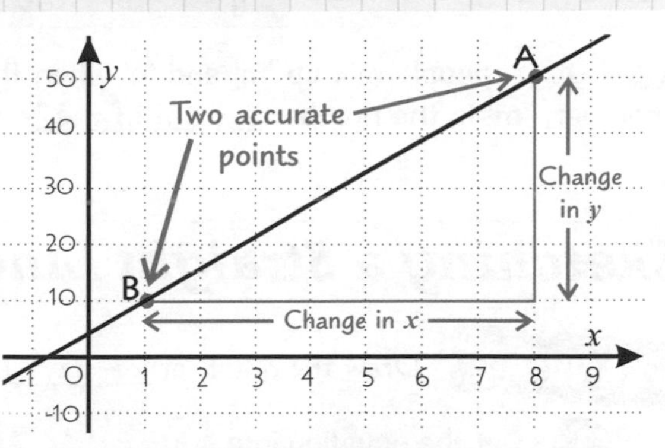

1. Find two accurate points and complete the triangle.

Both points should be in the upper right quadrant if possible (to keep all the numbers positive).

Two points that can be read accurately are:

Point A: (8, 50) Point B: (1, 10)

2. Find the change in y and the change in x.

Change in $y = 50 - 10 = 40$

Change in $x = 8 - 1 = 7$

Make sure you subtract the x-coordinates the SAME WAY ROUND as you do the y-coordinates. E.g. y-coord. of pt A − y-coord. of pt B and x-coord. of pt A − x-coord. of pt B

3. LEARN this formula, and use it:

$$\text{GRADIENT} = \frac{\text{CHANGE IN Y}}{\text{CHANGE IN X}}$$

Gradient $= \frac{40}{7} = 5.71$ (2 d.p.)

4. Check the sign's right.

If it slopes uphill left → right (↗) then it's positive.

If it slopes downhill left → right (↘) then it's negative.

As the graph goes uphill, the gradient is positive.
So the gradient is 5.71 (not -5.71)

If you subtracted the coordinates the right way round, the sign should be correct. If it's not, go back and check your working.

Step 4 catches a lot of folks out in exams. It's easy to divide, get a nice positive number and breathe a sigh of relief. You've got to check that sign.

Finding gradients is often an uphill battle...

Learn the four steps for finding a gradient then have a bash at this Practice Exam Question. Take care — you might not be able to pick two points with nice, positive coordinates. Fun times ahoy.

Q1 Find the gradient of the line shown on the right. [2 marks] (C)

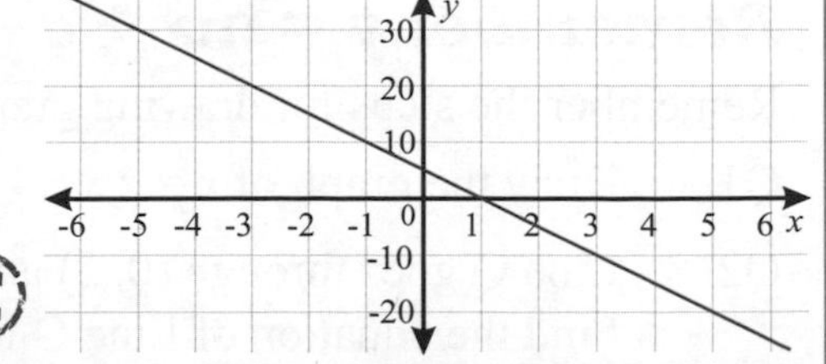

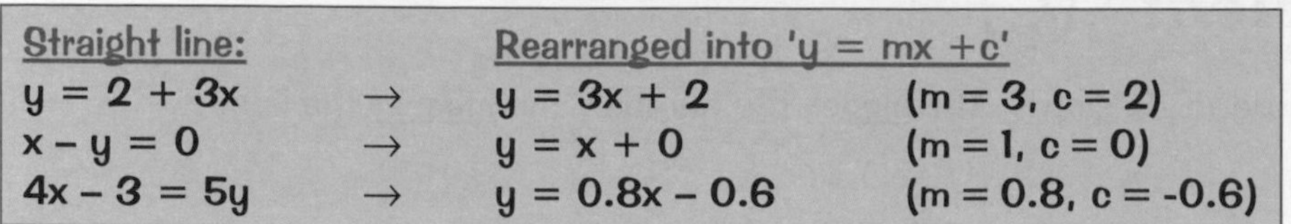

"y = mx + c"

Using '$y = mx + c$' is perhaps the 'proper' way of dealing with straight-line equations, and it's a nice trick if you can do it. The first thing you have to do though is rearrange the equation into the standard format like this:

Straight line:		Rearranged into '$y = mx + c$'	
$y = 2 + 3x$	→	$y = 3x + 2$	($m = 3$, $c = 2$)
$x - y = 0$	→	$y = x + 0$	($m = 1$, $c = 0$)
$4x - 3 = 5y$	→	$y = 0.8x - 0.6$	($m = 0.8$, $c = -0.6$)

REMEMBER:
'm' = gradient of the line.
'c' = 'y-intercept' (where it hits the y-axis)

WATCH OUT: people mix up 'm' and 'c' when they get something like $y = 5 + 2x$. Remember, 'm' is the number in front of the 'x' and 'c' is the number on its own.

Sketching a Straight Line using y = mx + c

EXAMPLE: Draw the graph of $y - 2x = 1$.

1. Get the equation into the form $y = mx + c$. $y - 2x = 1 \rightarrow y = 2x + 1$
2. Put a dot on the y-axis at the value of c. 'c' = 1, so put a dot here.
3. Go along 1 unit and up or down by m. Make another dot, then repeat this step a few times in both directions. Go 1 along and 2 up because 'm' = +2. (If 'm' was −2, you'd go down.)
4. When you have 4 or 5 dots, draw a straight line through them.
5. Finally check that the gradient looks right. A gradient of +2 should be quite steep and uphill left to right — which it is, so it looks OK.

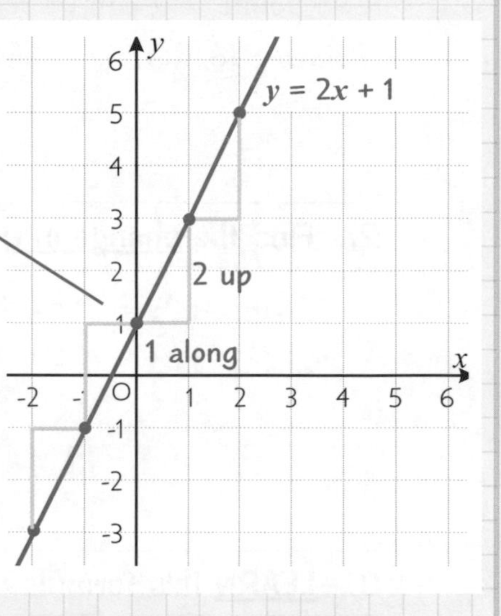

Finding the Equation of a Straight-Line Graph

This is the reverse process and is easier.

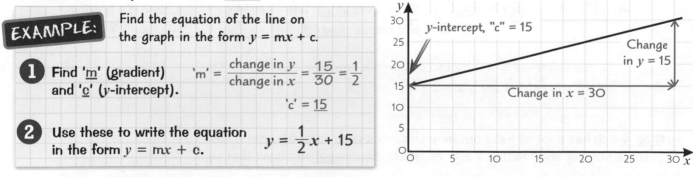

EXAMPLE: Find the equation of the line on the graph in the form $y = mx + c$.

1. Find 'm' (gradient) and 'c' (y-intercept). $'m' = \frac{\text{change in } y}{\text{change in } x} = \frac{15}{30} = \frac{1}{2}$ 'c' = 15
2. Use these to write the equation in the form $y = mx + c$. $y = \frac{1}{2}x + 15$

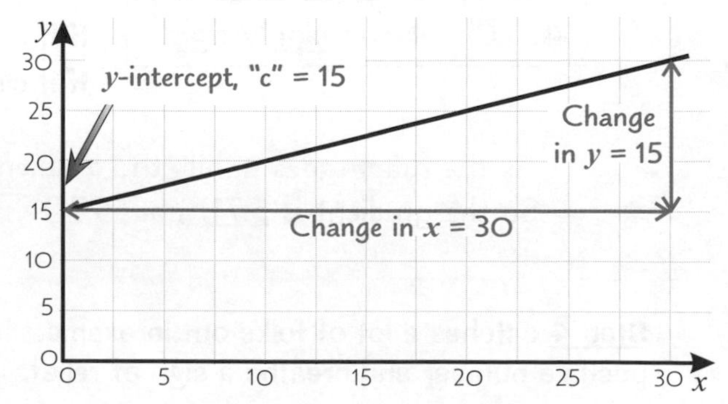

Remember y = mx + c — it'll keep you on the straight and narrow...

Remember the steps for drawing graphs and finding the equations. And try these questions.

Q1 Draw the graph of $x = 2y + 4$ for values of x between –4 and 4. [3 marks] (C)

Q2 Line Q goes through (0, 5) and (4, 7).
Find the equation of Line Q in the form $y = mx + c$. [3 marks] (C)

Parallel and Perpendicular Lines

You've just seen how to write the <u>equation of a straight line</u>. Well, you also have to be able to write the equation of a line that's <u>parallel</u> or <u>perpendicular</u> to the straight line you're given. The fun just never ends.

Parallel Lines Have the Same Gradient (B)

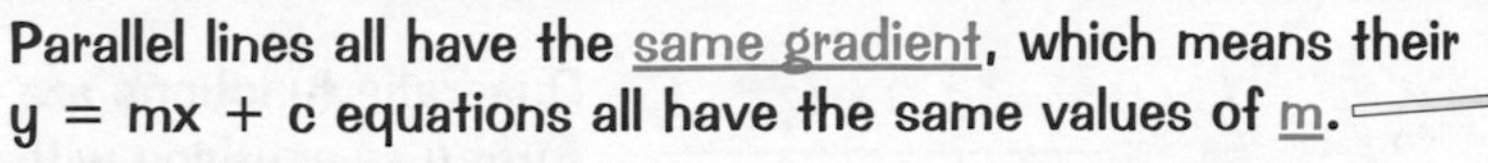

Parallel lines all have the <u>same gradient</u>, which means their y = mx + c equations all have the same values of <u>m</u>.
So the lines: y = 2x + 3, y = 2x and y = 2x – 4 are all parallel.

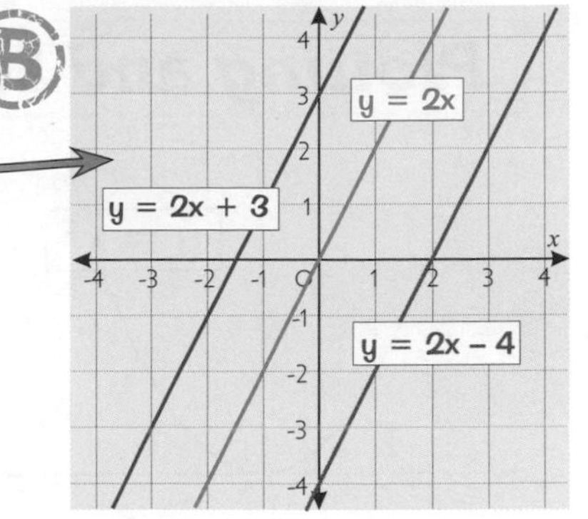

EXAMPLE: Line J has a gradient of <u>–0.25</u>. Find the equation of Line K, which is <u>parallel</u> to Line J and passes through point (2, 3).

1) Lines J and K are <u>parallel</u> so their <u>gradients</u> are the same $\Rightarrow$ m = –0.25
2) $y = -0.25x + c$
3) $x = 2, y = 3$
   $3 = (-0.25 \times 2) + c \Rightarrow 3 = -0.5 + c$
   $c = 3 + 0.5 = 3.5$
4) $y = -0.25x + 3.5$

1) First find the <u>'m' value</u> for Line K.
2) Substitute the value for 'm' into <u>y = mx + c</u> to give you the 'equation so far'.
3) Substitute the <u>x and y values</u> for the given point on Line K and solve for <u>'c'</u>.
4) Write out the <u>full equation</u>.

Perpendicular Line Gradients (A)

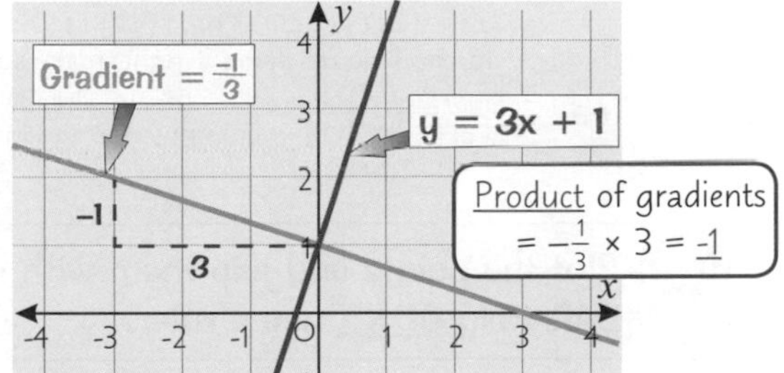

If the gradient of the first line is m, the gradient of the other line will be $-\frac{1}{m}$, because $m \times -\frac{1}{m} = -1$.

EXAMPLE: Lines A and B are <u>perpendicular</u> and intersect at <u>(3, 3)</u>.
If Line A has the equation <u>$3y - x = 6$</u>, what is the equation of Line B?

Find <u>'m'</u> (the gradient) for Line A.	$3y - x = 6 \Rightarrow 3y = x + 6$ $\Rightarrow y = \frac{1}{3}x + 2$, so $m_A = \frac{1}{3}$
Find the 'm' value for the <u>perpendicular</u> line (Line B).	$m_B = -\frac{1}{m_A} = -1 \div \frac{1}{3} = -3$
Put this into y = mx + c to give the 'equation so far'.	$y = -3x + c$
Put in the <u>x and y values</u> of the point and solve for <u>'c'</u>.	$x = 3, y = 3$ gives: $3 = (-3 \times 3) + c$ $\Rightarrow 3 = -9 + c \Rightarrow c = 12$
Write out the full equation.	$y = -3x + 12$

This stuff is a way to get one over on the examiners (well –1 actually)...

So basically, use one gradient to find the other, then use the known x and y values to work out c.

Q1 Find the equation of the line parallel to $2x + 2y = 3$ which passes through the point (1, 4). Give your answer in the form $y = mx + c$. [3 marks] (B)

Q2 Show that the lines $y + 5x = 2$ and $5y = x + 3$ are perpendicular. [3 marks] (A)

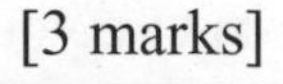

Quadratic Graphs

Quadratic functions can sound pretty darn impressive — "What did you do in Maths today, dear?", "Drawing the graphs of quadratic functions and solving the resulting quadratic equation graphically." Like wow. Seriously.

Plotting and Solving Quadratics

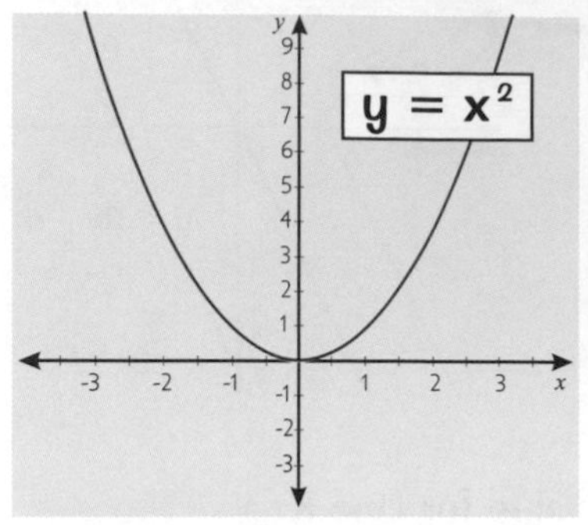

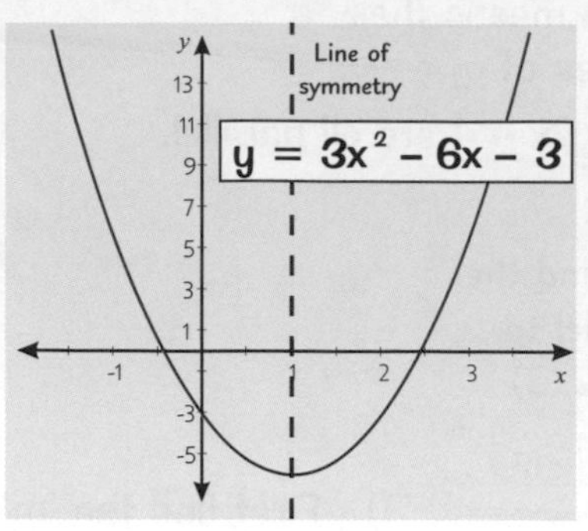

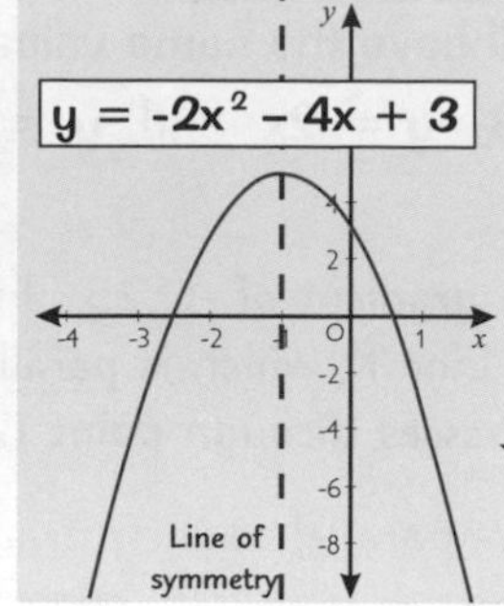

Quadratic functions are of the form <u>y = anything with x^2</u> (but not higher powers of x).

x^2 graphs all have the same <u>symmetrical</u> bucket shape.

If the x^2 bit has a '–' in front of it then the bucket is <u>upside down</u>.

EXAMPLE: Complete the table of values for the equation $y = x^2 + 2x - 3$ and then draw the graph.

x	-5	-4	-3	-2	-1	0	1	2	3
y	12	5	0	-3	-4	-3	0	5	12

1 Work out each <u>y-value</u> by <u>substituting</u> the corresponding <u>x-value</u> into the equation.

$y = (-5)^2 + (2 \times -5) - 3$
$= 25 - 10 - 3 = 12$

$y = (2)^2 + (2 \times 2) - 3$
$= 4 + 4 - 3 = 5$

To check you're doing it right, make sure you can <u>reproduce the y-values</u> they've already given you.

2 Plot the points and join them with a <u>completely smooth curve</u>. Definitely <u>DON'T</u> use a ruler.

<u>NEVER EVER</u> let one point drag your line off in some ridiculous direction. When a graph is generated from an equation, you never get spikes or lumps — only <u>MISTAKES</u>.

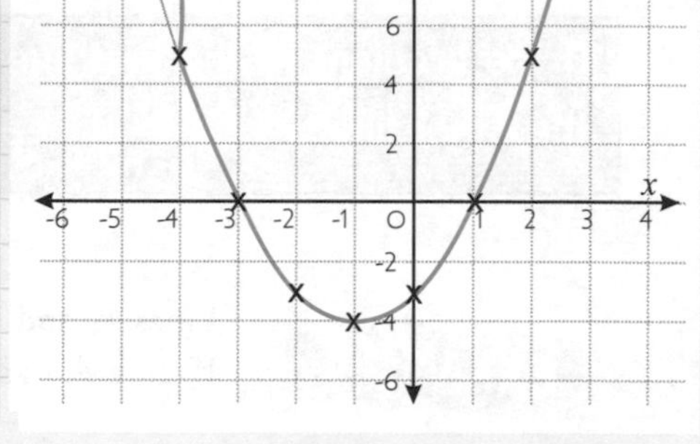

Solving Quadratic Equations

EXAMPLE: Use the graph of $y = x^2 + 2x - 3$ to solve the equation $x^2 + 2x - 3 = 0$.

The equation $x^2 + 2x - 3 = 0$ is what you get when you put <u>$y = 0$</u> into the graph's equation, $y = x^2 + 2x - 3$.

So to <u>solve</u> the equation, all you do is <u>read the x-values</u> where $y = 0$, i.e. where it crosses the x-axis.

So the solutions are <u>$x = -3$</u> and <u>$x = 1$</u>.

Quadratic equations usually have <u>2 solutions</u>.

Now celebrate the only way graphs know how: line dancing.

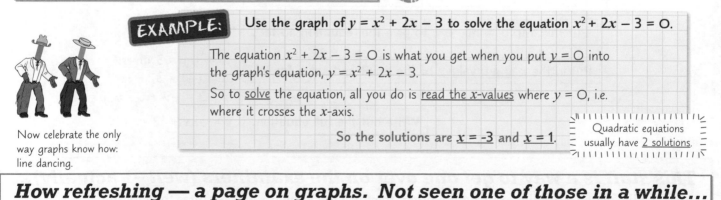

How refreshing — a page on graphs. Not seen one of those in a while...

You know the deal by now — learn what's on this page, then treat yourself to answering the question below.

Q1 a) Draw the graph of $y = x^2 - 4x - 1$ for values of x between –2 and 6. [4 marks] (C)

b) Use your graph to estimate the solutions to $5 = x^2 - 4x - 1$. [1 mark]

Harder Graphs

Graphs come in all sorts of shapes, sizes and wiggles — here are the first of four more types you need to know:

x^3 Graphs: $y = ax^3 + bx^2 + cx + d$ (b, c and/or d can be zero)

All x^3 graphs have a <u>wiggle</u> in the middle — sometimes it's a flat wiggle, sometimes it's more pronounced. $-x^3$ graphs always go down from <u>top left</u>, $+x^3$ ones go up from <u>bottom left</u>.

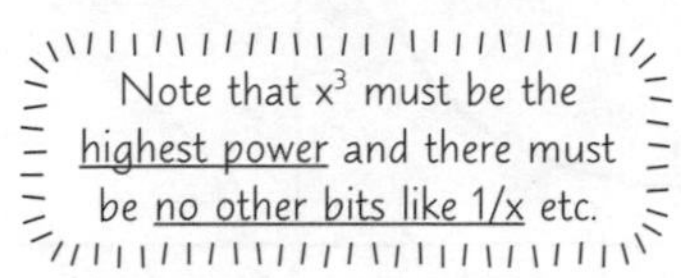

Note that x³ must be the highest power and there must be no other bits like 1/x etc.

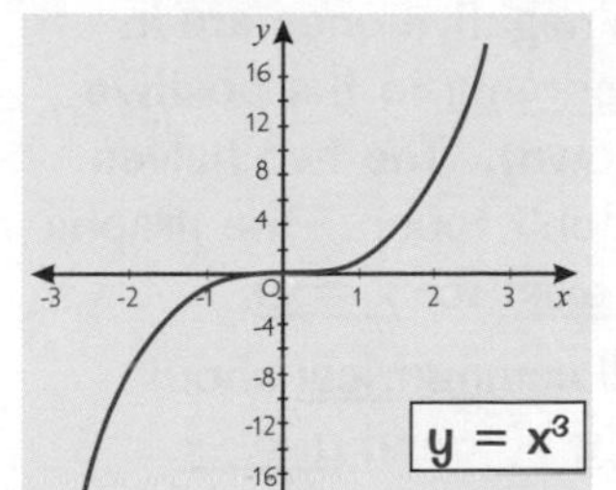

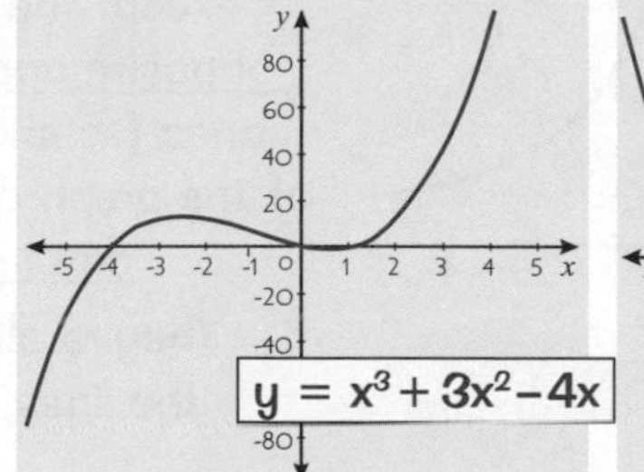

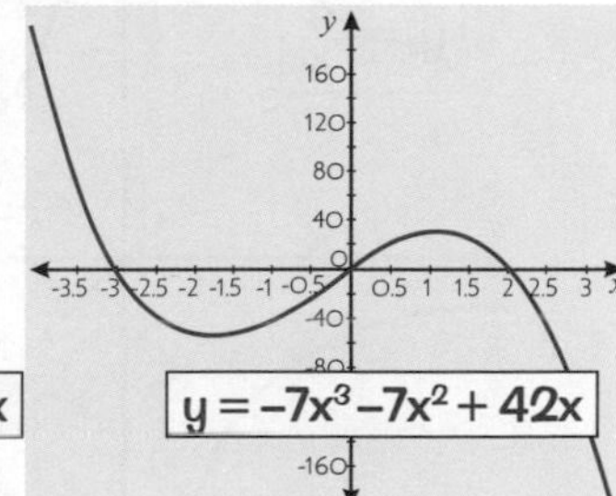

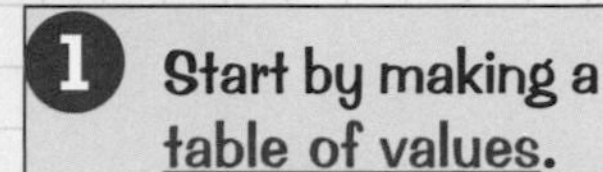

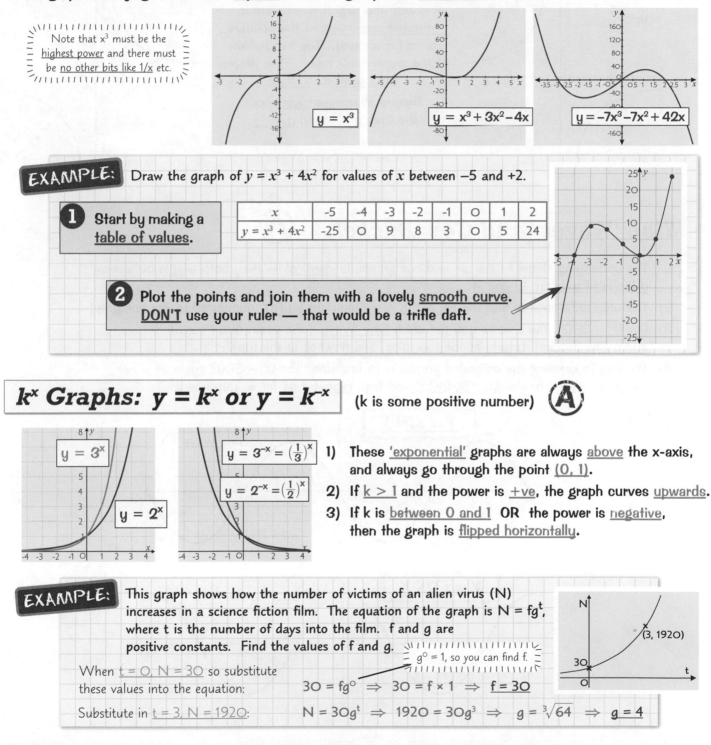

Draw the graph of $y = x^3 + 4x^2$ for values of x between –5 and +2.

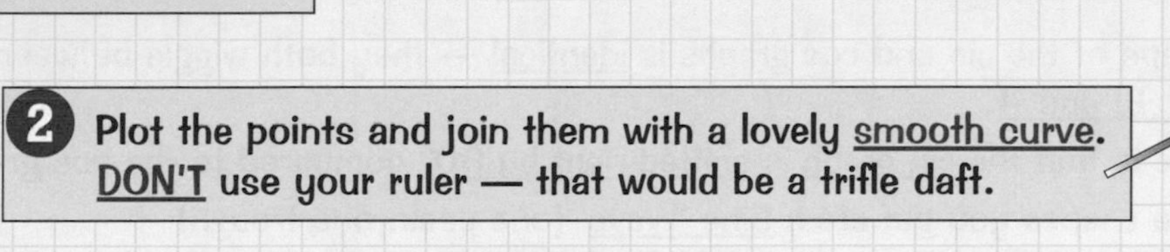

1. Start by making a <u>table of values</u>.

x	-5	-4	-3	-2	-1	0	1	2
$y = x^3 + 4x^2$	-25	0	9	8	3	0	5	24

2. Plot the points and join them with a lovely <u>smooth curve</u>. <u>DON'T</u> use your ruler — that would be a trifle daft.

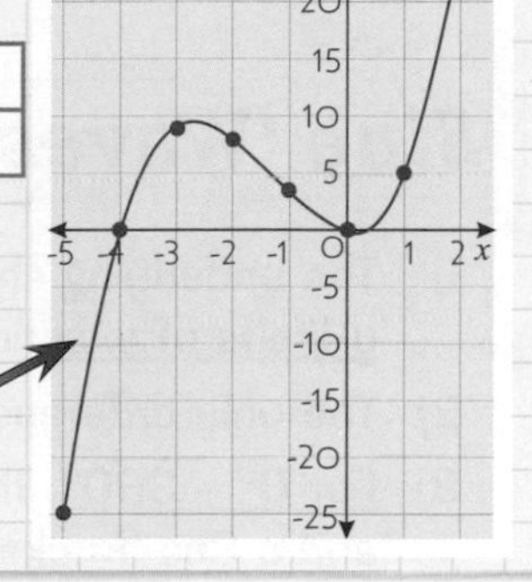

k^x Graphs: $y = k^x$ or $y = k^{-x}$ (k is some positive number)

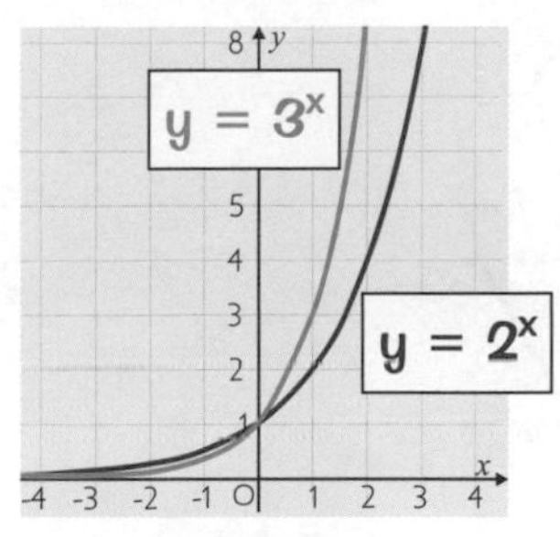

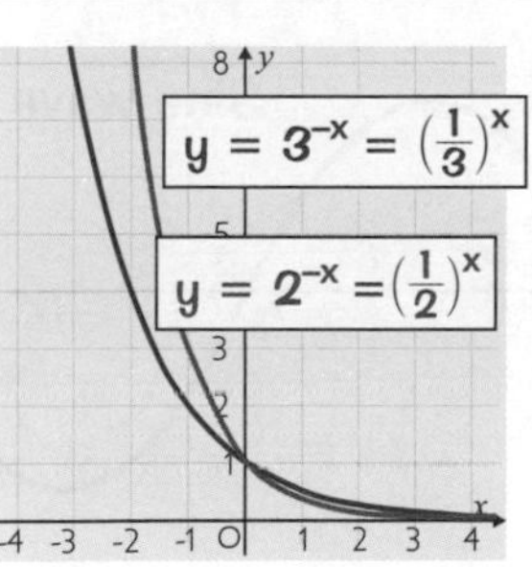

1) These <u>'exponential'</u> graphs are always <u>above</u> the x-axis, and always go through the point <u>(0, 1)</u>.
2) If <u>$k > 1$</u> and the power is <u>+ve</u>, the graph curves <u>upwards</u>.
3) If k is <u>between 0 and 1</u> OR the power is <u>negative</u>, then the graph is <u>flipped horizontally</u>.

EXAMPLE: This graph shows how the number of victims of an alien virus (N) increases in a science fiction film. The equation of the graph is $N = fg^t$, where t is the number of days into the film. f and g are positive constants. Find the values of f and g.

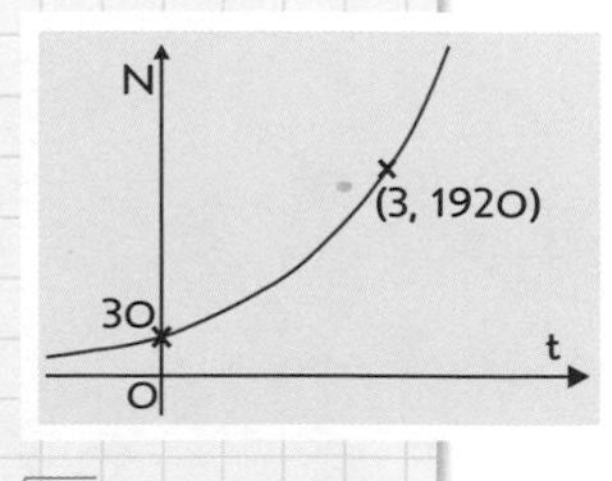

When <u>t = 0, N = 30</u> so substitute these values into the equation: $30 = fg^0 \Rightarrow 30 = f \times 1 \Rightarrow \underline{f = 30}$

Substitute in <u>t = 3, N = 1920</u>: $N = 30g^t \Rightarrow 1920 = 30g^3 \Rightarrow g = \sqrt[3]{64} \Rightarrow \underline{g = 4}$

Phew — that page could seriously drive you round the k^x

Learn what type of graph you get from each sort of equation. Then try this Exam Practice Question.

Q1 a) Complete this table for $y = x^3 - 2x + 1$. [2 marks]

x	–2	–1	0	1	2
y				0	

b) Hence sketch the graph of $y = x^3 - 2x + 1$. [2 marks]

Harder Graphs

Here are the final graph types you need to be able to sketch. Knowing what you're aiming for really helps.

1/x (Reciprocal) Graphs: y = A/x or xy = A 

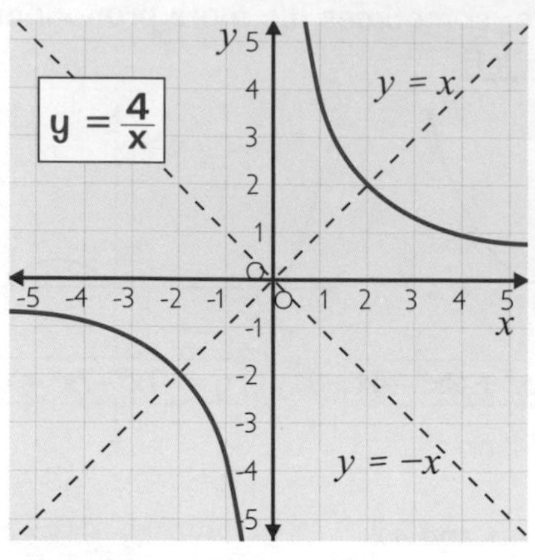

These are <u>all the same basic shape</u>, except the negative ones are in <u>opposite quadrants</u> to the positive ones (as shown). The two halves of the graph don't touch. The graphs <u>don't exist</u> for <u>$x = 0$</u>.

They're all <u>symmetrical</u> about the lines <u>$y = x$</u> and <u>$y = -x$</u>.

(You get this type of graph with inverse proportion — see p45)

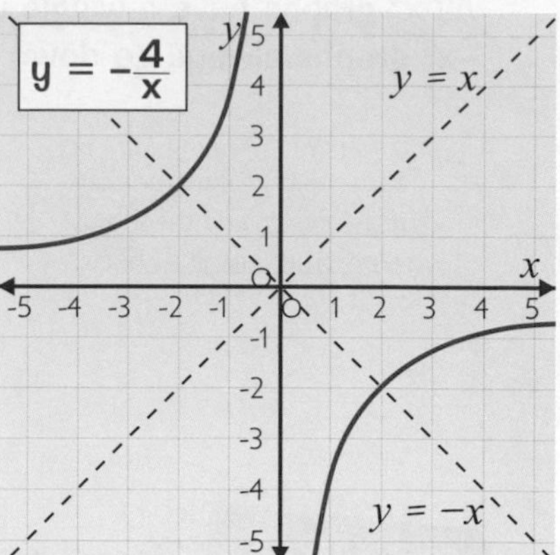

Sine 'Waves' and Cos 'Buckets' 

1) The underlying shape of the sin and cos graphs is <u>identical</u> — they both wiggle between <u>y-limits of exactly +1 and -1</u>.
2) The only difference is that the <u>sin graph</u> is <u>shifted right by 90°</u> compared to the cos graph.
3) <u>For 0° – 360°</u>, the shapes you get are a <u>Sine 'Wave'</u> (one peak, one trough) and a <u>Cos 'Bucket'</u> (starts at the top, dips, and finishes at the top).
4) The key to drawing the extended graphs is to first draw the 0° – 360° cycle of either the <u>Sine 'WAVE'</u> or the <u>Cos 'BUCKET'</u> and then <u>repeat it</u> as far as you need.

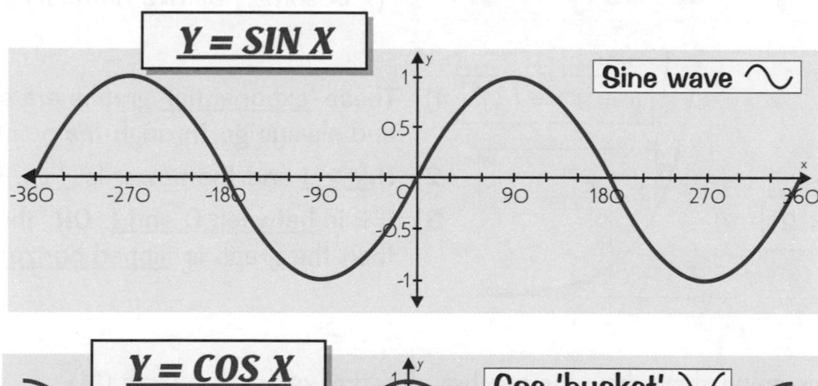

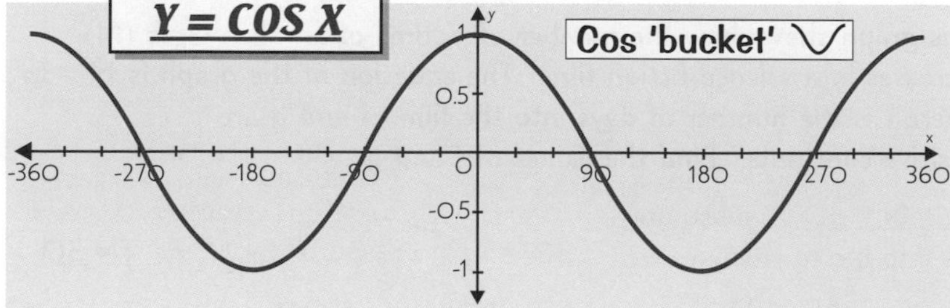

The sine wave and the cos bucket — a great day out at the beach...

You could be asked to plot any of these graphs. Use a table of values (see p50) but make sure you find enough points to get the shape right. That's another reason for learning the basic shapes of the graphs.

Q1 a) Sketch the graph of $y = \cos x$ for values of x between –360° and 360°. [4 marks] 

b) Give the values of x (to the nearest degree) where $\cos x = 0.67$, for $-360° < x < 0°$. [1 mark] 

Graph Transformations

Don't be put off by function notation involving f(x). It doesn't mean anything complicated, it's just a fancy way of saying "an equation in x". In other words "y = f(x)" just means "y = some totally mundane equation in x, which we won't tell you, we'll just call it f(x) instead to see how many of you get in a flap about it".

Learn These Five Types of Graph Transformation

In a question on transforming graphs they will either use function notation or they'll use a known function instead. There aren't many different types of graph transformations so just learn them and be done with it.

1) y-Stretch: $y = k \times f(x)$

This is where the original graph is stretched parallel to the y-axis by multiplying the whole function by a number, i.e. y = f(x) becomes y = kf(x) (where k = 2 or 5 etc.). If k is less than 1, then the graph is squashed down in the y-direction instead:

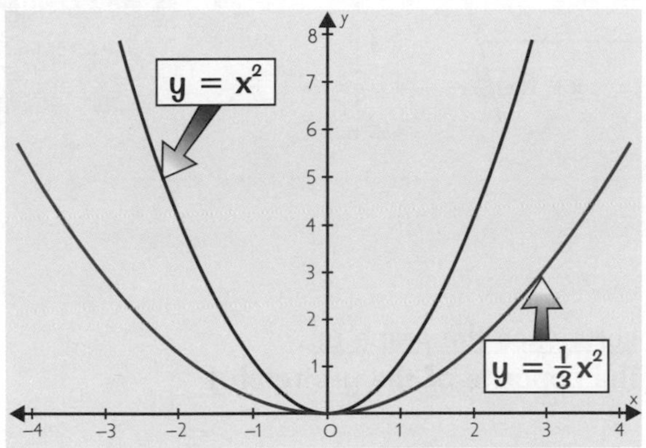

This graph shows y = f(x) and $y = \frac{1}{3}f(x)$ ($y = x^2$ and $y = \frac{1}{3}x^2$)

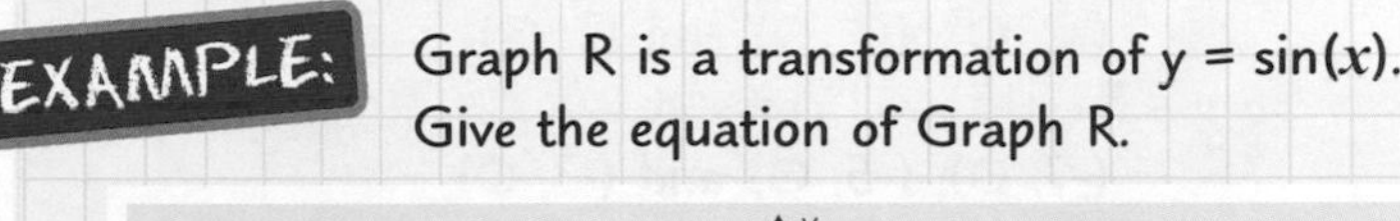

Graph R is a transformation of $y = \sin(x)$. Give the equation of Graph R.

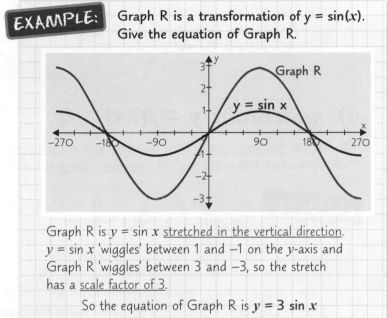

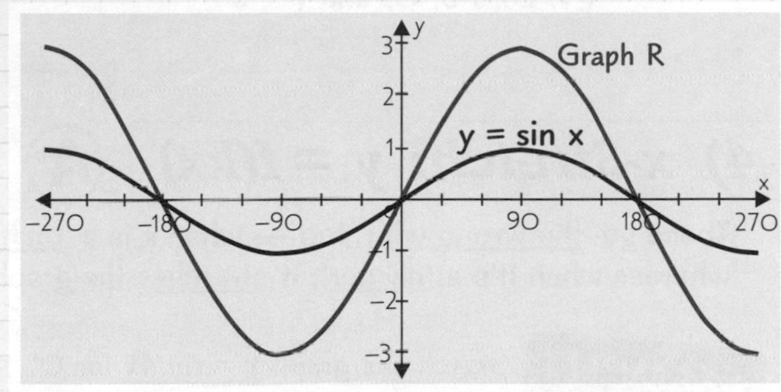

Graph R is $y = \sin x$ stretched in the vertical direction. $y = \sin x$ 'wiggles' between 1 and –1 on the y-axis and Graph R 'wiggles' between 3 and –3, so the stretch has a scale factor of 3.

So the equation of Graph R is $y = 3 \sin x$

2) y-Shift: $y = f(x) + a$

This is where the whole graph is slid up or down the y-axis with no distortion, and is achieved by simply adding a number onto the end of the equation: y = f(x) + a.

This shows y = f(x) and y = f(x) – 3 i.e. y = sin x and y = (sin x) – 3

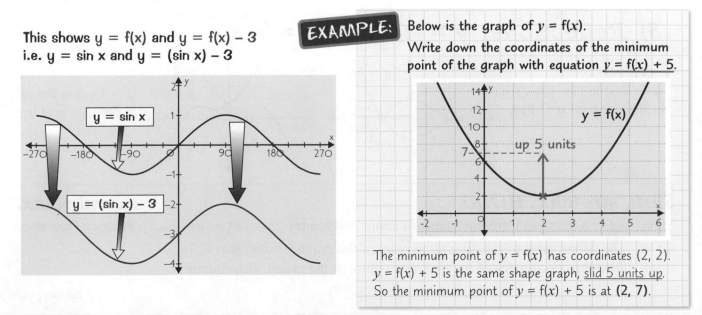

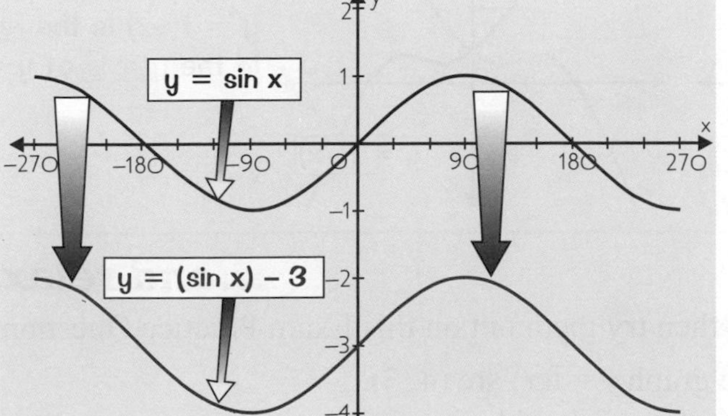

Below is the graph of $y = f(x)$. Write down the coordinates of the minimum point of the graph with equation $y = f(x) + 5$.

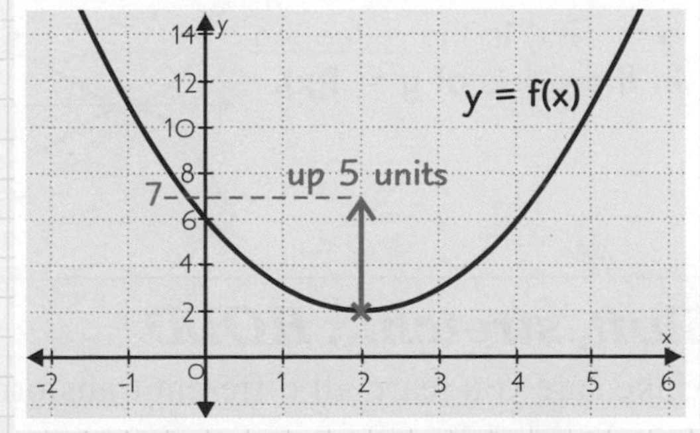

The minimum point of $y = f(x)$ has coordinates (2, 2). $y = f(x) + 5$ is the same shape graph, slid 5 units up. So the minimum point of $y = f(x) + 5$ is at **(2, 7)**.

Graph Transformations

Keep going, this is the last page on graph transformations. Promise.

3) x-Shift: y = f(x – a)

This is where the whole graph slides to the left or right and it only happens when you replace 'x' everywhere in the equation with 'x – a'. These are tricky because they go 'the wrong way'. If you want to go from y = f(x) to y = f(x – a) you must move the whole graph a distance 'a' in the positive x-direction → (and vice versa).

EXAMPLE: The graph $y = f(x)$ is shown. Give the coordinates of the points where the graph $y = f(x + 5)$ will cross the x-axis.

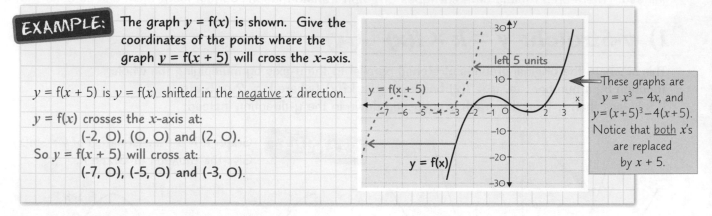

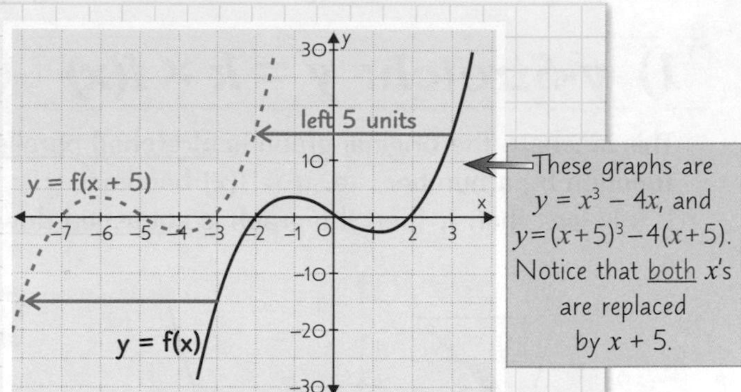

These graphs are $y = x^3 - 4x$, and $y = (x+5)^3 - 4(x+5)$. Notice that both x's are replaced by $x + 5$.

$y = f(x + 5)$ is $y = f(x)$ shifted in the negative x direction.

$y = f(x)$ crosses the x-axis at:
(-2, 0), (0, 0) and (2, 0).

So $y = f(x + 5)$ will cross at:
(-7, 0), (-5, 0) and (-3, 0).

4) x-Stretch: y = f(kx)

These go 'the wrong way' too — when k is a 'multiplier' it scrunches the graph up, whereas when it's a 'divider', it stretches the graph out. (The opposite of the y-stretch.)

EXAMPLE: Sketch the graph $y = \sin 4x$ for $0° \le x \le 360°$. The graph of $y = \sin x$ is shown.

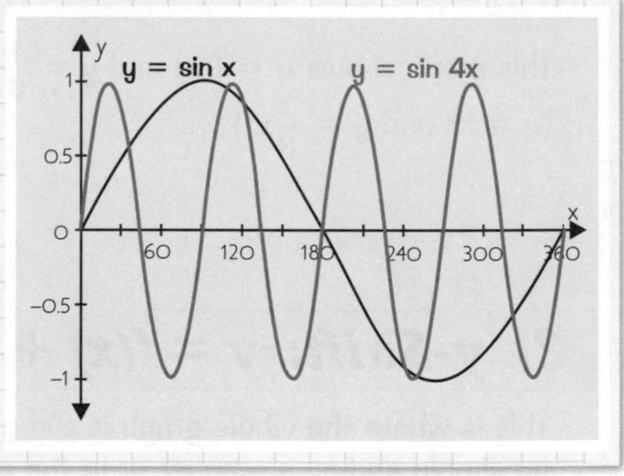

$y = \sin 4x$ has a multiplier of 4, so its graph will be 4 times as squashed up as $y = \sin x$.

There is one cycle of up and down on the $y = \sin x$ graph, so you can fit four cycles of the $y = \sin 4x$ graph in the same space.

Remember, if k is a divider, then the graph spreads out. So if the squashed-up graph was the original, $y = f(x)$, then the more spread out one would be $y = f(\frac{x}{4})$.

5) Reflections: y = –f(x) and y = f(–x)

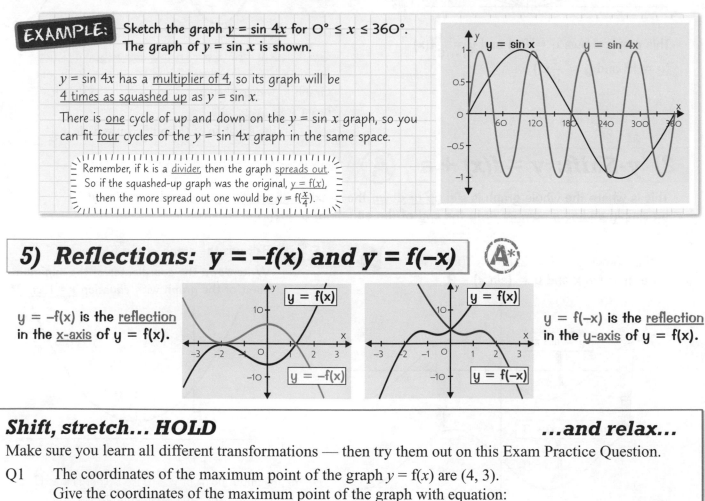

y = –f(x) is the reflection in the x-axis of y = f(x).

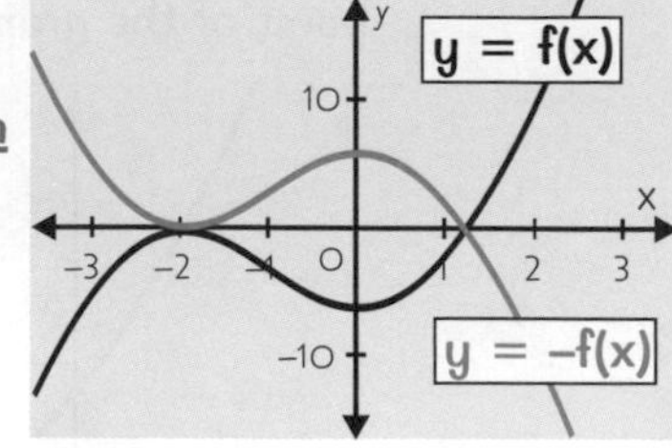

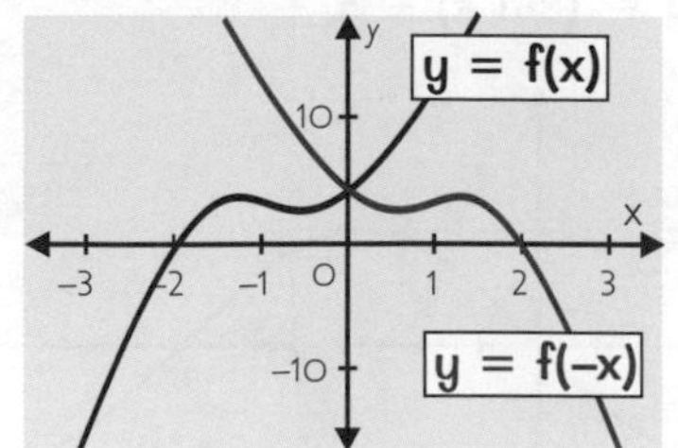

y = f(–x) is the reflection in the y-axis of y = f(x).

Shift, stretch... HOLD ...and relax...

Make sure you learn all different transformations — then try them out on this Exam Practice Question.

Q1 The coordinates of the maximum point of the graph $y = f(x)$ are (4, 3).
Give the coordinates of the maximum point of the graph with equation:
a) $y = f(\frac{x}{2})$ b) $y = f(x) - 4$ c) $y = 0.5\,f(x)$ [3 marks]

Real-Life Graphs

Now and then, graphs mean something more interesting than just $y = x^3 + 4x^2 - 6x + 4$...

Graphs Can Show Billing Structures

Many bills are made up of two charges — a fixed charge and a cost per unit. E.g. You might pay £11 each month for your phone line, and then be charged 3p for each minute of calls you make. This type of thing is perfect exam question fodder.

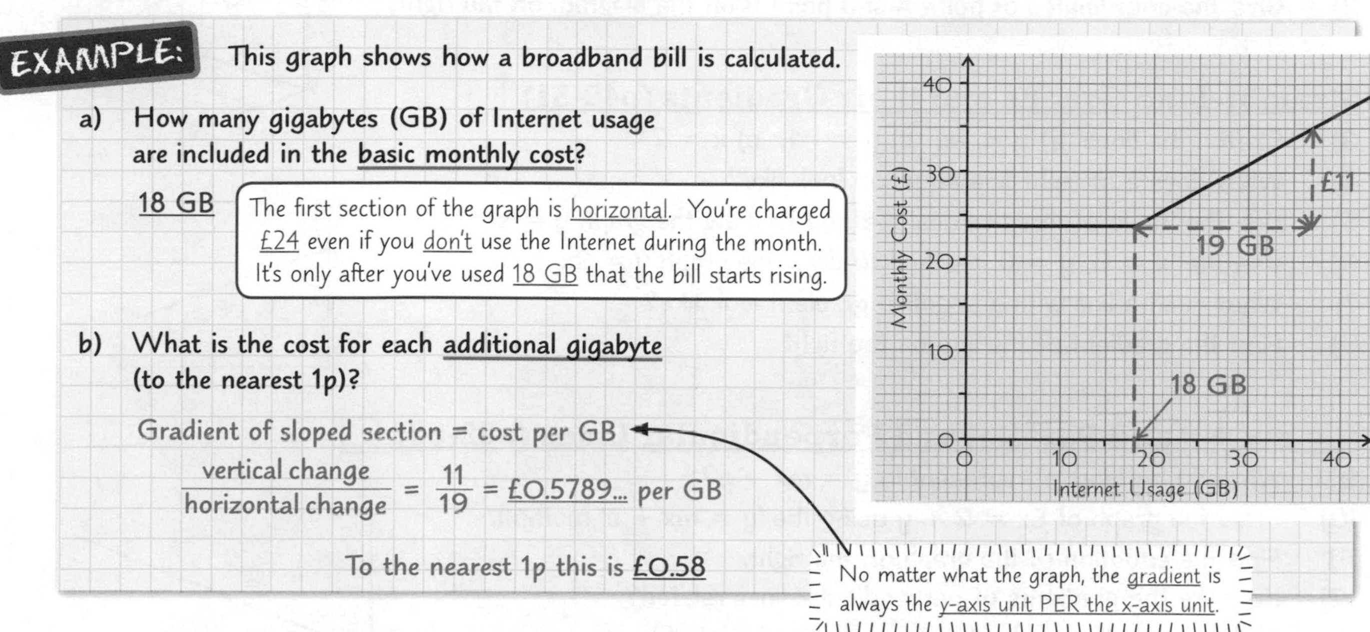

EXAMPLE: This graph shows how a broadband bill is calculated.

a) How many gigabytes (GB) of Internet usage are included in the basic monthly cost?

18 GB

The first section of the graph is horizontal. You're charged £24 even if you don't use the Internet during the month. It's only after you've used 18 GB that the bill starts rising.

b) What is the cost for each additional gigabyte (to the nearest 1p)?

Gradient of sloped section = cost per GB

$$\frac{\text{vertical change}}{\text{horizontal change}} = \frac{11}{19} = £0.5789... \text{ per GB}$$

To the nearest 1p this is £0.58

No matter what the graph, the gradient is always the y-axis unit PER the x-axis unit.

Graphs Can Show Changes with Time

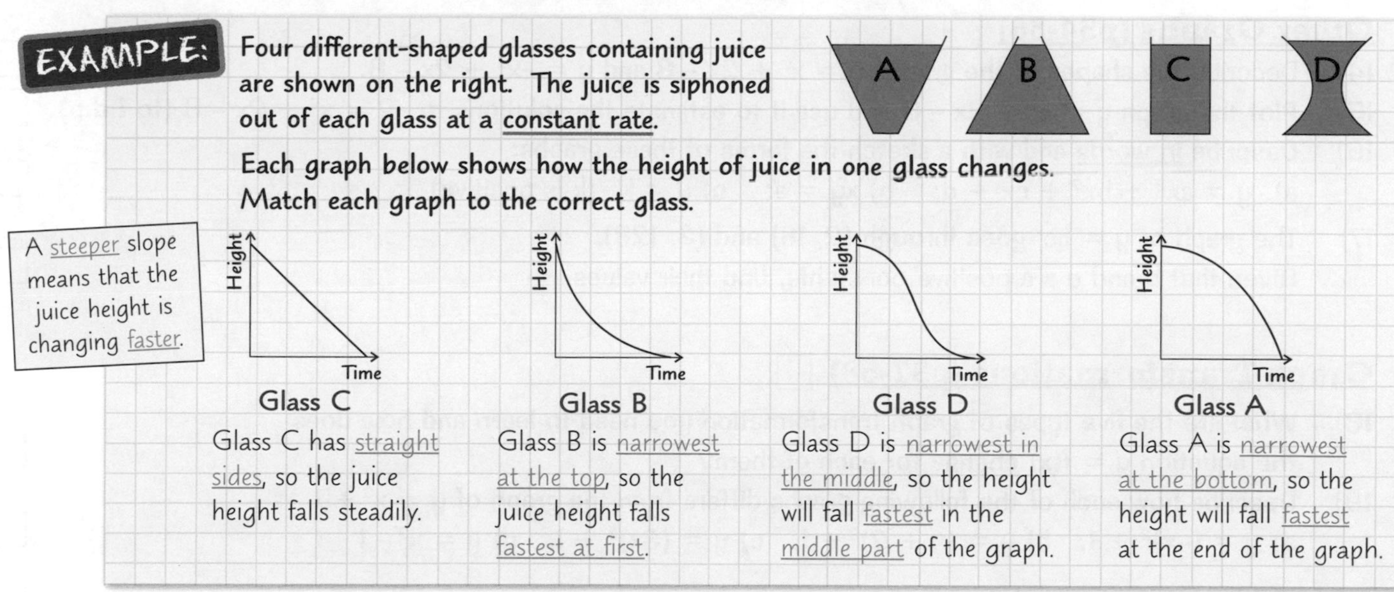

EXAMPLE: Four different-shaped glasses containing juice are shown on the right. The juice is siphoned out of each glass at a constant rate.

Each graph below shows how the height of juice in one glass changes. Match each graph to the correct glass.

A steeper slope means that the juice height is changing faster.

Glass C — Glass C has straight sides, so the juice height falls steadily.

Glass B — Glass B is narrowest at the top, so the juice height falls fastest at first.

Glass D — Glass D is narrowest in the middle, so the height will fall fastest in the middle part of the graph.

Glass A — Glass A is narrowest at the bottom, so the height will fall fastest at the end of the graph.

Exam marks per unit of brainpower...

Distance-time graphs and unit conversion graphs are real-life graphs too — see p80 and p82.

Q1 A taxi charges a minimum fare of £4.50, which includes the first three miles.
It then charges 80p for each additional mile.
Draw a graph to show the cost of journeys of up to 10 miles. [4 marks]

Revision Questions for Section Three

Well, that wraps up Section Three — time to put yourself to the test and find out how much you really know.

- Try these questions and tick off each one when you get it right.
- When you've done all the questions for a topic and are completely happy with it, tick off the topic.

X, Y and Z Coordinates (p48)

1) Find the midpoint of a line segment with end points (–2, 3) and (7, –4).
2) Give the coordinates of point A and point B on the diagram on the right.

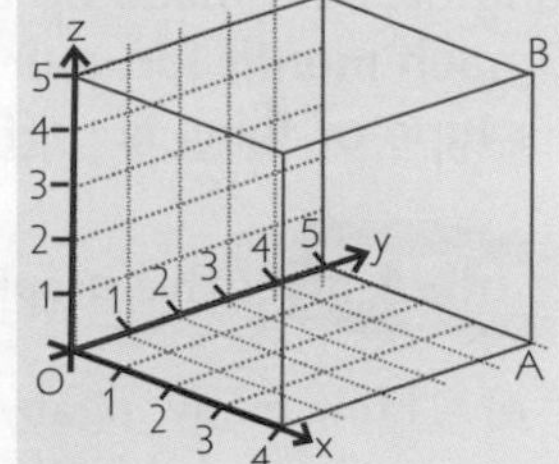

Straight-Line Graphs and their Gradients (p49-51)

3) Sketch the lines a) $y = -x$, b) $y = -4$, c) $x = 2$
4) What does a straight-line equation look like?
5) Use the 'table of three values' method to draw the graph $y = x + \frac{1}{10}x$
6) Use the '$x = 0$, $y = 0$' method to draw the graph $y = 3x + 5$.
7) What does a line with a negative gradient look like?
8) Find the gradient of the line on the right.

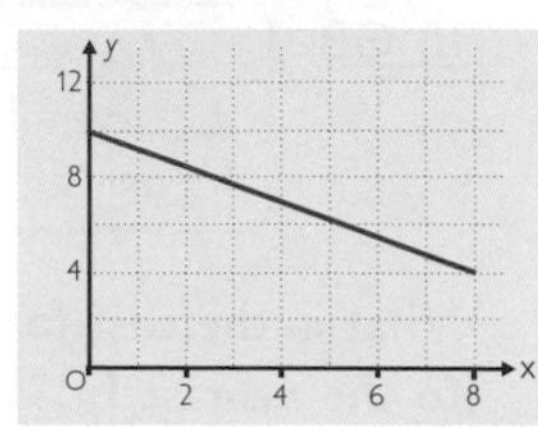

y = mx + c and Parallel and Perpendicular Lines (p52-53)

9) What do 'm' and 'c' represent in $y = mx + c$?
10) Draw the graph of $5x = 2 + y$ using the '$y = mx + c$' method.
11) Find the equation of the graph on the right.
12) How are the gradients of perpendicular lines related? And how are the gradients of parallel lines related?
13) Find the equation of the line passing through (4, 2) which is perpendicular to $y = 2x - 1$.

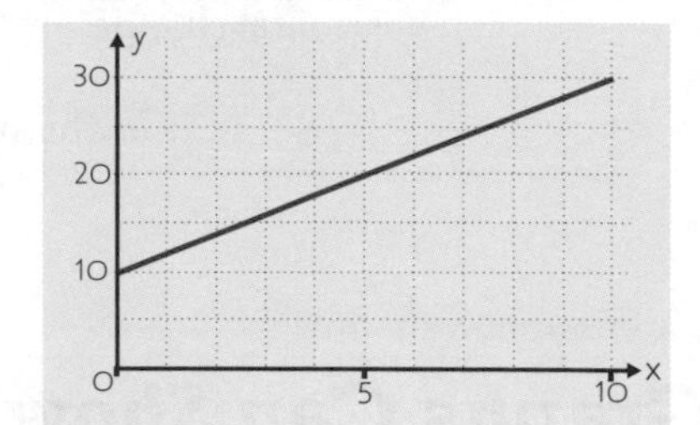

Other Graphs (p54-56)

14) Describe the shapes of the graphs $y = x^2 + 2x - 8$ and $y = -x^2 + 2x - 8$.
15) Plot the graph $y = x^2 + 2x - 8$ and use it to estimate the solutions to $-2 = x^2 + 2x - 8$ (to 1 d.p).
16) Describe in words and with a sketch the forms of these graphs:
a) $y = ax^3 + bx^2 + cx + d$; b) $xy = a$; c) $y = k^x$ (k is positive)
17) The graph of $y = bc^x$ goes through (2, 16) and (3, 128).
Given that b and c are positive constants, find their values.

Graph Transformations (p57-58)

18) What are the five types of graph transformation you need to learn and how does the equation $y = f(x)$ change for each of them?
19) Describe how each of the following graphs differs from the graph of $y = x^3 + 1$
a) $y = (-x)^3 + 1$, b) $y = (x + 2)^3 + 1$, c) $y = (3x)^3 + 1$, d) $y = x^3 - 1$

Real-Life Graphs (p59)

20) Sweets'R'Yum sells chocolate drops. They charge 90p per 100 g for the first kg, then 60p per 100 g after that. Sketch a graph to show the cost of buying up to 3 kg of chocolate drops.

Geometry

If you know all these rules thoroughly, you'll at least have a fighting chance of working out problems with lines and angles. If you don't — you've no chance. Sorry to break it to you like that.

6 Simple Rules — that's all

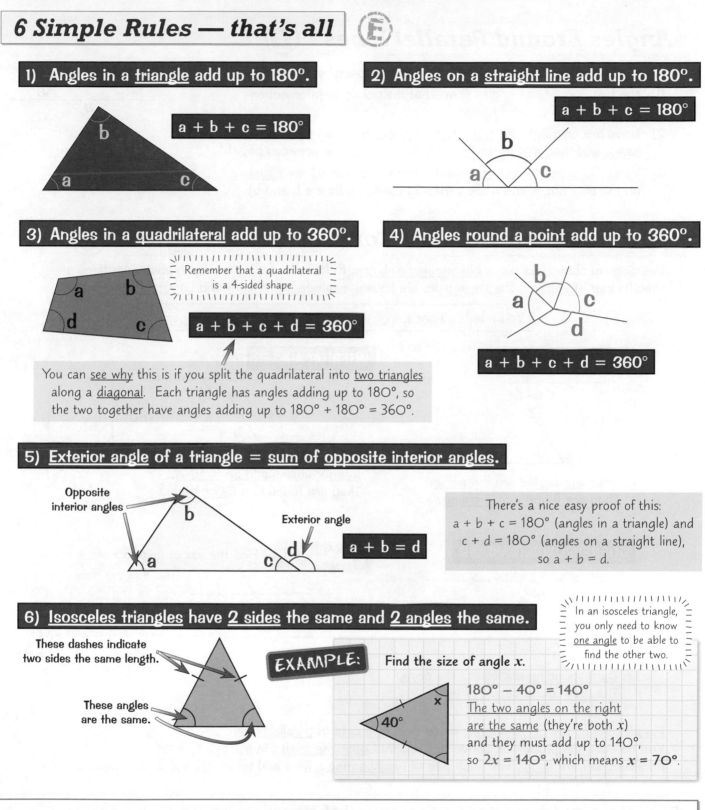

1) Angles in a triangle add up to 180°.

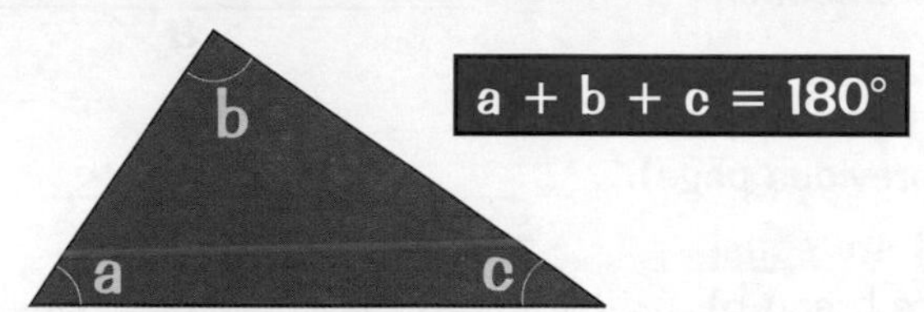

2) Angles on a straight line add up to 180°.

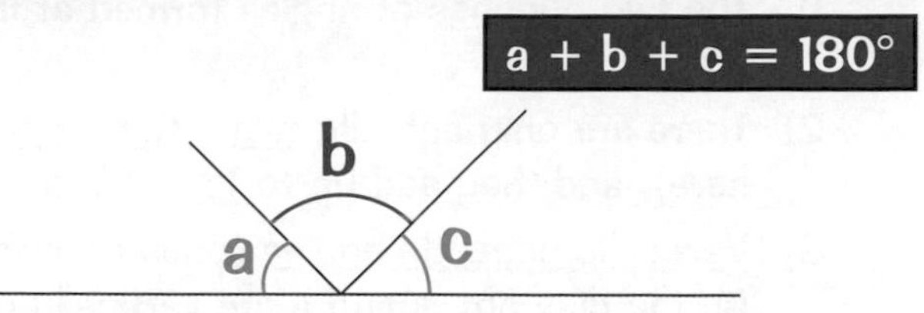

3) Angles in a quadrilateral add up to 360°.

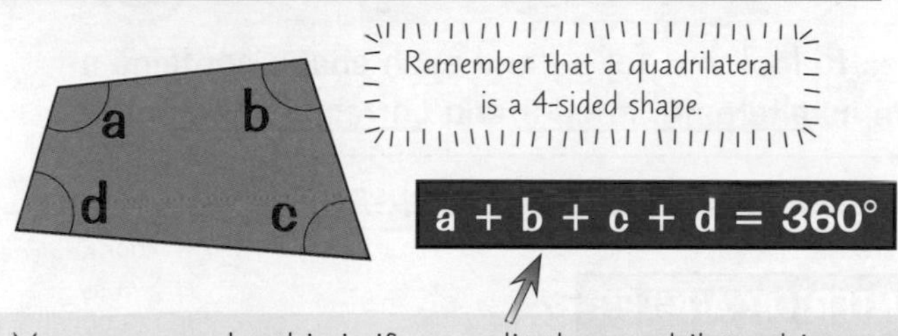

You can see why this is if you split the quadrilateral into two triangles along a diagonal. Each triangle has angles adding up to 180°, so the two together have angles adding up to 180° + 180° = 360°.

4) Angles round a point add up to 360°.

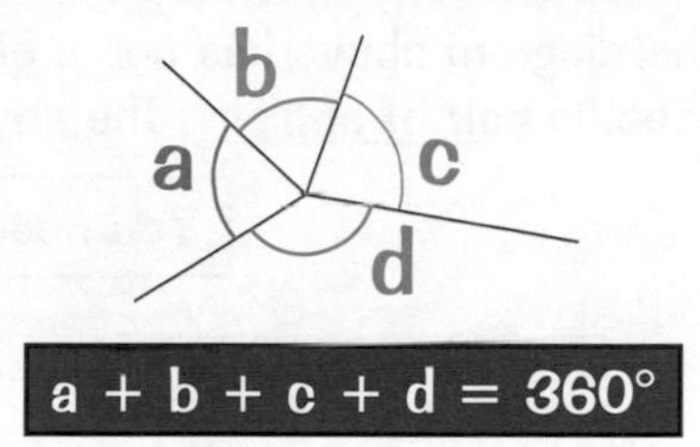

5) Exterior angle of a triangle = sum of opposite interior angles.

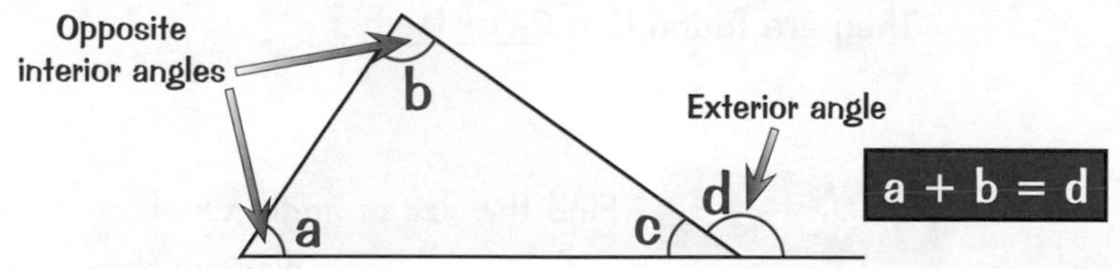

There's a nice easy proof of this:
a + b + c = 180° (angles in a triangle) and
c + d = 180° (angles on a straight line),
so a + b = d.

6) Isosceles triangles have 2 sides the same and 2 angles the same.

In an isosceles triangle, you only need to know one angle to be able to find the other two.

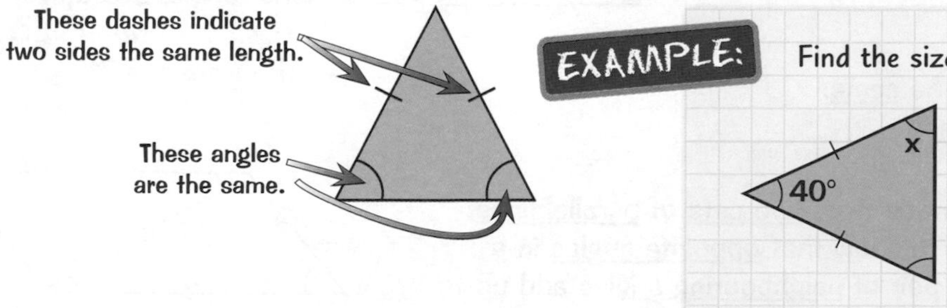

Find the size of angle x.

180° – 40° = 140°
The two angles on the right are the same (they're both x) and they must add up to 140°, so $2x = 140°$, which means $x = 70°$.

Heaven must be missing an angle...

All the basic facts are pretty easy really, but examiners like to combine them in questions to confuse you. There are some examples of these on p63, but have a go at this one as a warm-up.

Q1 Find the size of the angle marked x. [2 marks]

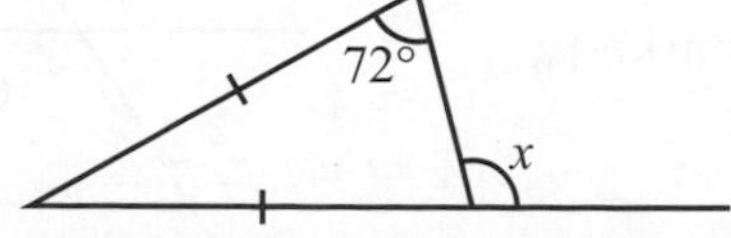

Parallel Lines

Parallel lines are quite straightforward really. (They're also quite straight. And parallel.) There are a few rules you need to learn — make sure you don't get them mixed up.

Angles Around Parallel Lines 

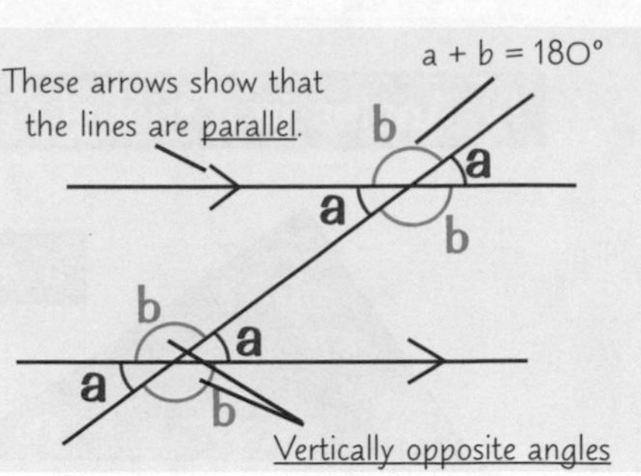

When a line crosses two parallel lines, it forms special sets of angles.

1) The two bunches of angles formed at the points of intersection are the same.
2) There are only actually two different angles involved (labelled a and b here), and they add up to 180° (from rule 2 on the previous page).
3) Vertically opposite angles (ones opposite each other) are equal (in the diagram, a and a are vertically opposite, as are b and b).

Alternate, Interior and Corresponding Angles 

The diagram above has some characteristic shapes to look out for — and each shape contains a specific pair of angles. The angle pairs are known as alternate, interior and corresponding angles.

You need to spot the characteristic Z, C, U and F shapes:

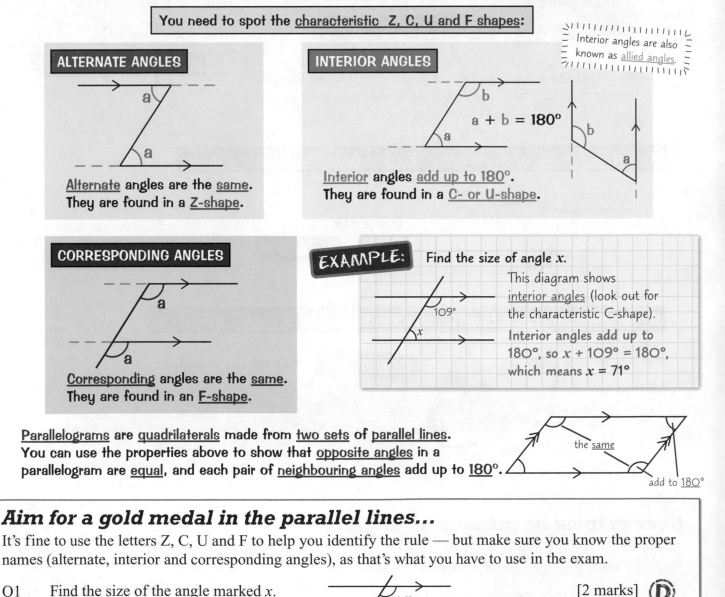

Interior angles are also known as allied angles.

ALTERNATE ANGLES

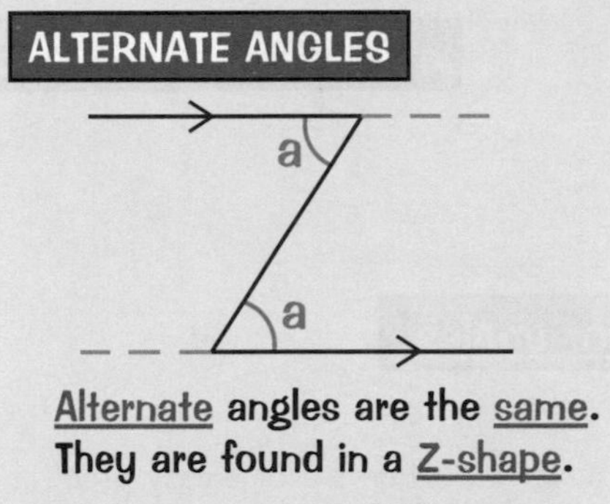

Alternate angles are the same. They are found in a Z-shape.

INTERIOR ANGLES

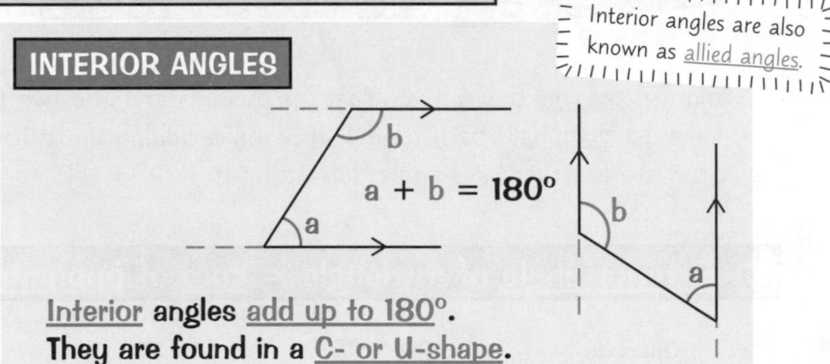

Interior angles add up to 180°. They are found in a C- or U-shape.

CORRESPONDING ANGLES

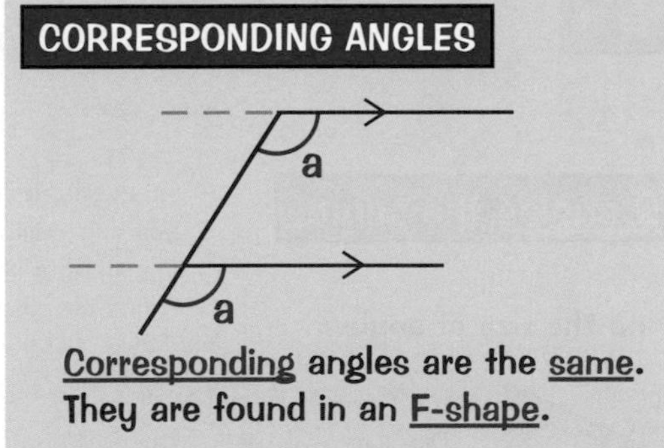

Corresponding angles are the same. They are found in an F-shape.

EXAMPLE: Find the size of angle x.

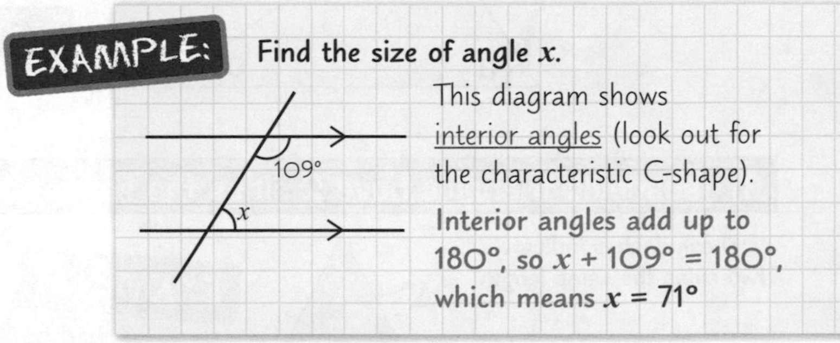

This diagram shows interior angles (look out for the characteristic C-shape).

Interior angles add up to 180°, so $x + 109° = 180°$, which means $x = 71°$

Parallelograms are quadrilaterals made from two sets of parallel lines. You can use the properties above to show that opposite angles in a parallelogram are equal, and each pair of neighbouring angles add up to 180°.

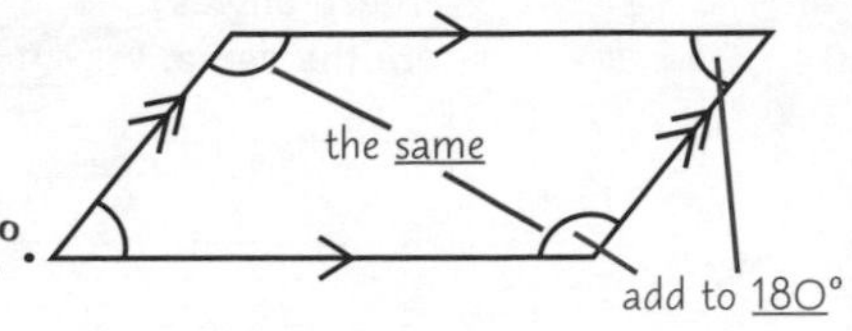

Aim for a gold medal in the parallel lines...

It's fine to use the letters Z, C, U and F to help you identify the rule — but make sure you know the proper names (alternate, interior and corresponding angles), as that's what you have to use in the exam.

Q1 Find the size of the angle marked x. [2 marks] 

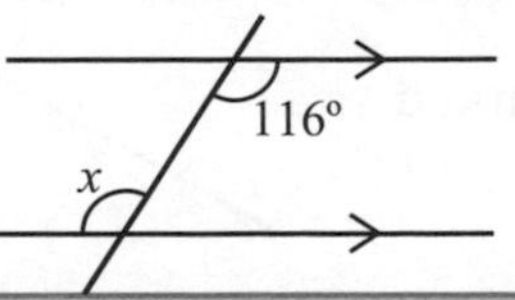

Geometry Problems

My biggest geometry problem is that I have to do geometry problems in the first place. *Sigh* Ah well, best get practising — these problems aren't going to solve themselves.

Try Out All The Rules One By One

1) Don't concentrate too much on the angle you have been asked to find. The best method is to find ALL the angles in whatever order they become obvious.
2) Don't sit there waiting for inspiration to hit you. It's all too easy to find yourself staring at a geometry problem and getting nowhere. The method is this:

GO THROUGH ALL THE RULES OF GEOMETRY (including PARALLEL LINES), ONE BY ONE, and apply each of them in turn in as many ways as possible — one of them is bound to work.

3) Before we get going, there's one bit of notation you need to be familiar with — three-letter angle notation. It's not hard — if you get an angle written as ∠ABC (or just ABC), it's the angle formed at letter B (it's always the middle letter).

B Angle ∠ABC

A C

EXAMPLE: Find the size of angles x and y.

Write down everything you know (or can work out) about the shape:

Triangle ABD is isosceles, so ∠BAD = ∠ABD = 76°.

That means ∠ADB = 180° – 76° – 76° = 28°.

∠ADC is a right angle (= 90°), so angle x = 90° – 28° = 62°

ABCD is a quadrilateral, so all the angles add up to 360°. 76° + 90° + y + 72° = 360°, so y = 360° – 76° – 90° – 72° = 122°

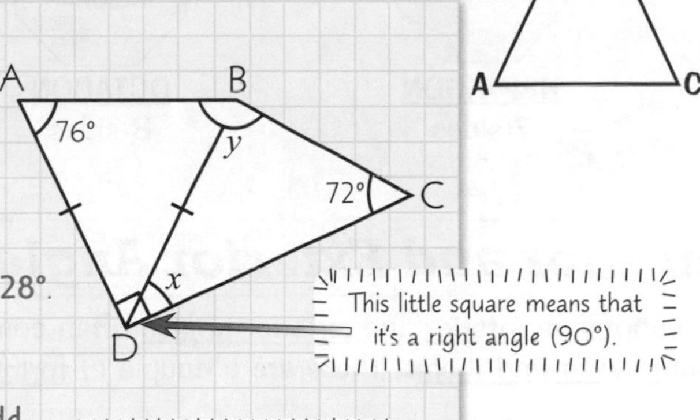

You could have worked out angle y before angle x.

EXAMPLE: Find all the missing angles in the diagram below.

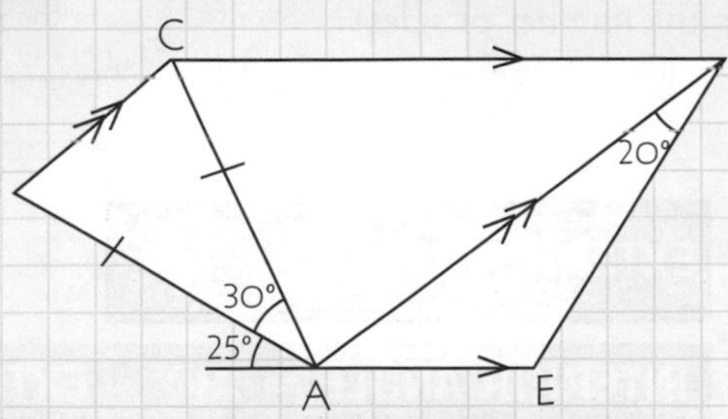

1) Triangle ABC is isosceles, so...
   ∠ABC = ∠ACB = $\frac{(180° - 30°)}{2}$ = 75°
2) BC and AD are parallel, BCAD is a Z-shape, so...
   ∠ACB and ∠CAD are alternate angles.
   As ∠ACB = 75° then ∠CAD = 75° too
3) Angles on a straight line means...
   ∠EAD = 180° – 25° – 30° – 75° = 50°
4) AE and CD are parallel so ∠ADC = ∠EAD = 50° also.
5) Angles in triangle ACD add up to 180° so ∠ACD = 180° – 75° – 50° = 55°
6) Angles in triangle ADE add up to 180° so ∠AED = 180° – 50° – 20° = 110°

Missing: angle x. If found, please return to Amy...

Geometry problems often look a lot worse than they are — don't panic, just write down everything you can work out. Watch out for hidden parallel lines and isosceles triangles — they can help you work out angles.

Q1 Find the size of angle x.

[3 marks] D

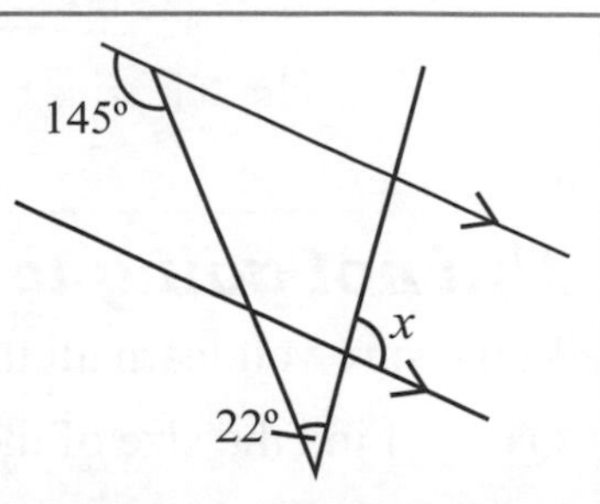

Polygons

A polygon is a many-sided shape, and can be regular or irregular. A regular polygon is one where all the sides and angles are the same (in an irregular polygon, the sides and angles are different).

Regular Polygons

You need to be familiar with the first few regular polygons — ones with up to 10 sides. You need to know their names and how many sides they have (remember that all the sides and angles in a regular polygon are the same).

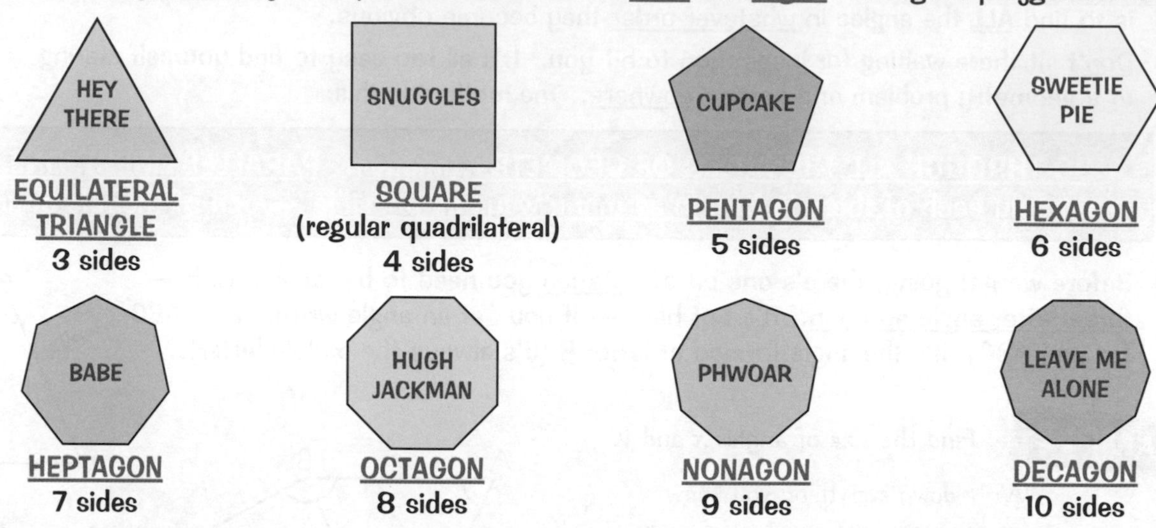

Interior and Exterior Angles

Questions on interior and exterior angles often come up in exams — so you need to know what they are and how to find them. There are a couple of formulas you need to learn as well.

For ANY POLYGON (regular or irregular):

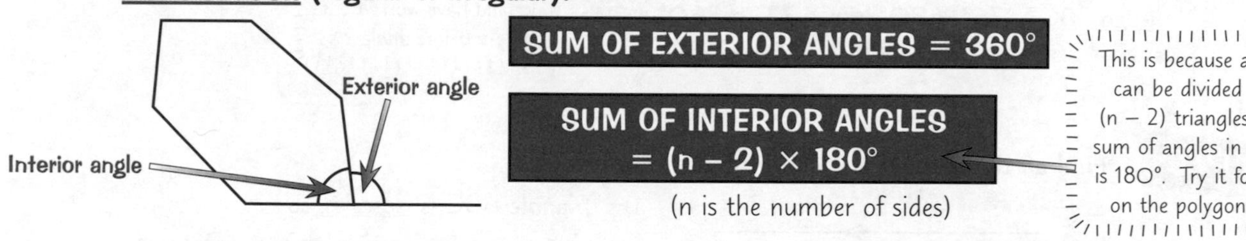

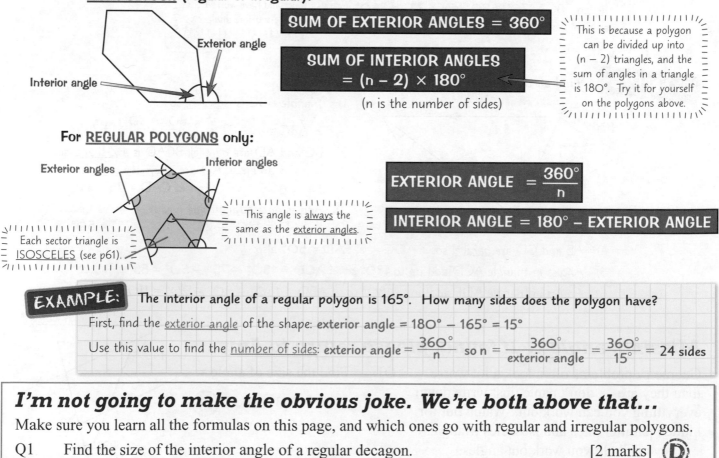

SUM OF EXTERIOR ANGLES = 360°

SUM OF INTERIOR ANGLES = (n – 2) × 180°

(n is the number of sides)

This is because a polygon can be divided up into (n – 2) triangles, and the sum of angles in a triangle is 180°. Try it for yourself on the polygons above.

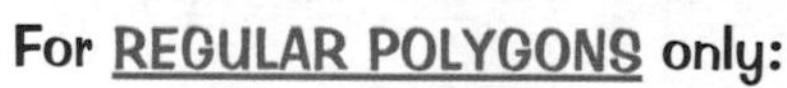

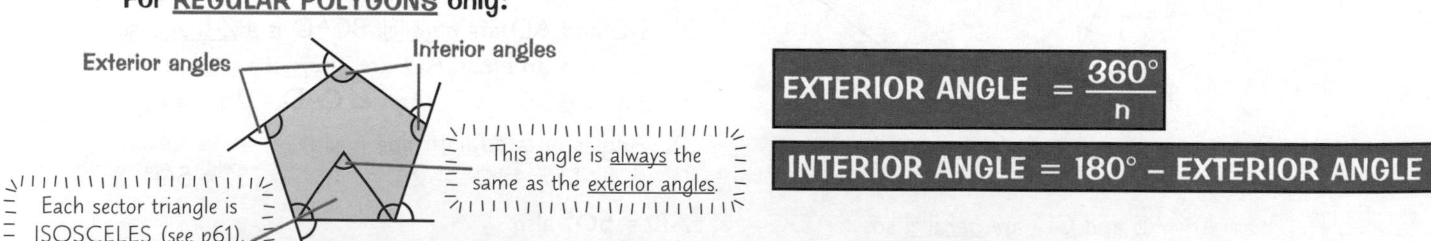

$$\text{EXTERIOR ANGLE} = \frac{360°}{n}$$

INTERIOR ANGLE = 180° – EXTERIOR ANGLE

EXAMPLE: The interior angle of a regular polygon is 165°. How many sides does the polygon have?

First, find the exterior angle of the shape: exterior angle = 180° – 165° = 15°

Use this value to find the number of sides: exterior angle $= \frac{360°}{n}$ so $n = \frac{360°}{\text{exterior angle}} = \frac{360°}{15°} = 24$ sides

I'm not going to make the obvious joke. We're both above that...

Make sure you learn all the formulas on this page, and which ones go with regular and irregular polygons.

Q1 Find the size of the interior angle of a regular decagon. [2 marks]

Aww man, this was gonna be my big break an' everythin'.

Symmetry

After you've finished this page, remind me that I need to pop out and buy a pint of milk. There'll be none left for my breakfast otherwise. Anyway, sorry, symmetry... Right, yes, there are TWO types of symmetry:

Symmetry

1) LINE SYMMETRY

This is where you draw one or more MIRROR LINES across a shape and both sides will fold exactly together. A regular polygon (see previous page) has the same number of lines of symmetry as the number of sides.

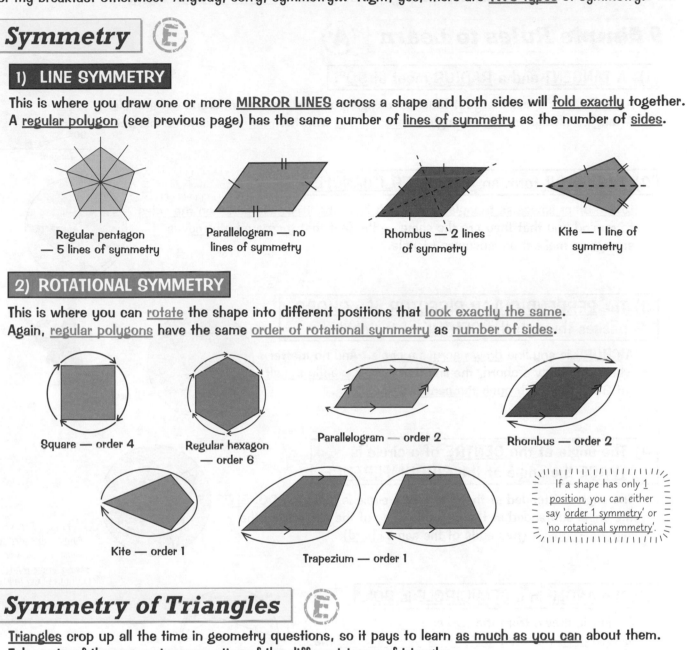

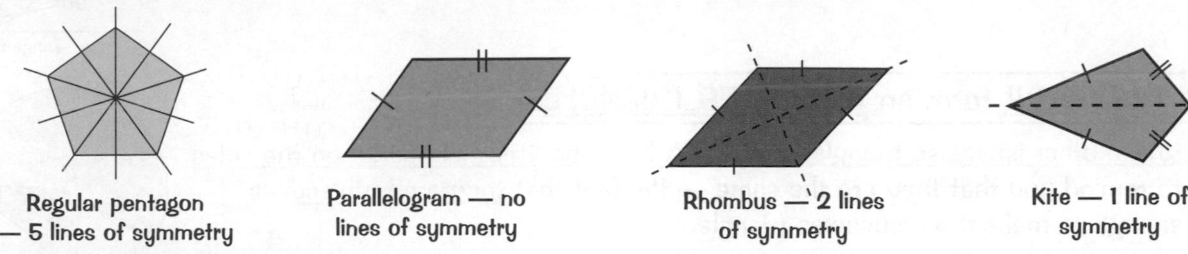

2) ROTATIONAL SYMMETRY

This is where you can rotate the shape into different positions that look exactly the same. Again, regular polygons have the same order of rotational symmetry as number of sides.

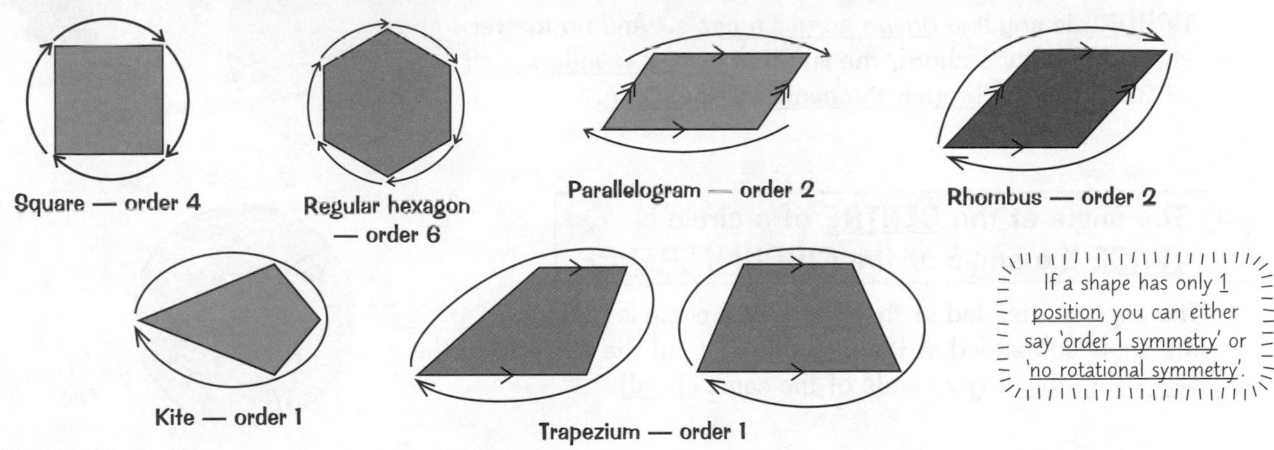

Symmetry of Triangles

Triangles crop up all the time in geometry questions, so it pays to learn as much as you can about them. Take note of the symmetry properties of the different types of triangle:

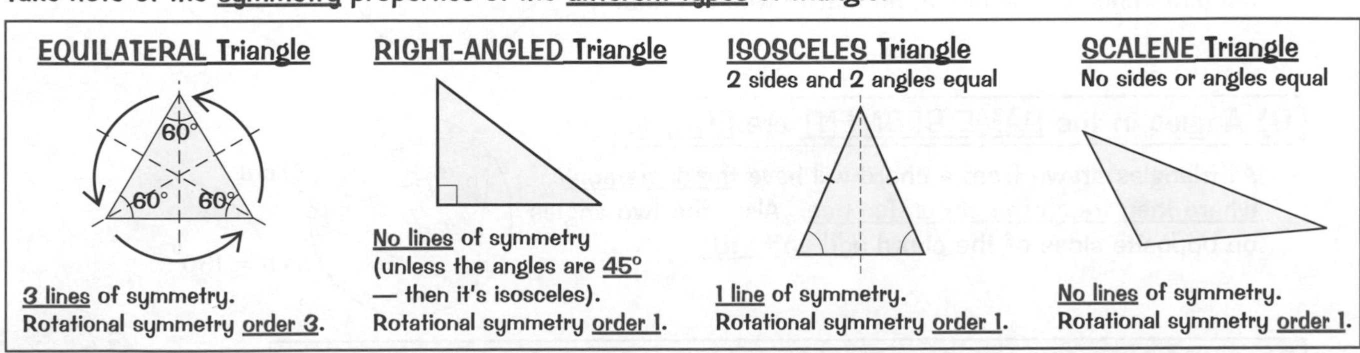

EQUILATERAL Triangle
60°
60°
60°
3 lines of symmetry.
Rotational symmetry order 3.
RIGHT-ANGLED Triangle
No lines of symmetry (unless the angles are 45° — then it's isosceles).
Rotational symmetry order 1.
ISOSCELES Triangle
2 sides and 2 angles equal
1 line of symmetry.
Rotational symmetry order 1.
SCALENE Triangle
No sides or angles equal
No lines of symmetry.
Rotational symmetry order 1.

A regular heptagon has 7 lines of symmetry...

This is a nice easy page. Cover it up and try this Exam Practice Question:

Q1 Write down the number of lines of reflection symmetry and order of rotational symmetry of:
a) an equilateral triangle b) a parallelogram c) a regular octagon [3 marks]

Circle Geometry

After lulling you into a false sense of security with a nice easy page on symmetry, it's time to plunge you into the depths of mathematical peril with a 2-page extravaganza on circle theorems. Sorry.

9 ~~Simple~~ Rules to Learn

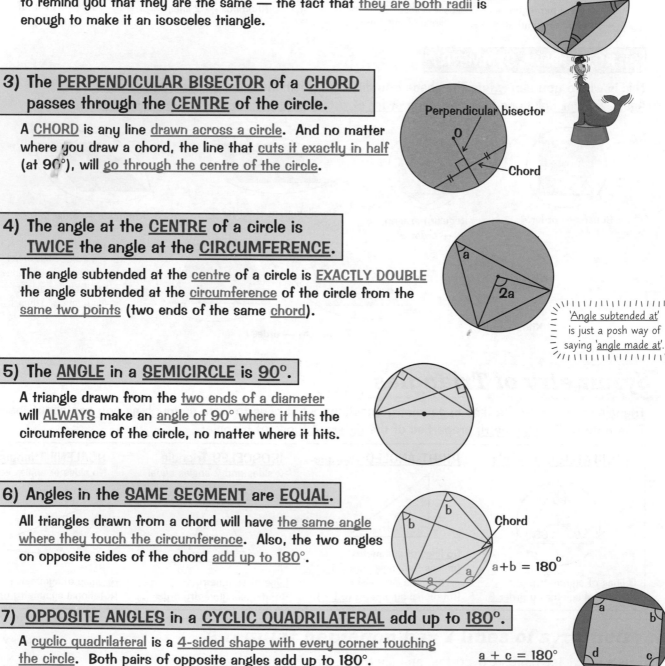

1) A TANGENT and a RADIUS meet at 90°.

A TANGENT is a line that just touches a single point on the circumference of a circle. A tangent always makes an angle of exactly 90° with the radius it meets at this point.

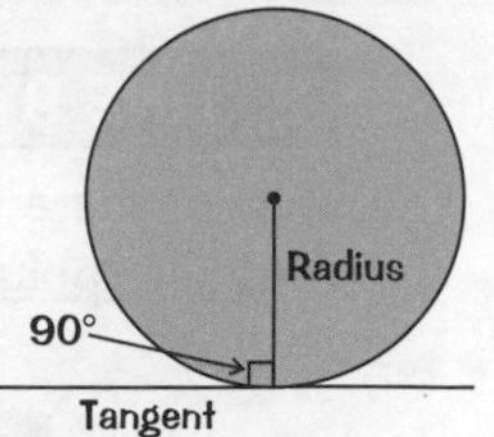

2) TWO RADII form an ISOSCELES TRIANGLE.

Radii is the plural of radius.

Unlike other isosceles triangles they don't have the little tick marks on the sides to remind you that they are the same — the fact that they are both radii is enough to make it an isosceles triangle.

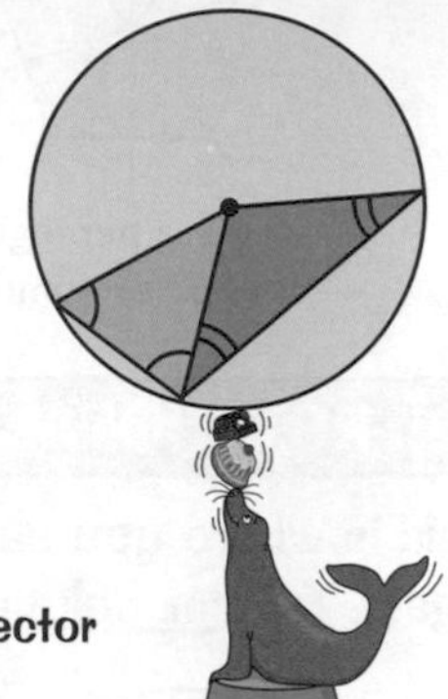

3) The PERPENDICULAR BISECTOR of a CHORD passes through the CENTRE of the circle.

A CHORD is any line drawn across a circle. And no matter where you draw a chord, the line that cuts it exactly in half (at 90°), will go through the centre of the circle.

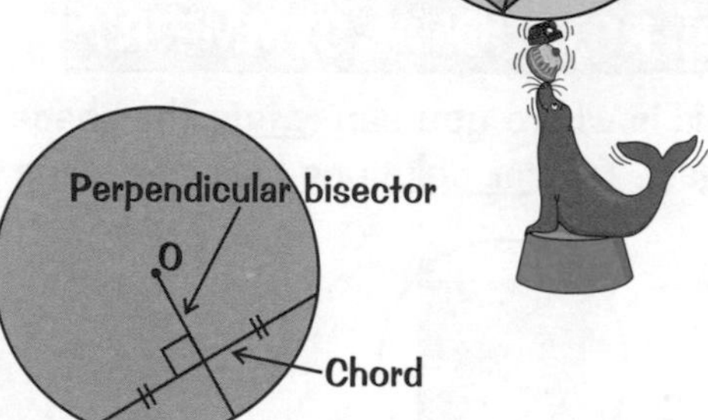

4) The angle at the CENTRE of a circle is TWICE the angle at the CIRCUMFERENCE.

The angle subtended at the centre of a circle is EXACTLY DOUBLE the angle subtended at the circumference of the circle from the same two points (two ends of the same chord).

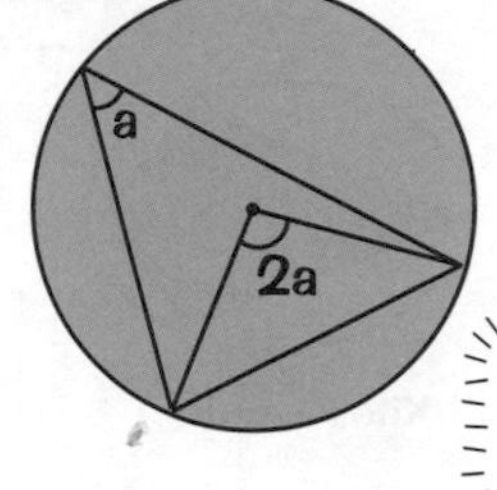

'Angle subtended at' is just a posh way of saying 'angle made at'.

5) The ANGLE in a SEMICIRCLE is 90°.

A triangle drawn from the two ends of a diameter will ALWAYS make an angle of 90° where it hits the circumference of the circle, no matter where it hits.

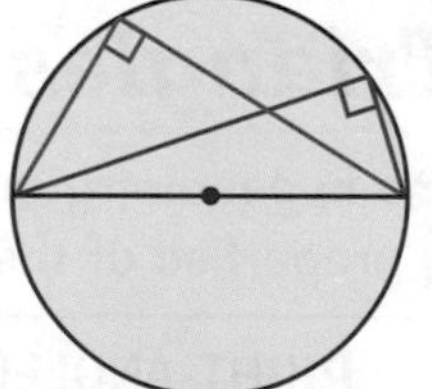

6) Angles in the SAME SEGMENT are EQUAL.

All triangles drawn from a chord will have the same angle where they touch the circumference. Also, the two angles on opposite sides of the chord add up to 180°.

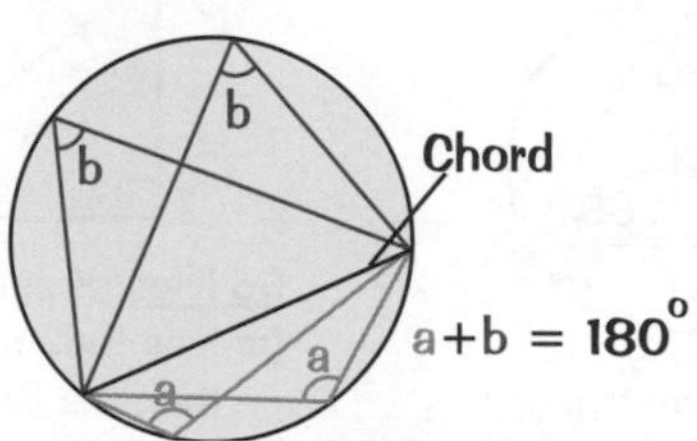

7) OPPOSITE ANGLES in a CYCLIC QUADRILATERAL add up to 180°.

A cyclic quadrilateral is a 4-sided shape with every corner touching the circle. Both pairs of opposite angles add up to 180°.

$a + c = 180°$

$b + d = 180°$

What? No Exam Practice Questions? I feel cheated.

Circle Geometry

More circle theorems? But I've had enough. Can't I go home now?

Final 2 Rules to Learn

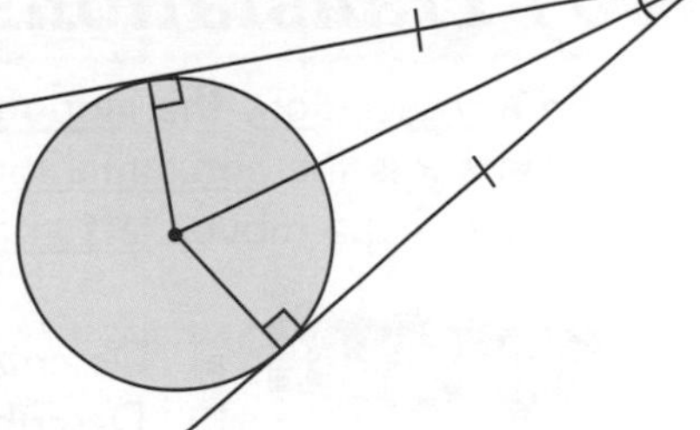

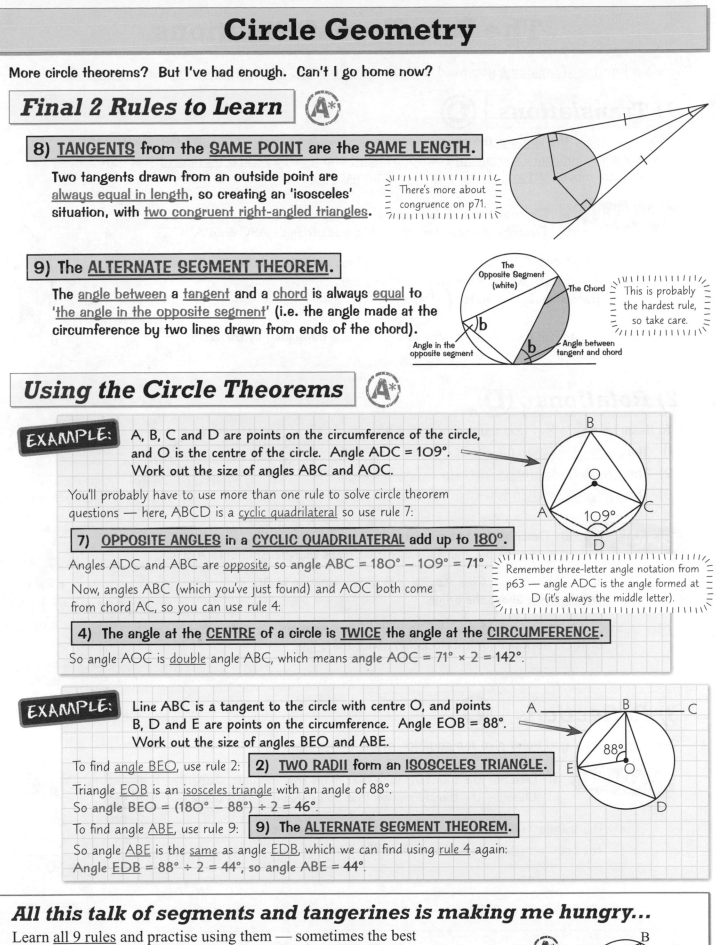

8) TANGENTS from the SAME POINT are the SAME LENGTH.

Two tangents drawn from an outside point are always equal in length, so creating an 'isosceles' situation, with two congruent right-angled triangles.

There's more about congruence on p71.

9) The ALTERNATE SEGMENT THEOREM.

The angle between a tangent and a chord is always equal to 'the angle in the opposite segment' (i.e. the angle made at the circumference by two lines drawn from ends of the chord).

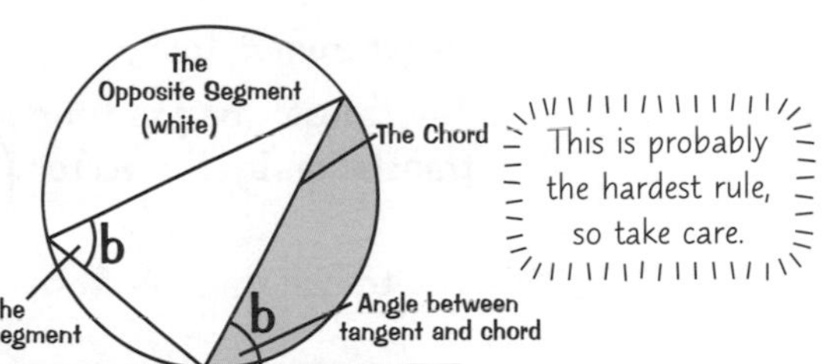

Using the Circle Theorems

EXAMPLE: A, B, C and D are points on the circumference of the circle, and O is the centre of the circle. Angle ADC = 109°. Work out the size of angles ABC and AOC.

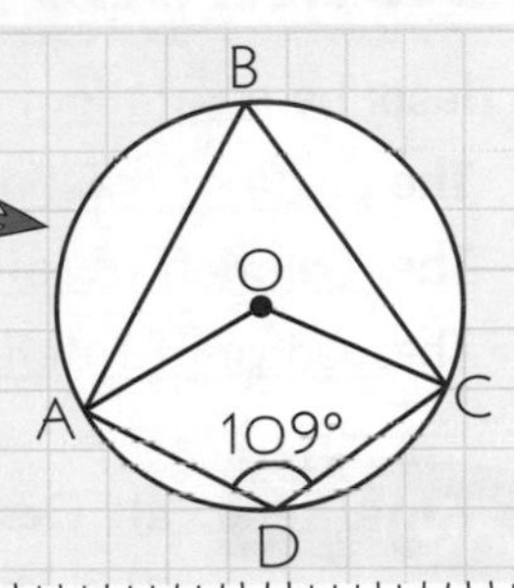

You'll probably have to use more than one rule to solve circle theorem questions — here, ABCD is a cyclic quadrilateral so use rule 7:

7) OPPOSITE ANGLES in a CYCLIC QUADRILATERAL add up to 180°.

Angles ADC and ABC are opposite, so **angle ABC = 180° – 109° = 71°**.

Remember three-letter angle notation from p63 — angle ADC is the angle formed at D (it's always the middle letter).

Now, angles ABC (which you've just found) and AOC both come from chord AC, so you can use rule 4:

4) The angle at the CENTRE of a circle is TWICE the angle at the CIRCUMFERENCE.

So angle AOC is double angle ABC, which means **angle AOC = 71° × 2 = 142°**.

EXAMPLE: Line ABC is a tangent to the circle with centre O, and points B, D and E are points on the circumference. Angle EOB = 88°. Work out the size of angles BEO and ABE.

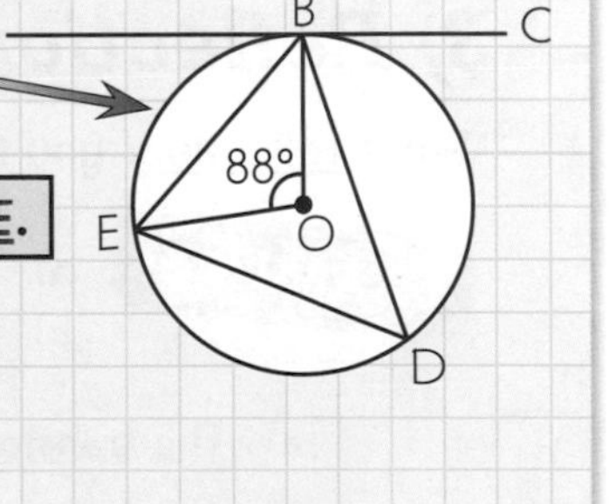

To find angle BEO, use rule 2: **2) TWO RADII form an ISOSCELES TRIANGLE.**

Triangle EOB is an isosceles triangle with an angle of 88°.
So **angle BEO = (180° – 88°) ÷ 2 = 46°**.

To find angle ABE, use rule 9: **9) The ALTERNATE SEGMENT THEOREM.**

So angle ABE is the same as angle EDB, which we can find using rule 4 again:
Angle EDB = 88° ÷ 2 = 44°, so angle ABE = 44°.

All this talk of segments and tangerines is making me hungry...

Learn all 9 rules and practise using them — sometimes the best approach is to try different rules until you find one that works.

Q1 A, B, C and D are points on the circumference of the circle with centre O. The line EF is a tangent to the circle, and touches the circle at D. Angle ADE is 63°. Find the size of angles ABD and ACD. [2 marks]

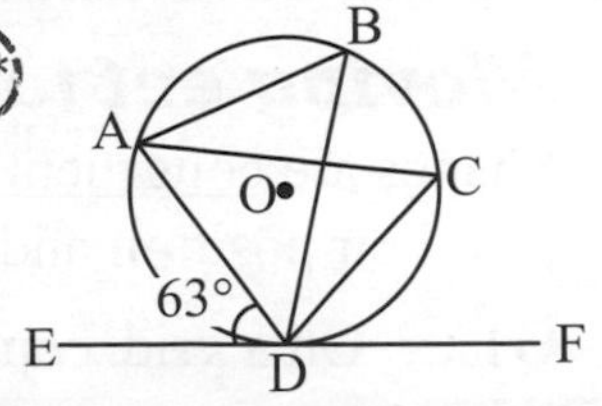

The Four Transformations

There are four transformations you need to know — translation, rotation, reflection and enlargement.

1) Translations

In a translation, the amount the shape moves by is given as a vector (see p97-98) written $\begin{pmatrix} x \\ y \end{pmatrix}$ — where x is the horizontal movement (i.e. to the right) and y is the vertical movement (i.e. up). If the shape moves left and down, x and y will be negative.

EXAMPLE:

a) Describe the transformation that maps triangle ABC onto A'B'C'.

b) Describe the transformation that maps triangle ABC onto A''B''C''.

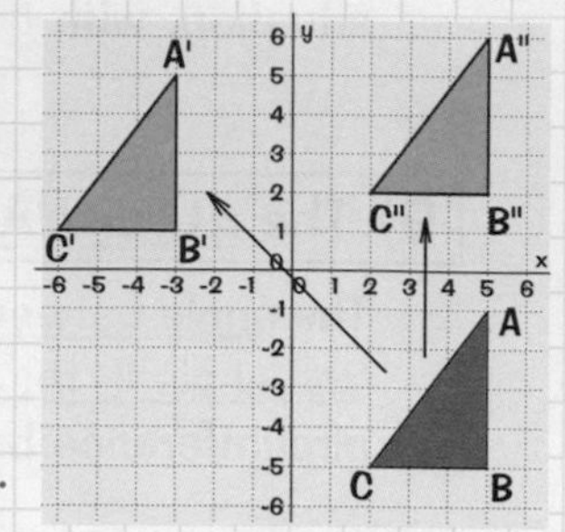

a) To get from A to A', you need to move 8 units left and 6 units up, so...
The transformation from ABC to A'B'C' is a translation by the vector $\begin{pmatrix} -8 \\ 6 \end{pmatrix}$.

b) The transformation from ABC to A''B''C'' is a translation by the vector $\begin{pmatrix} 0 \\ 7 \end{pmatrix}$.

2) Rotations

To describe a rotation, you must give 3 details:

1) The angle of rotation (usually 90° or 180°).
2) The direction of rotation (clockwise or anticlockwise).
3) The centre of rotation (often, but not always, the origin).

For a rotation of 180°, it doesn't matter whether you go clockwise or anticlockwise.

EXAMPLE:

a) Describe the transformation that maps triangle ABC onto A'B'C'.

b) Describe the transformation that maps triangle ABC onto A''B''C''.

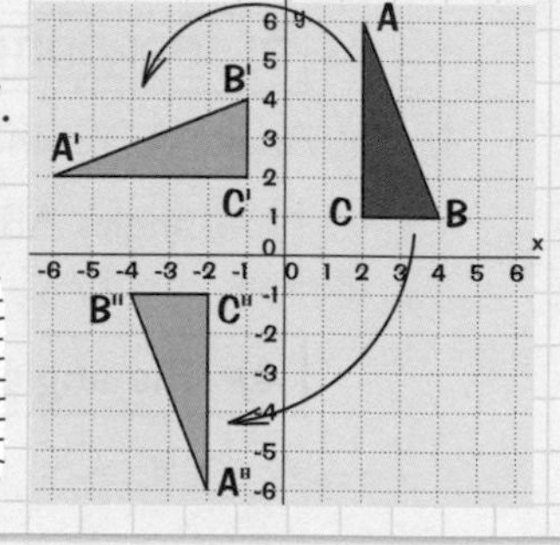

a) The transformation from ABC to A'B'C' is a rotation of 90° anticlockwise about the origin.

b) The transformation from ABC to A''B''C'' is a rotation of 180° clockwise (or anticlockwise) about the origin.

If it helps, you can use tracing paper to help you find the centre of rotation.

3) Reflections

For a reflection, you must give the equation of the mirror line.

EXAMPLE:

a) Describe the transformation that maps shape A onto shape B.

b) Describe the transformation that maps shape A onto shape C.

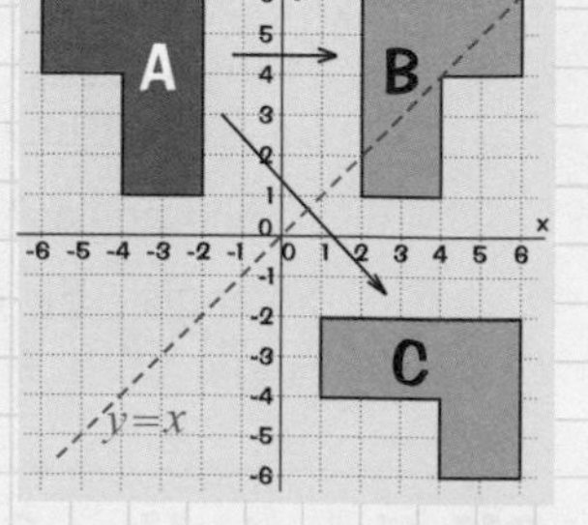

a) The transformation from A to B is a reflection in the y-axis.

b) The transformation from A to C is a reflection in the line $y = x$.

Moving eet to ze left — a perfect translation...

Shapes are congruent under translation, reflection and rotation because their size and shape don't change, just their position and orientation (congruence is on p71). Now have a go at this question:

Q1 On a grid, copy shape A above and rotate it 90° clockwise about the point (–1, –1). [2 marks]

The Four Transformations

One more transformation coming up — enlargements. They're the trickiest, but also the most fun (honest).

4) Enlargements 

For an enlargement, you must specify:

1) **The scale factor.**
2) **The centre of enlargement.**

$$\text{scale factor} = \frac{\text{new length}}{\text{old length}}$$

EXAMPLE:

a) Describe the transformation that maps triangle A onto triangle B.
b) Describe the transformation that maps triangle B onto triangle A.

a) Use the formula above to find the scale factor (just choose one side): scale factor $= \frac{6}{3} = 2$

For the centre of enlargement, draw lines that go through corresponding vertices of both shapes and see where they cross.

So the transformation from A to B is an enlargement of scale factor 2, centre (2, 6)

b) Using a similar method, **scale factor** $= \frac{3}{6} = \frac{1}{2}$ and the centre of enlargement is the same as before, **so the transformation from B to A is an enlargement of scale factor $\frac{1}{2}$, centre (2, 6)**

If the scale factor is less than 1, the shape will get smaller (A is smaller than B). There's more on this below.

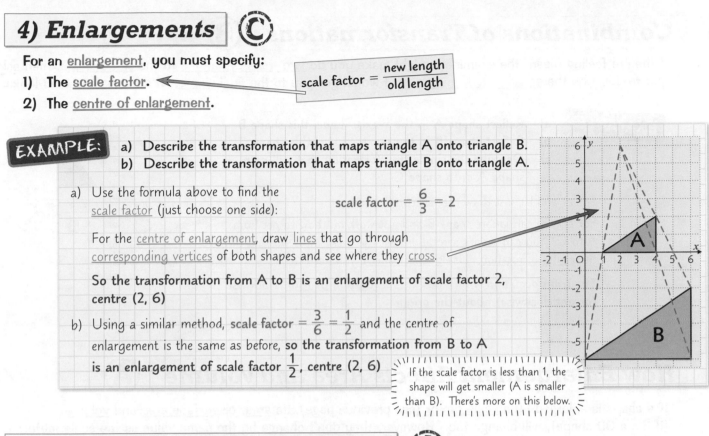

Scale Factors — Four Key Facts 

1) **If the scale factor is bigger than 1 the shape gets bigger.**
2) **If the scale factor is smaller than 1 (e.g. ½) it gets smaller.**
3) **If the scale factor is negative then the shape pops out the other side of the enlargement centre. If the scale factor is -1, it's exactly the same as a rotation of 180°.**
4) **The scale factor also tells you the relative distance of old points and new points from the centre of enlargement — this is very useful for drawing an enlargement, because you can use it to trace out the positions of the new points.**

EXAMPLE:

Enlarge shape A below by a scale factor of -3, centre (1, 1). Label the transformed shape B.

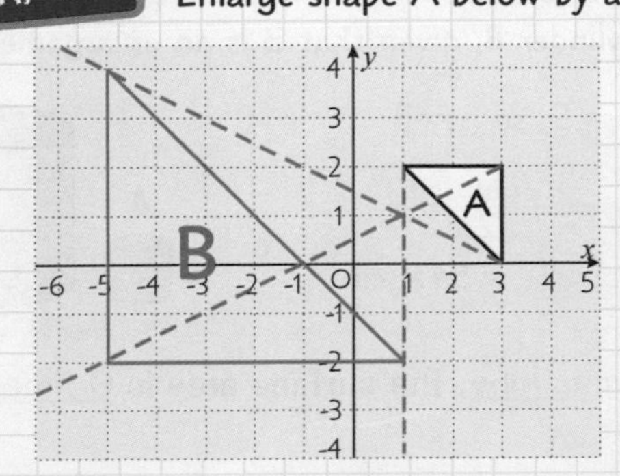

1) First, draw lines going through (1, 1) from each vertex of shape A.
2) Then, multiply the distance from each vertex to the centre of enlargement by 3, and measure this distance coming out the other side of the centre of enlargement.
So on shape A, vertex (3, 2) is 2 right and 1 up from (1, 1) — so the corresponding point on shape B will be 6 left and 3 down from (1, 1). Do this for every point.
3) Join the points you've drawn to form shape **B**.

Scale factors — they're enough to put the fear of cod into you...

Shapes are similar under enlargement — the position and the size change, but the angles and ratios of the sides don't (see p72). Make sure you learn the 4 key facts about scale factors.

Q1 On a grid, draw triangle A with vertices (2, 1), (4, 1) and (4, 3), then enlarge it by a scale factor of –2 about point (1, 1). [3 marks] 

More Transformation Stuff

Just one more page on transformations, and then you're done. With transformations anyway, not with Maths.

Combinations of Transformations

If they're feeling mean, the examiners might make you do two transformations to the same shape, then ask you to describe the single transformation that would get you to the final shape. It's not as bad as it looks.

a) Reflect shape A in the x-axis. Label this shape B.
b) Reflect shape B in the y-axis. Label this shape C.
c) Describe the single transformation that will map shape A onto shape C.

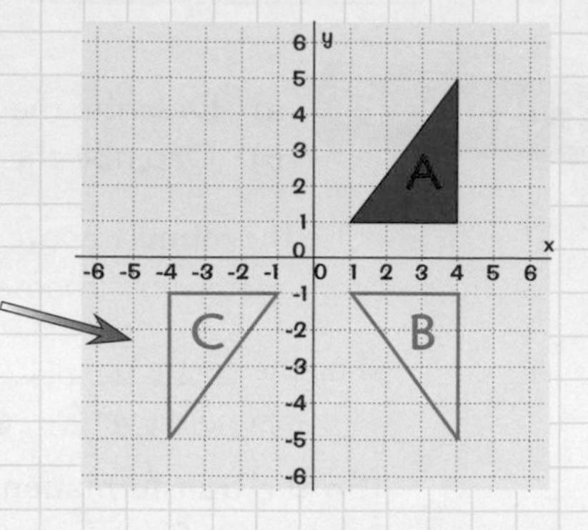

For a) and b), just draw the reflections.

For c), you can ignore shape B and just work out how to get from A to C. You can see it's a rotation, but the tricky bit is working out the centre of rotation. Use tracing paper if you need to.

The transformation from A to C is a rotation of 180° clockwise (or anticlockwise) about the origin.

How Enlargement Affects Area and Volume

If a shape is enlarged by a scale factor (see previous page), its area, or surface area and volume (if it's a 3D shape), will change too. However, they don't change by the same value as the scale factor:

For a SCALE FACTOR n:

The **SIDES** are n times bigger
The **AREAS** are n^2 times bigger
The **VOLUMES** are n^3 times bigger

And:

$$n = \frac{\text{new length}}{\text{old length}} \qquad n^2 = \frac{\text{new area}}{\text{old area}}$$

$$n^3 = \frac{\text{new volume}}{\text{old volume}}$$

So if the scale factor is 2, the lengths are twice as long, the area is 2^2 = 4 times as big, and the volume is 2^3 = 8 times as big.

There's more on areas on p74-76 and volumes on p77-78.

EXAMPLE:

Cylinder A has surface area 6π cm², and cylinder B has surface area 54π cm². The volume of cylinder A is 2π cm³. Find the volume of cylinder B, given that B is an enlargement of A.

First, work out the scale factor, n: $n^2 = \frac{\text{Area B}}{\text{Area A}} = \frac{54\pi}{6\pi} = 9$, so $n = 3$

Use this in the volume formula: $n^3 = \frac{\text{Volume B}}{\text{Volume A}} \Rightarrow 3^3 = \frac{\text{Volume B}}{2\pi}$

$\Rightarrow$ Volume of B $= 2\pi \times 27 = 54\pi$ cm³

A B

This shows that if the scale factor is 3, lengths are 3 times as long, the surface area is 9 times as big and the volume is 27 times as big.

Twice as much learning, 4 times better results, 8 times more fun...

Make sure you don't get the scale factors mixed up — try them out on this Exam Practice Question.

Q1 There are 3 stacking dolls in a set. The dolls are mathematically similar and have heights of 5 cm, 10 cm and 15 cm. The surface area of the middle doll is 80 cm², and the volume of the largest doll is 216 cm³. Find the surface area and volume of the smallest doll. [4 marks]

A

Congruent Shapes

Congruence is another ridiculous maths word which sounds really complicated when it's not. If two shapes are congruent, they are simply the same — the same size and the same shape. That's all it is. They can however be reflected or rotated.

CONGRUENT — same size, same shape

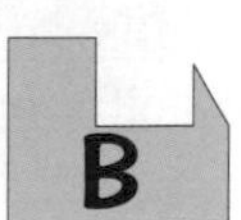

Proving Triangles are Congruent

To prove that two triangles are congruent, you have to show that one of the conditions below holds true:

1) **SSS** three sides are the same
2) **AAS** two angles and a corresponding side match up
3) **SAS** two sides and the angle between them match up
4) **RHS** a right angle, the hypotenuse and one other side all match up

The hypotenuse is the longest side of a right-angled triangle — the one opposite the right angle.

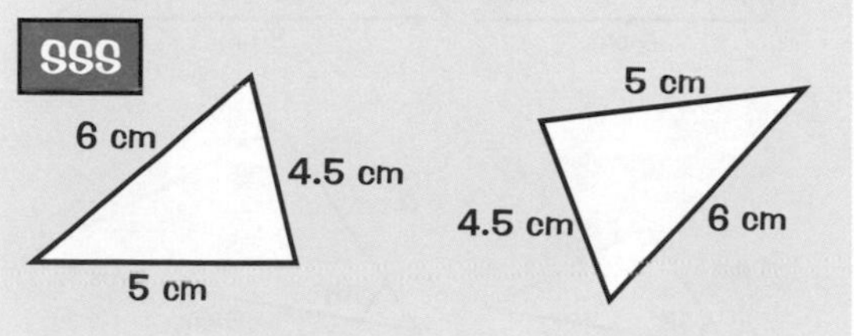

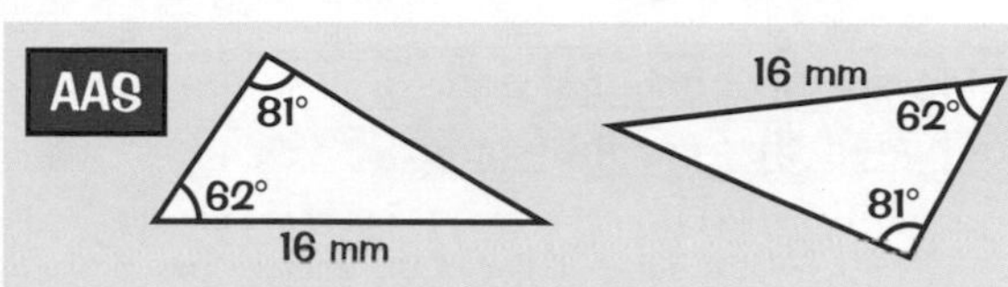

Make sure the sides match up — here, the side is opposite the 81° angle.

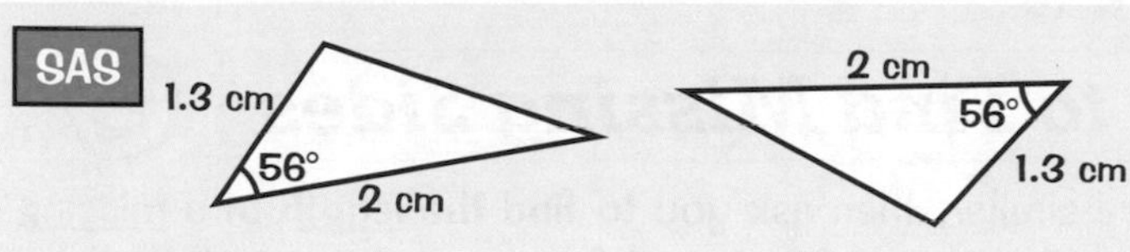

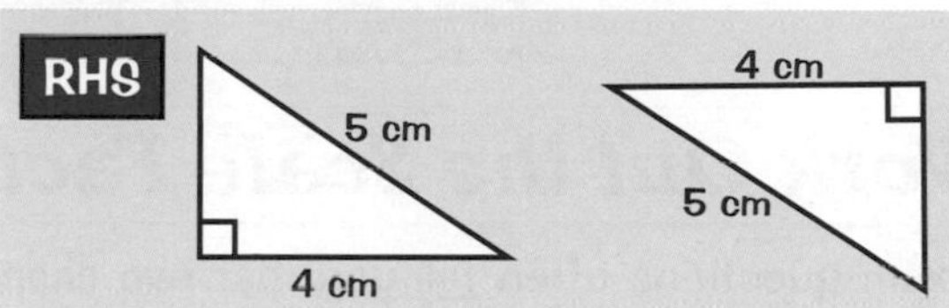

Work Out All the Sides and Angles You Can Find

The best approach to proving two triangles are congruent is to write down everything you can find out, then see which condition they fit. Watch out for things like parallel lines (p62) and circle theorems (p66-67).

EXAMPLE: XY and YZ are tangents to the circle with centre O, and touch the circle at points X and Z respectively. Prove that the triangles OXY and OYZ are congruent.

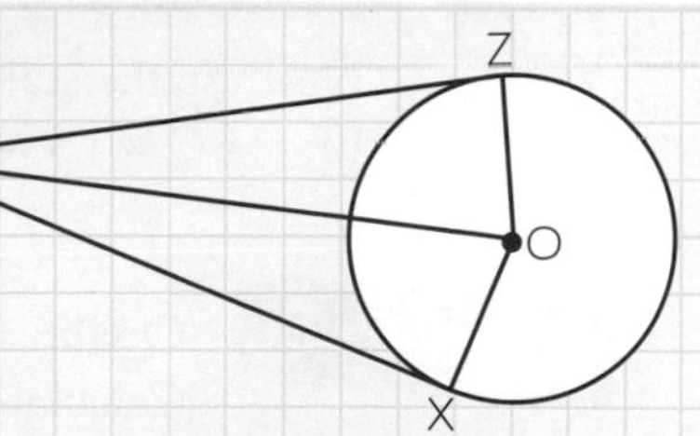

Write down what you know (you're going to have to use circle theorems):

- Sides OX and OZ are the same length (as they're both radii).
- Both triangles have a right angle (OXY and OZY) as a tangent meets a radius at 90°.
- OY is the hypotenuse of each triangle.

So the condition RHS holds, as there is a right angle, the hypotenuses are the same and one other side of each triangle (OX and OZ) are the same. **RHS holds, so OXY and OYZ are congruent triangles.**

SAS? More like SOS...

Learn all 4 conditions and make sure you know how to use them to prove that triangles are congruent. Then have a go at this Exam Practice Question:

Q1 Prove that triangles ABD and BCD are congruent. [3 marks] (B)

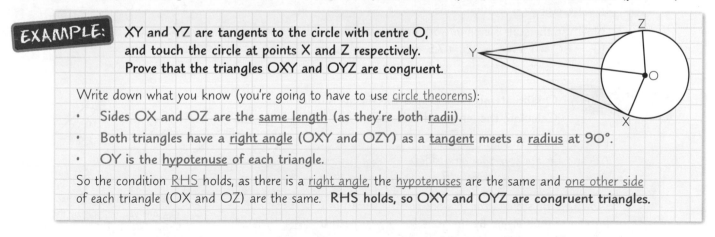

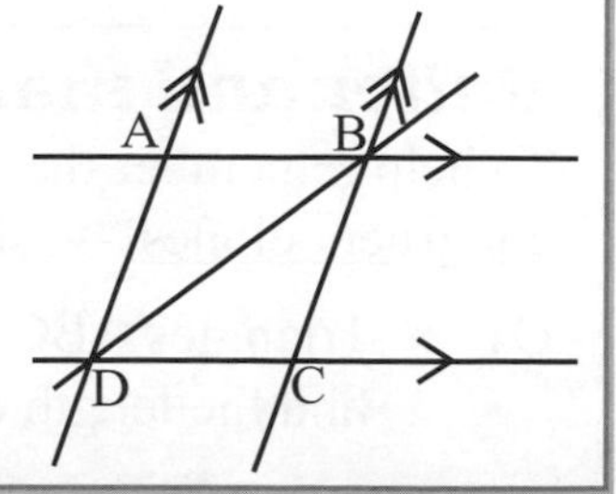

Similar Shapes

Similar shapes are exactly the same shape, but can be different sizes (they can also be rotated or reflected).

SIMILAR — same shape, different size

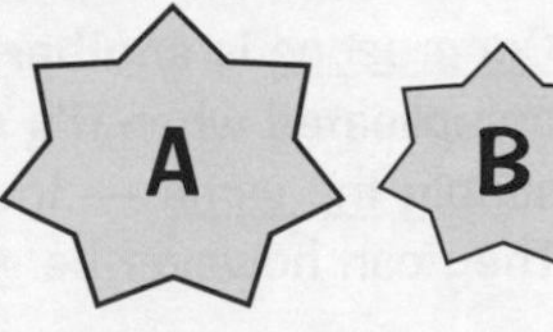

Similar Shapes Have the Same Angles

Generally, for two shapes to be similar, all the angles must match and the sides must be proportional. But for triangles, there are three special conditions — if any one of these is true, you know they're similar.

Two triangles are similar if:

1) **All the angles match up** i.e. the angles in one triangle are the same as the other.
2) **All three sides are proportional** i.e. if one side is twice as long as the corresponding side in the other triangle, all the sides are twice as long as the corresponding sides.
3) **Any two sides are proportional and the angle between them is the same.**

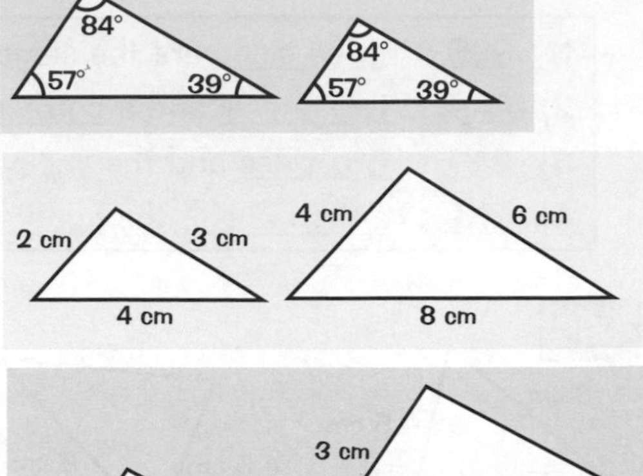

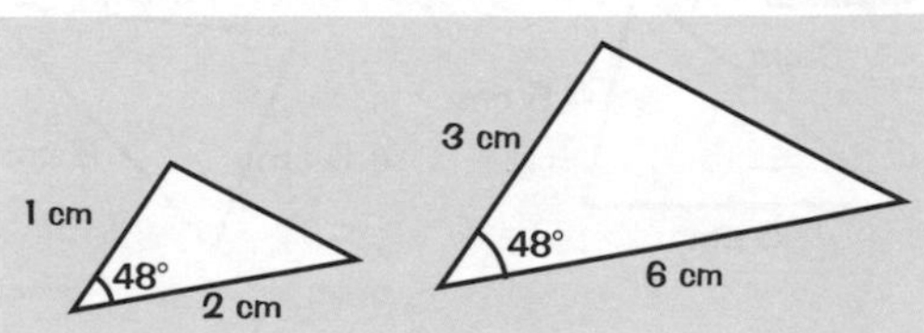

Watch out — if one of the triangles has been rotated or flipped over, it might look as if they're not similar, but don't be fooled.

Work Out the Scale Factor to Find Missing Sides

Exam questions often tell you that two shapes are similar, then ask you to find the length of a missing side. You need to find the scale factor (remember enlargements — p69) to get from one shape to the other.

EXAMPLE: Triangles ABC and DEF are similar. Calculate the length of side EF.

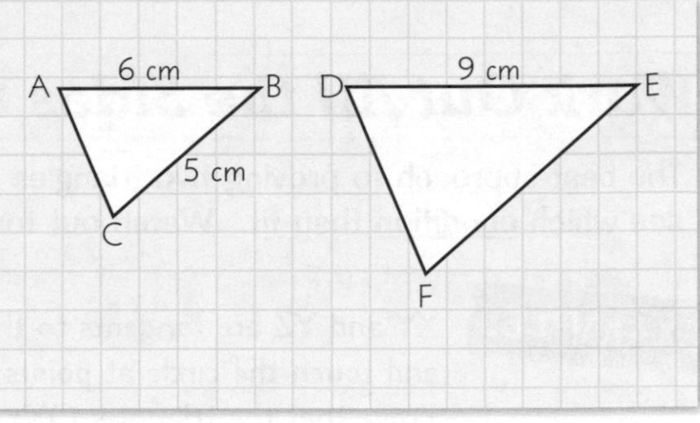

The first thing to do is to work out the scale factor: AB and DE are corresponding sides:

scale factor $= \frac{DE}{AB} = \frac{9}{6} = 1.5$.

Now use the scale factor to work out the length of EF:

EF = BC × 1.5 = 5 × 1.5 = 7.5 cm

EXAMPLE: Quadrilaterals ABCD and EFGH are similar. Calculate the length of side CD.

First, work out the scale factor: AB and EF are corresponding sides,

so the scale factor is $\frac{EF}{AB} = \frac{7}{4} = 1.75$.

Now use the scale factor to work out the length of CD (this time, instead of multiplying by the scale factor, you need to divide by it):

CD = GH ÷ 1.75 = 10.5 ÷ 1.75 = 6 cm

To get from the original shape to the new one you multiply by the scale factor — so to get from EFGH to ABCD you divide by the scale factor.

Butter and margarine — similar products...

To help remember the difference between similarity and congruence, think 'similar siblings, congruent clones' — siblings are alike but not the same, clones are identical.

Q1 Triangles ABC and ADE are similar. Given that AB = 20 cm, find the length of AD. [2 marks]

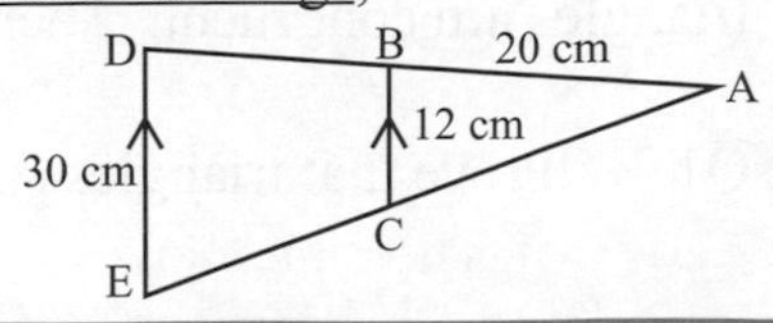

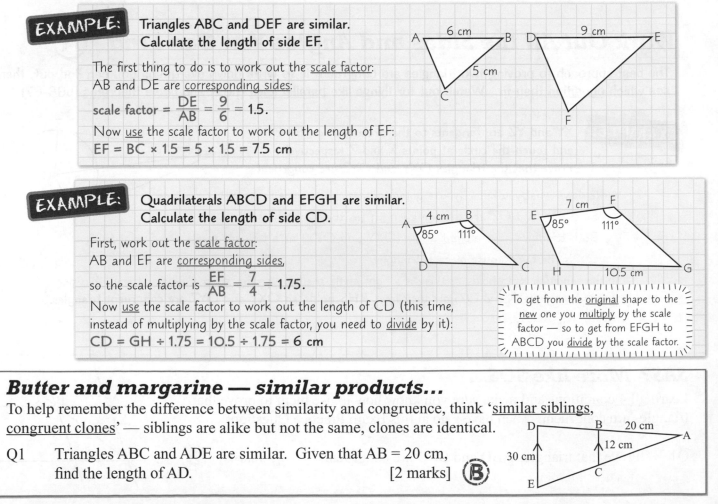

Projections

<u>Projections</u> are just different <u>views</u> of a 3D solid shape — looking at it from the <u>front</u>, the <u>side</u> and the <u>top</u>.

The Three Different Projections

There are three different types of projections — <u>front elevations</u>, <u>side elevations</u> and <u>plans</u> (elevation is just another word for projection).

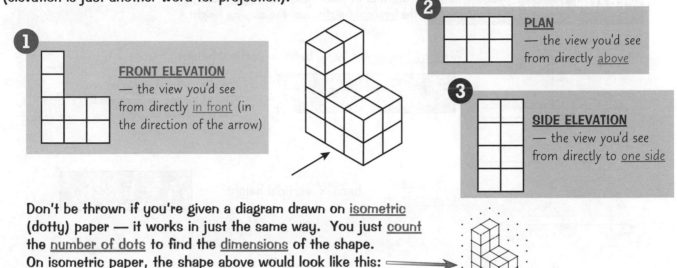

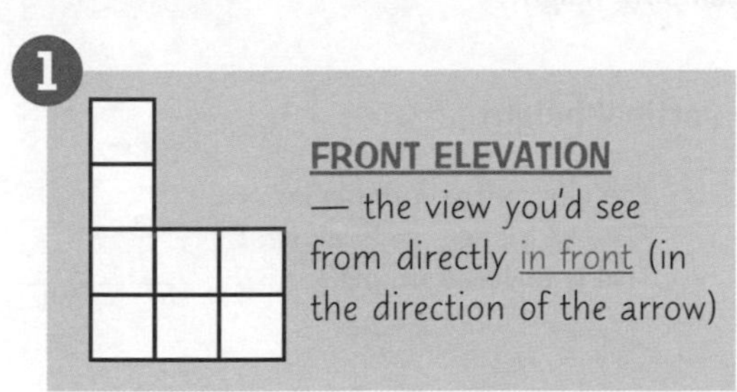

1
FRONT ELEVATION
— the view you'd see from directly in front (in the direction of the arrow)

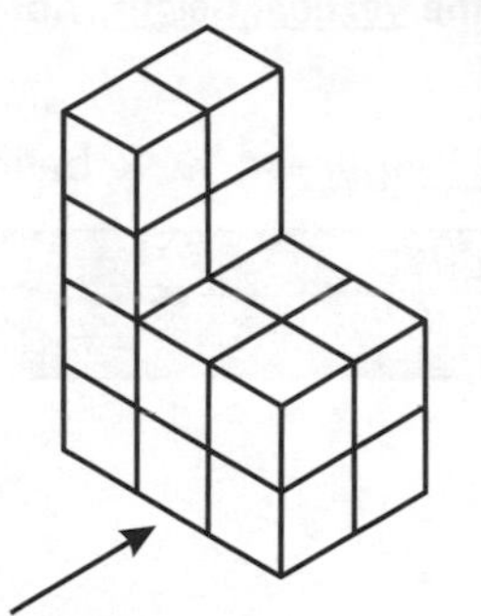

2 **PLAN** — the view you'd see from directly <u>above</u>

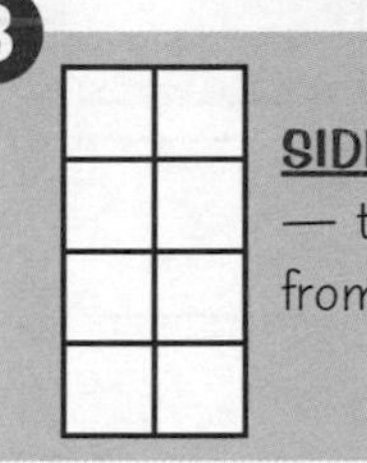

3 **SIDE ELEVATION** — the view you'd see from directly to <u>one side</u>

Don't be thrown if you're given a diagram drawn on <u>isometric</u> (dotty) paper — it works in just the same way. You just <u>count</u> the <u>number of dots</u> to find the <u>dimensions</u> of the shape. On isometric paper, the shape above would look like this:

Decide If You Need To Draw Projections To Scale

1) You'll often be given <u>grid paper</u> to draw your projections on — so make sure you get the <u>dimensions</u> right.
2) However, if you're asked to <u>sketch</u> a projection (and not given grid paper), it <u>doesn't</u> have to be perfectly to <u>scale</u>.
3) <u>Lines</u> on the projection should show any changes in <u>depth</u>.

A question might give you a plan and elevation and ask you to sketch the solid... Just piece together the original shape from the info given. And <u>check</u> your final sketch against the question.

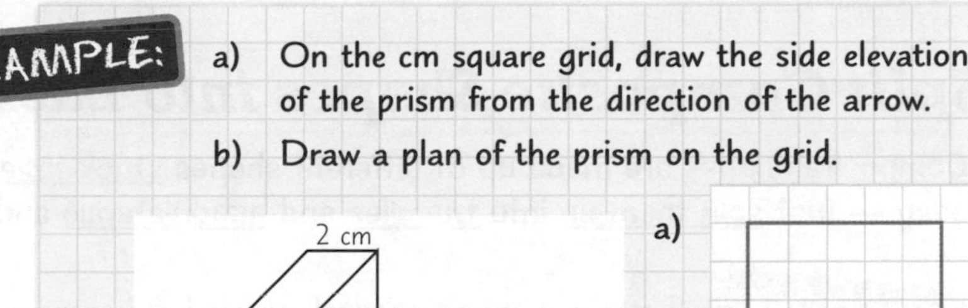

EXAMPLE:
a) On the cm square grid, draw the side elevation of the prism from the direction of the arrow.
b) Draw a plan of the prism on the grid.

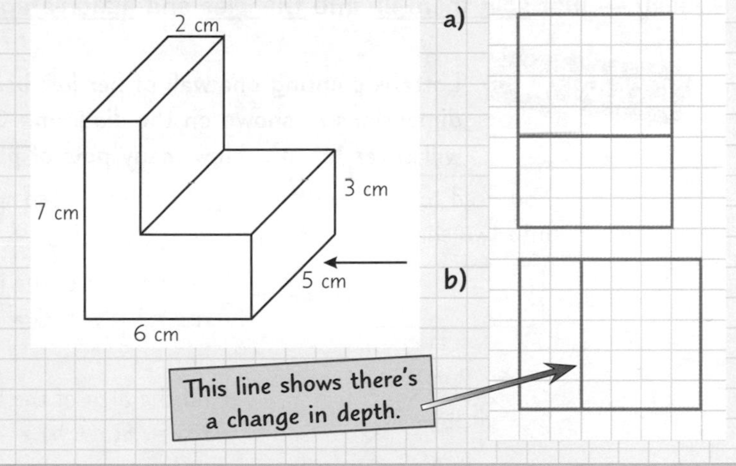

Projections — enough to send you dotty...

Projection questions aren't too bad — just take your time and sketch the diagrams carefully. Watch out for questions on isometric paper — they might look confusing, but they can actually be easier than other questions.

Q1 For the shape on the right, sketch: (D)
a) The front elevation (from the direction of the arrow), [1 mark]
b) The side elevation, [1 mark]
c) The plan view. [1 mark]

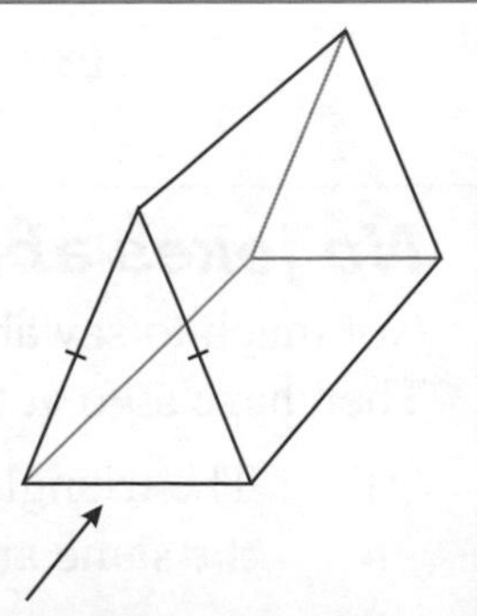

Areas

Be warned — there are lots of area formulas coming up on the next two pages for you to learn. By the way, I'm assuming that you know the formulas for the area of a rectangle ($A = l \times w$) and the area of a square ($A = l^2$).

Areas of Triangles and Quadrilaterals

LEARN these Formulas:

Note that in each case the height must be the vertical height, not the sloping height.

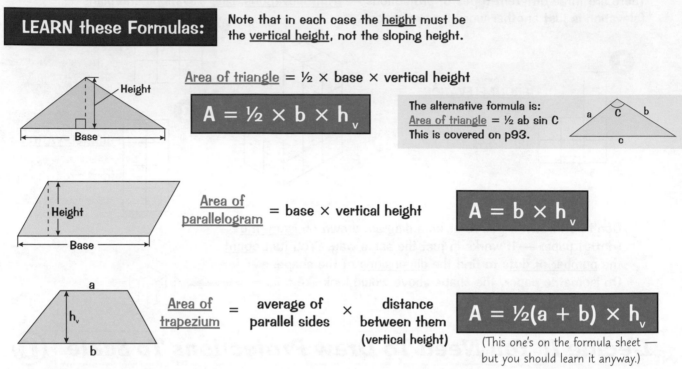

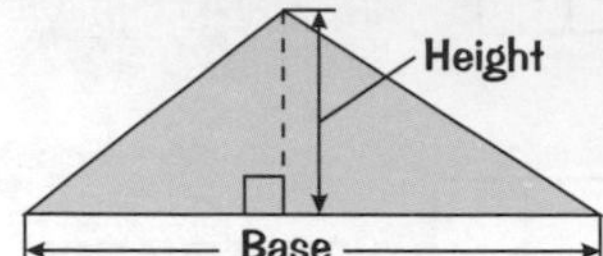

Area of triangle = ½ × base × vertical height

$$A = \frac{1}{2} \times b \times h_v$$

The alternative formula is:
Area of triangle = ½ ab sin C
This is covered on p93.

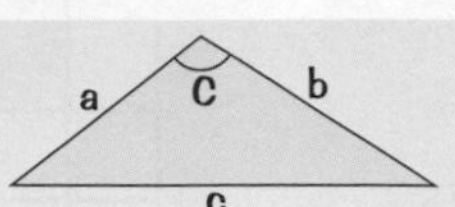

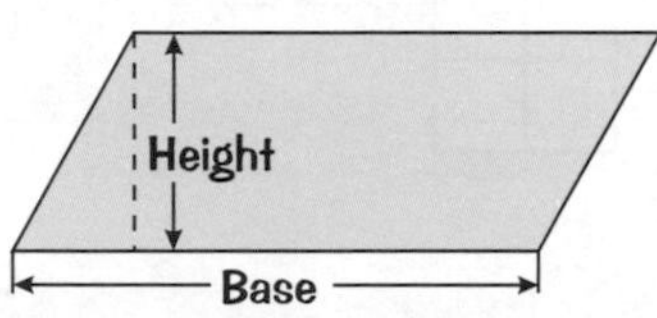

Area of parallelogram = base × vertical height

$$A = b \times h_v$$

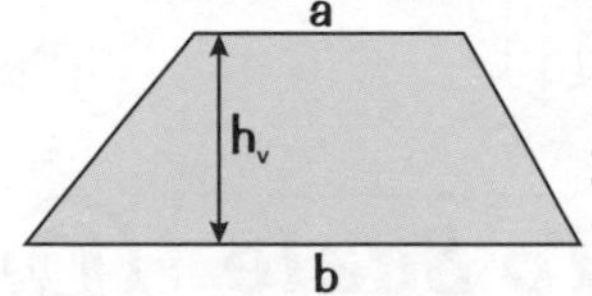

Area of trapezium = average of parallel sides × distance between them (vertical height)

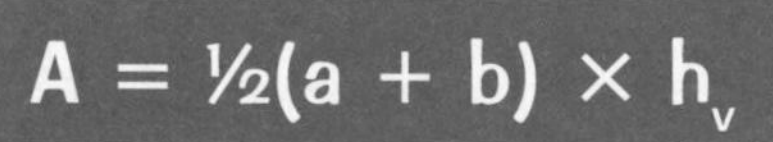

$$A = \frac{1}{2}(a + b) \times h_v$$

(This one's on the formula sheet — but you should learn it anyway.)

Split Composite Shapes into Easier Shapes

Composite shapes are made up of different shapes stuck together. Finding their area is actually dead easy — just split them up into triangles and quadrilaterals and work out the area of each bit.

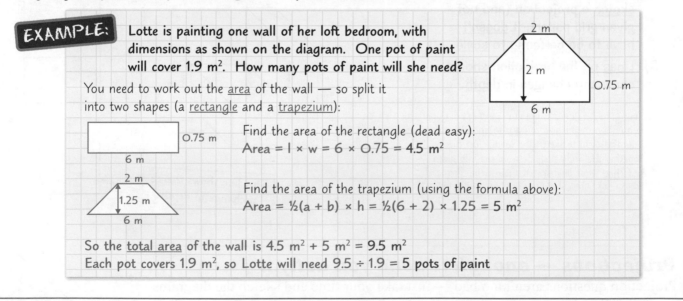

EXAMPLE: Lotte is painting one wall of her loft bedroom, with dimensions as shown on the diagram. One pot of paint will cover 1.9 m². How many pots of paint will she need?

You need to work out the area of the wall — so split it into two shapes (a rectangle and a trapezium):

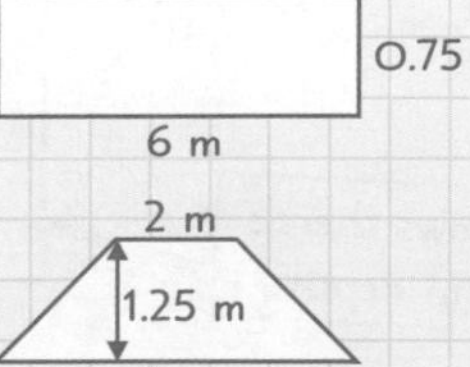

Find the area of the rectangle (dead easy):
Area = $l \times w = 6 \times 0.75 = 4.5$ m²

Find the area of the trapezium (using the formula above):
Area = $\frac{1}{2}(a + b) \times h = \frac{1}{2}(6 + 2) \times 1.25 = 5$ m²

So the total area of the wall is 4.5 m² + 5 m² = 9.5 m²
Each pot covers 1.9 m², so Lotte will need 9.5 ÷ 1.9 = **5 pots of paint**

No jokes about my vertical height please...

Not much to say about this page really — LEARN the formulas and practise using them. Then have a go at this Exam Practice Question.

Q1 The triangle and rectangle shown on the right have the same area. Find the value of x. [4 marks]

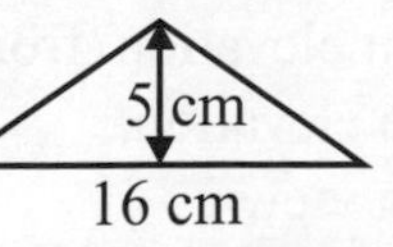

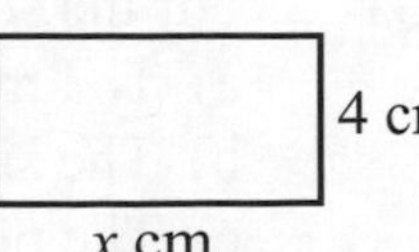

Areas

Yes, I thought I could detect some groaning when you realised that this is another page of formulas. You know the drill...

LEARN these Formulas

Area and Circumference of Circles 

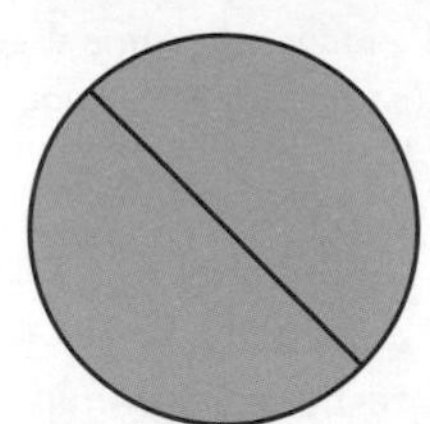

Area of circle = $\pi \times (\text{radius})^2$

Remember that the radius is half the diameter.

$$A = \pi r^2$$

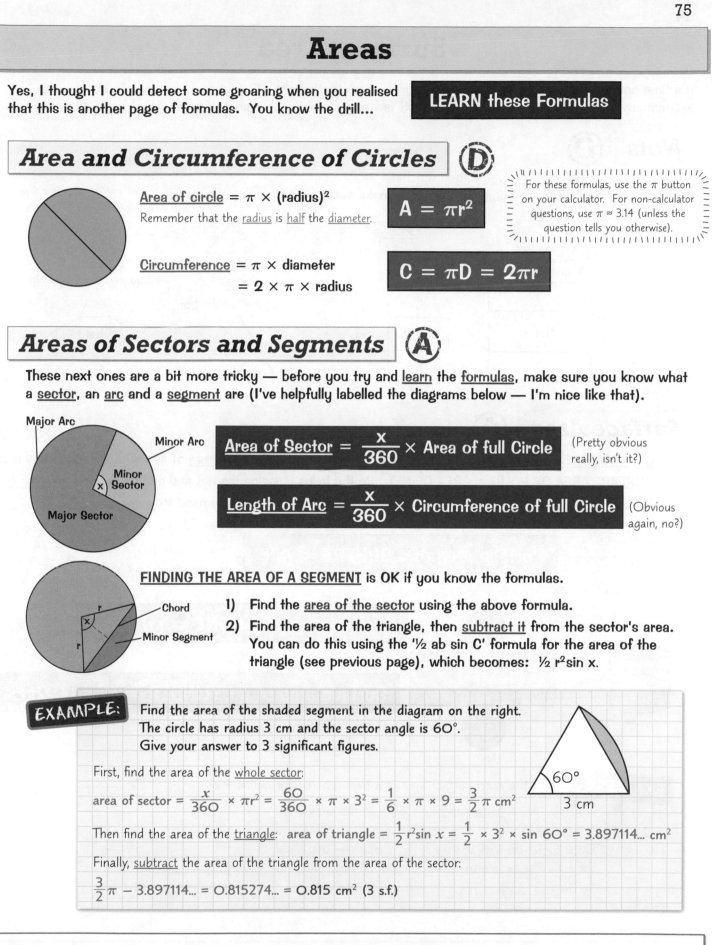

For these formulas, use the π button on your calculator. For non-calculator questions, use $\pi \approx 3.14$ (unless the question tells you otherwise).

Circumference = $\pi \times$ diameter
$= 2 \times \pi \times$ radius

$$C = \pi D = 2\pi r$$

Areas of Sectors and Segments 

These next ones are a bit more tricky — before you try and learn the formulas, make sure you know what a sector, an arc and a segment are (I've helpfully labelled the diagrams below — I'm nice like that).

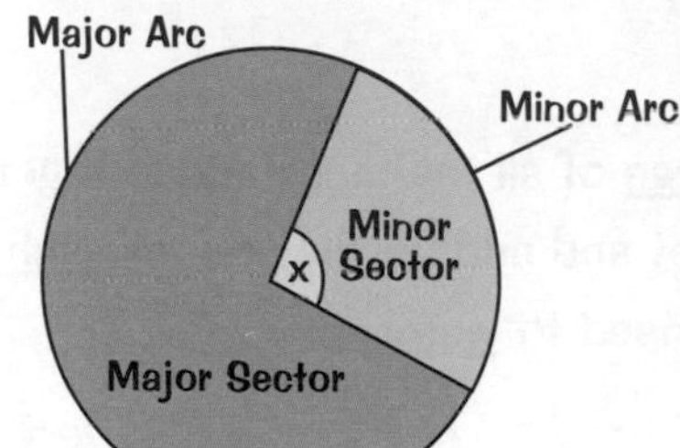

$$\text{Area of Sector} = \frac{x}{360} \times \text{Area of full Circle}$$

(Pretty obvious really, isn't it?)

$$\text{Length of Arc} = \frac{x}{360} \times \text{Circumference of full Circle}$$

(Obvious again, no?)

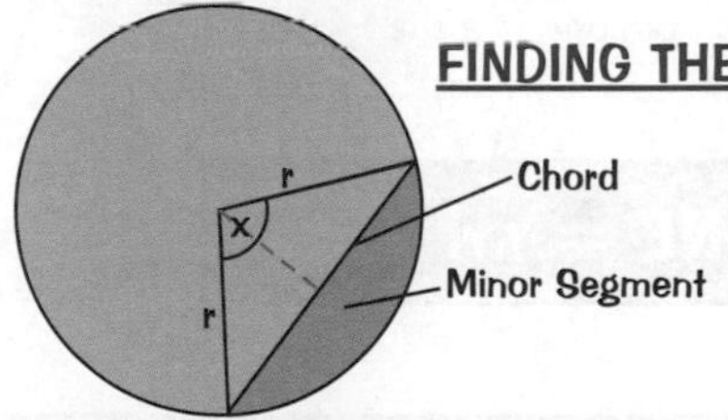

FINDING THE AREA OF A SEGMENT is OK if you know the formulas.

1) Find the area of the sector using the above formula.
2) Find the area of the triangle, then subtract it from the sector's area. You can do this using the '½ ab sin C' formula for the area of the triangle (see previous page), which becomes: $\frac{1}{2} r^2 \sin x$.

EXAMPLE: Find the area of the shaded segment in the diagram on the right. The circle has radius 3 cm and the sector angle is 60°. Give your answer to 3 significant figures.

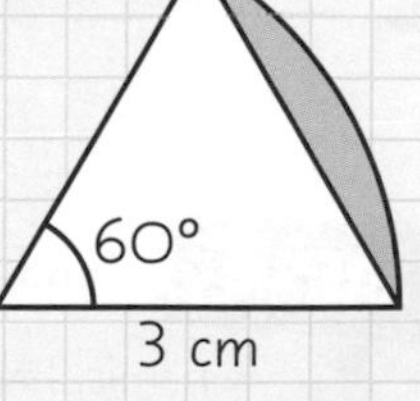

First, find the area of the whole sector:

$$\text{area of sector} = \frac{x}{360} \times \pi r^2 = \frac{60}{360} \times \pi \times 3^2 = \frac{1}{6} \times \pi \times 9 = \frac{3}{2}\pi \text{ cm}^2$$

Then find the area of the triangle: $\text{area of triangle} = \frac{1}{2} r^2 \sin x = \frac{1}{2} \times 3^2 \times \sin 60° = 3.897114... \text{ cm}^2$

Finally, subtract the area of the triangle from the area of the sector:

$$\frac{3}{2}\pi - 3.897114... = 0.815274... = 0.815 \text{ cm}^2 \text{ (3 s.f.)}$$

Pi r not square — pi are round. Pi are tasty...

Oo, one more thing — if you're asked to find the perimeter of a semicircle or quarter circle, don't forget to add on the straight edges too. It's an easy mistake to make, and it'll cost you marks.

Q1 For the sector on the right, find to 2 decimal places:
a) the area [2 marks] b) the arc length [2 marks] 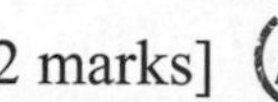

150°
8 cm

Surface Area

It's time now to move on to the next <u>dimension</u> — yep, that's right, <u>3D shapes</u>. I can hardly contain my excitement. If you do really well on these next few pages, we might even get on to <u>time travel</u>. Oooooo.

Nets

A <u>NET</u> is just a hollow <u>3D shape</u> folded out flat.
Here are the nets of some <u>common shapes</u> — make sure you can recognise them.

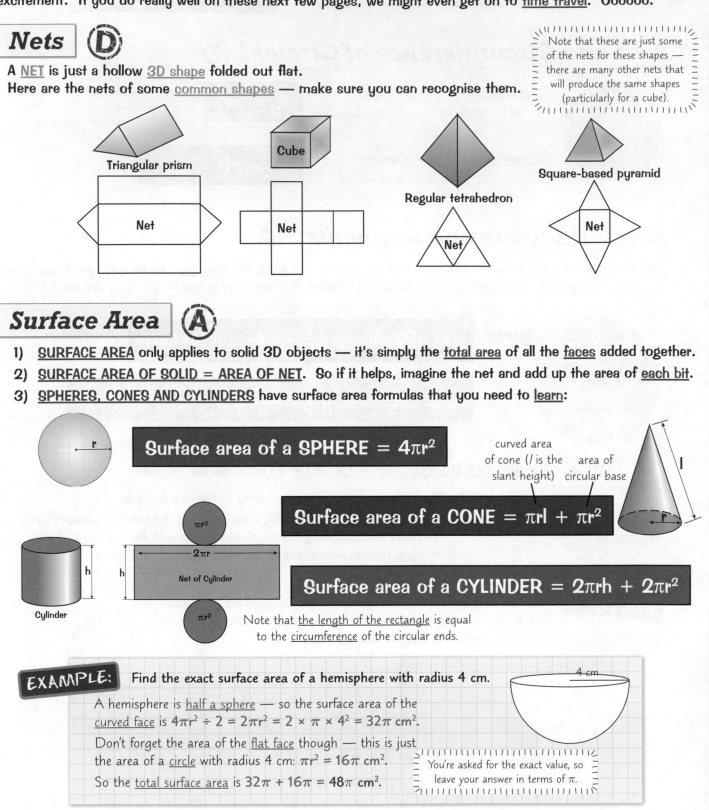

Note that these are just some of the nets for these shapes — there are many other nets that will produce the same shapes (particularly for a cube).

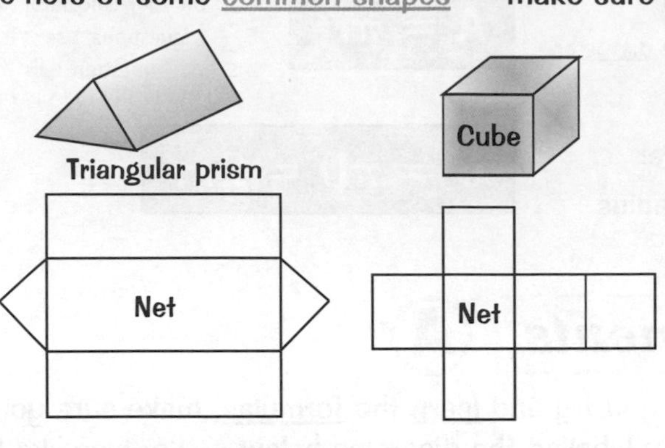

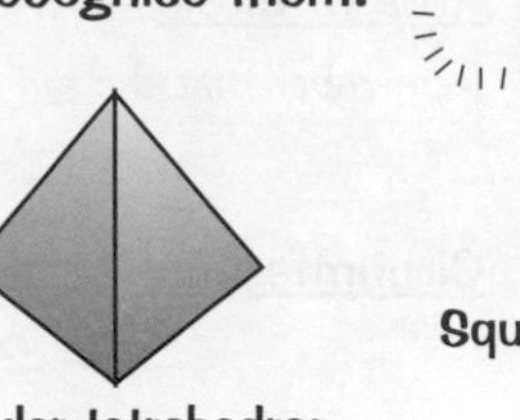

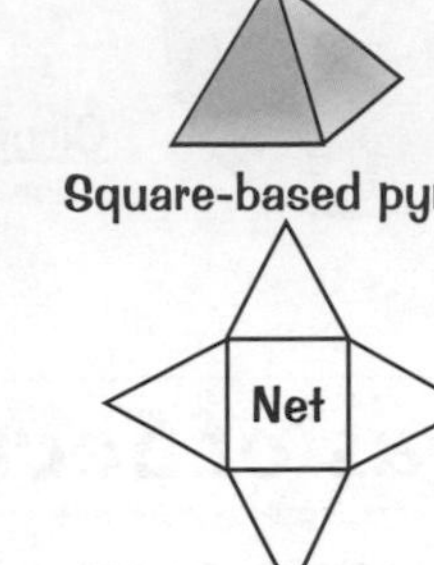

Surface Area

1) <u>SURFACE AREA</u> only applies to solid 3D objects — it's simply the <u>total area</u> of all the <u>faces</u> added together.
2) <u>SURFACE AREA OF SOLID = AREA OF NET</u>. So if it helps, imagine the net and add up the area of <u>each bit</u>.
3) <u>SPHERES, CONES AND CYLINDERS</u> have surface area formulas that you need to <u>learn</u>:

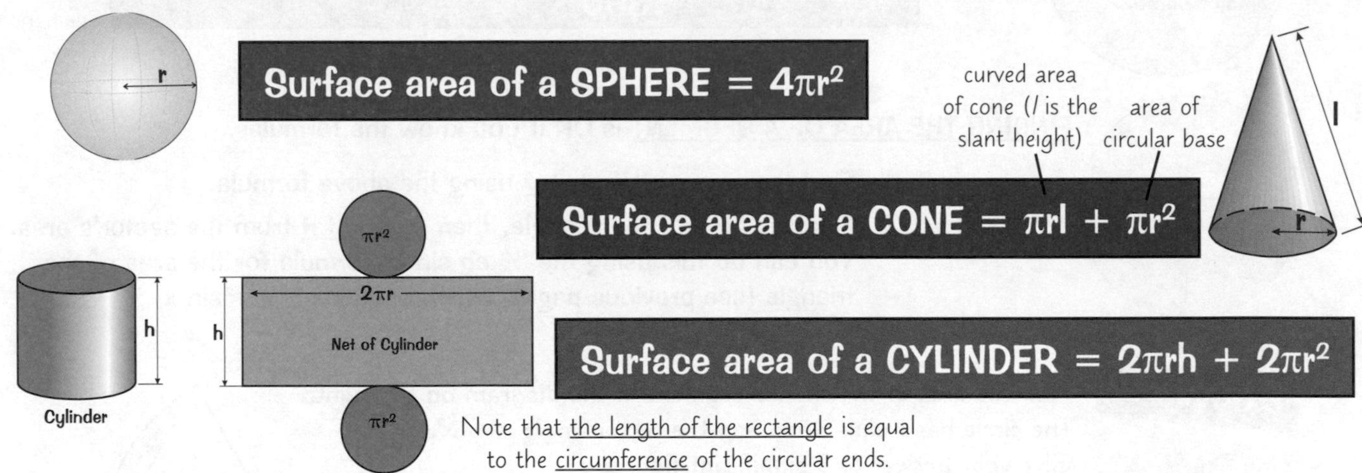

EXAMPLE: Find the exact surface area of a hemisphere with radius 4 cm.

A hemisphere is <u>half a sphere</u> — so the surface area of the <u>curved face</u> is $4\pi r^2 \div 2 = 2\pi r^2 = 2 \times \pi \times 4^2 = 32\pi \text{ cm}^2$.

Don't forget the area of the <u>flat face</u> though — this is just the area of a <u>circle</u> with radius 4 cm: $\pi r^2 = 16\pi \text{ cm}^2$.

So the <u>total surface area</u> is $32\pi + 16\pi = 48\pi \text{ cm}^2$.

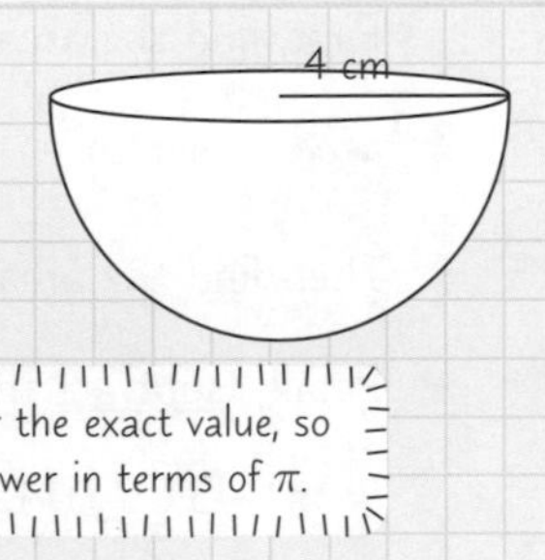

You're asked for the exact value, so leave your answer in terms of π.

Net yourself some extra marks...

Learn all the formulas on this page. Then have a go at this lovely Exam Practice Question:

Q1 The surface area of a cone with radius 5 cm is $125\pi \text{ cm}^2$.
Find the slant height, l, of the cone. [3 marks]

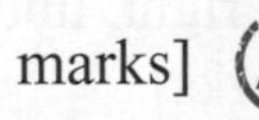

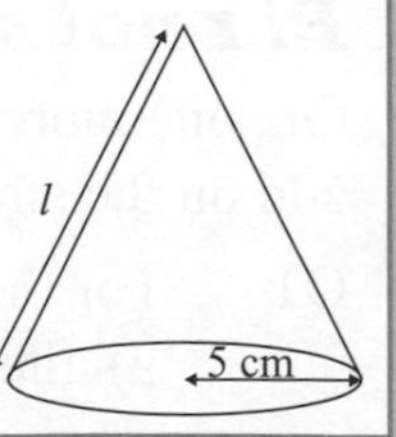

Volume

You might think you know some of this already, but I bet you don't know it all. There's only one thing for it...

LEARN these volume formulas... (Another word for volume is CAPACITY.)

Volumes of Cuboids

A cuboid is a rectangular block. Finding its volume is dead easy:

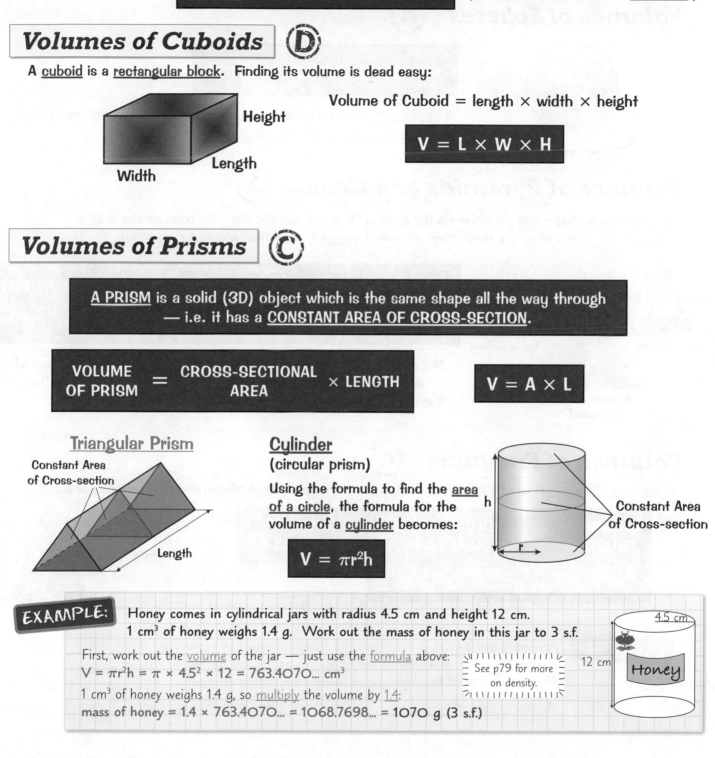

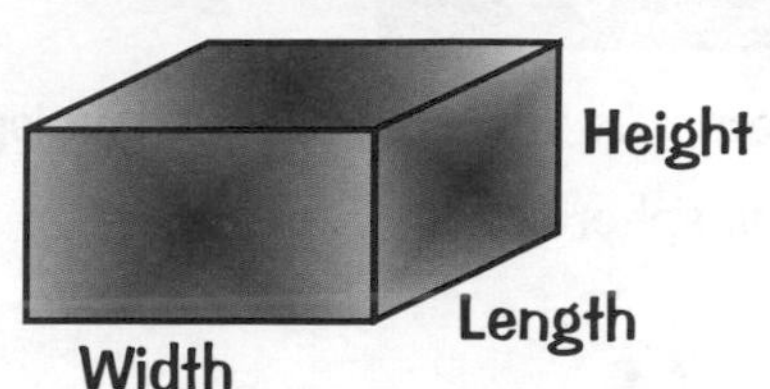

Volume of Cuboid = length × width × height

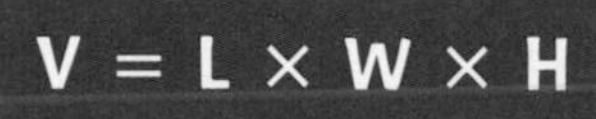

Volumes of Prisms

A PRISM is a solid (3D) object which is the same shape all the way through — i.e. it has a CONSTANT AREA OF CROSS-SECTION.

VOLUME OF PRISM = CROSS-SECTIONAL AREA × LENGTH

$$V = A \times L$$

Triangular Prism

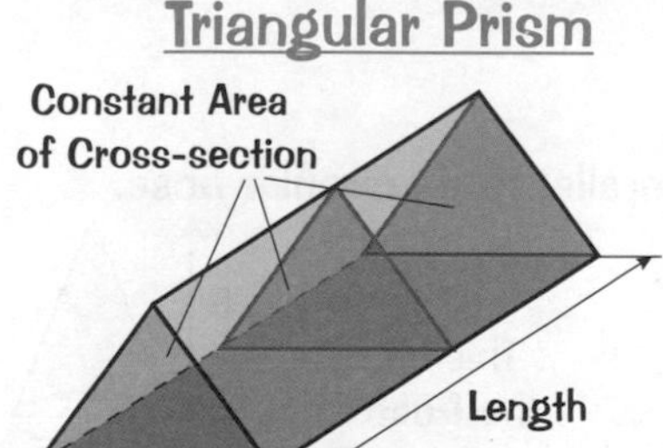

Cylinder

(circular prism)

Using the formula to find the area of a circle, the formula for the volume of a cylinder becomes:

$$V = \pi r^2 h$$

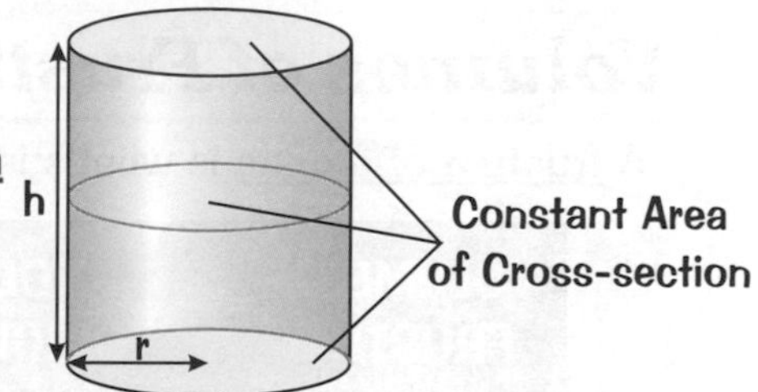

EXAMPLE: Honey comes in cylindrical jars with radius 4.5 cm and height 12 cm. 1 cm^3 of honey weighs 1.4 g. Work out the mass of honey in this jar to 3 s.f.

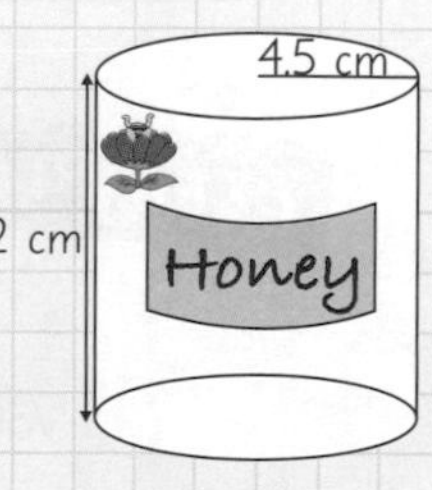

First, work out the volume of the jar — just use the formula above:
$V = \pi r^2 h = \pi \times 4.5^2 \times 12 = 763.4070...\ cm^3$

See p79 for more on density.

1 cm^3 of honey weighs 1.4 g, so multiply the volume by 1.4:
mass of honey = $1.4 \times 763.4070... = 1068.7698... = 1070$ g (3 s.f.)

Toblerones® — pretty good for prism food...

Learn the volume formulas on this page — and make sure you know the area formulas from p74-75 as well (you might need them to find the area of the cross-section of a prism).
Now try this question:

Q1 a) Find the volume of the triangular prism on the right. [2 marks]
b) The prism is made out of glass, which weighs 1.8 g per cm^3.
Work out the mass of the prism. [1 mark]

10 cm
12 cm
6 cm

Volume

This page has a great bonus — once you've learnt it you can amaze people by calculating the volume of their ice cream cones. Who says revision isn't fun. I love it. I take exams just for fun.

Volumes of Spheres

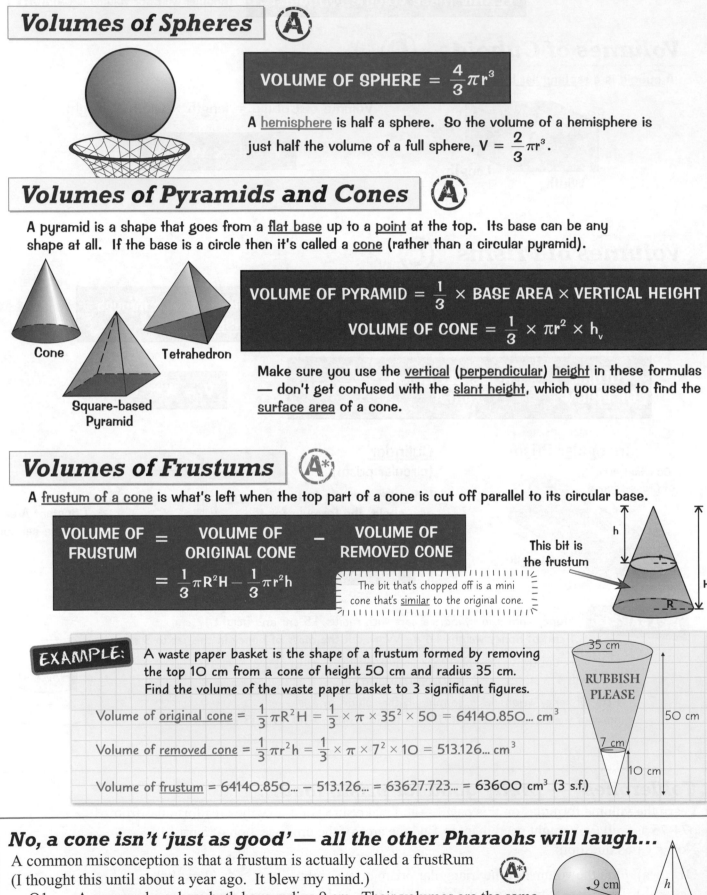

VOLUME OF SPHERE = $\frac{4}{3}\pi r^3$

A <u>hemisphere</u> is half a sphere. So the volume of a hemisphere is just half the volume of a full sphere, $V = \frac{2}{3}\pi r^3$.

Volumes of Pyramids and Cones

A pyramid is a shape that goes from a <u>flat base</u> up to a <u>point</u> at the top. Its base can be any shape at all. If the base is a circle then it's called a <u>cone</u> (rather than a circular pyramid).

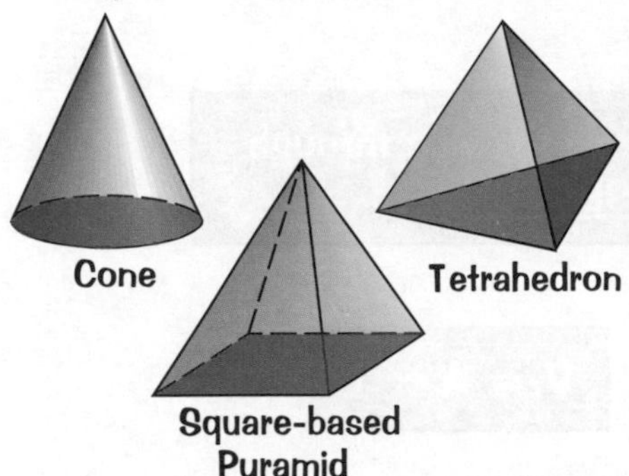

VOLUME OF PYRAMID = $\frac{1}{3}$ × BASE AREA × VERTICAL HEIGHT

VOLUME OF CONE = $\frac{1}{3} \times \pi r^2 \times h_v$

Make sure you use the <u>vertical</u> (<u>perpendicular</u>) <u>height</u> in these formulas — don't get confused with the <u>slant height</u>, which you used to find the <u>surface area</u> of a cone.

Volumes of Frustums

A <u>frustum of a cone</u> is what's left when the top part of a cone is cut off parallel to its circular base.

VOLUME OF FRUSTUM = VOLUME OF ORIGINAL CONE − VOLUME OF REMOVED CONE

$$= \frac{1}{3}\pi R^2 H - \frac{1}{3}\pi r^2 h$$

The bit that's chopped off is a mini cone that's <u>similar</u> to the original cone.

This bit is the frustum

EXAMPLE: A waste paper basket is the shape of a frustum formed by removing the top 10 cm from a cone of height 50 cm and radius 35 cm. Find the volume of the waste paper basket to 3 significant figures.

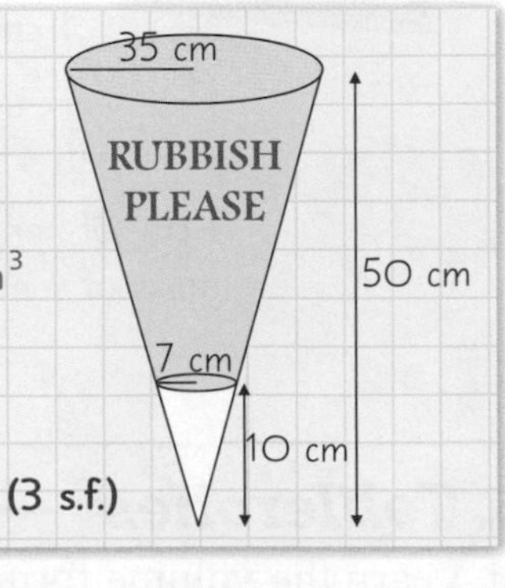

Volume of <u>original cone</u> = $\frac{1}{3}\pi R^2 H = \frac{1}{3} \times \pi \times 35^2 \times 50 = 64140.850...\ \text{cm}^3$

Volume of <u>removed cone</u> = $\frac{1}{3}\pi r^2 h = \frac{1}{3} \times \pi \times 7^2 \times 10 = 513.126...\ \text{cm}^3$

Volume of <u>frustum</u> = 64140.850... − 513.126... = 63627.723... = **63600 cm³ (3 s.f.)**

No, a cone isn't 'just as good' — all the other Pharaohs will laugh...

A common misconception is that a frustum is actually called a frustRum (I thought this until about a year ago. It blew my mind.)

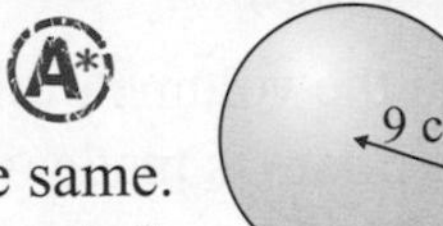

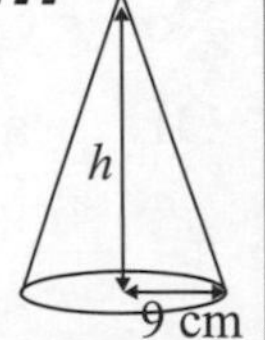

Q1 A cone and a sphere both have radius 9 cm. Their volumes are the same. Find the vertical height, h, of the cone. [4 marks]

There's no room left for time travel. Sorry.

Density and Speed

Density and speed. They're both a matter of learning the formulas, bunging the numbers in and watching the units.

Density = Mass ÷ Volume

Density is the mass per unit volume of a substance. It's usually measured in kg/m^3 or g/cm^3.

You might think this is physics, but density is specifically mentioned in the maths syllabus.

$$\text{DENSITY} = \frac{\text{MASS}}{\text{VOLUME}} \qquad \text{VOLUME} = \frac{\text{MASS}}{\text{DENSITY}} \qquad \text{MASS} = \text{DENSITY} \times \text{VOLUME}$$

A formula triangle is a mighty handy tool for remembering formulas like these. Here's the one for density. To remember the order of the letters in the formula triangle think D^MV or DiMoV (The Russian Agent).

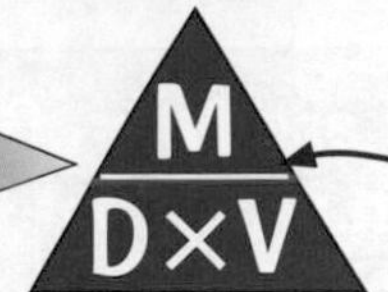

M
D×V

HOW DO YOU USE FORMULA TRIANGLES?

1) COVER UP the thing you want to find and WRITE DOWN what's left showing.
2) Now PUT IN THE VALUES for the other two things and WORK IT OUT.

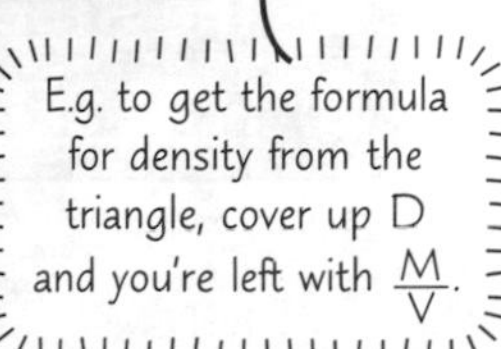

E.g. to get the formula for density from the triangle, cover up D and you're left with $\frac{M}{V}$.

EXAMPLE:

A giant 'Wunda-Choc' bar has a density of 1.3 g/cm^3. If the bar's volume is 1800 cm^3, what is the mass of the bar in kg?

First write down the formula: mass = density × volume

Put in the values and calculate: mass = 1.3 g/cm^3 × 1800 cm^3 = 2340 g

Check you've given the answer in the correct units: mass in kg = 2340 g ÷ 1000 = 2.34 kg

CHECK YOUR UNITS MATCH
If the density is in g/cm^3, the volume must be in cm^3 and you'll get a mass in g.

Speed = Distance ÷ Time

Speed is the distance travelled per unit time, e.g. the number of km per hour or metres per second.

$$\text{SPEED} = \frac{\text{DISTANCE}}{\text{TIME}} \qquad \text{TIME} = \frac{\text{DISTANCE}}{\text{SPEED}} \qquad \text{DISTANCE} = \text{SPEED} \times \text{TIME}$$

Here's the formula triangle for speed — this time, we have the words SaD Times to help you remember the order of the letters (S^DT). So if it's a question on speed, distance and time, just say SAD TIMES.

D
S×T

The units you get out of a formula DEPEND ENTIRELY on the units you put in. So, if you put a distance in cm and a time in seconds into the speed formula, the answer comes out in cm/second.

EXAMPLE: A car travels 9 miles at 36 miles per hour. How many minutes does it take?

Write down the formula, put in the values and calculate: $\text{time} = \frac{\text{distance}}{\text{speed}} = \frac{9 \text{ miles}}{36 \text{ mph}} = 0.25 \text{ hours}$

Convert the time from hours to minutes: 0.25 hours × 60 = 15 minutes

Formula triangles — it's all a big cover-up...

Write down the formula triangles from memory, then use them to generate the formulas in the boxes above.

Q1 A solid lead cone has a vertical height of 60 cm and a base radius of 20 cm. If the density of lead is 11.34 g/cm^3, find the mass of the cone in kilograms to 3 s.f. [5 marks] 

Q2 A camel gallops for 2 hours 15 minutes at an average speed of 45 km/h. Calculate the distance the camel travels. [2 marks] 

Distance-Time Graphs

Ah, what could be better than some nice D/T graphs? OK, so a picture of Jennifer Lawrence might be better. Or Hugh Jackman. But the section isn't called 'Hollywood Hotties' is it...

Distance-Time Graphs (D)

Distance-time graphs are pretty common in exams. They're not too bad once you get your head around them.

Just remember these 3 important points:

1) At any point, <u>GRADIENT = SPEED</u>, but watch out for the UNITS.
2) The <u>STEEPER</u> the graph, the <u>FASTER</u> it's going.
3) <u>FLAT SECTIONS</u> are where it is <u>STOPPED</u>.

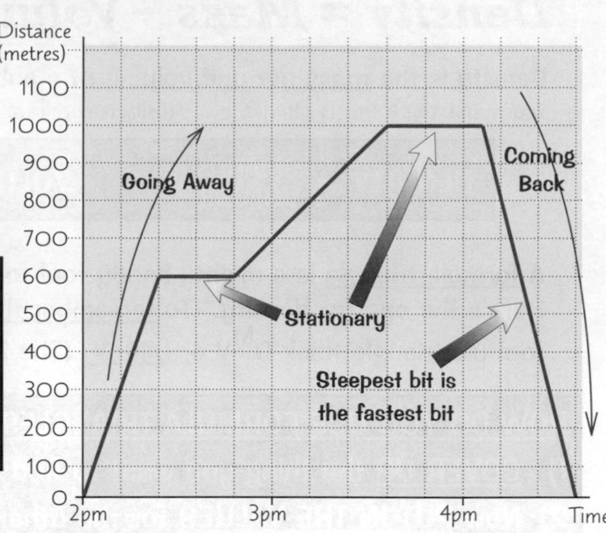

EXAMPLE: Henry went out for a ride on his bike. After a while he got a puncture and stopped to fix it. This graph shows the first part of Henry's journey.

a) **What time did Henry leave home?**

He left home at the point where the line starts. At <u>8:15</u>

b) **How far did Henry cycle before getting a puncture?**

The horizontal part of the graph is where Henry stopped. <u>12 km</u>

c) **What was Henry's speed before getting a puncture?**

Using the speed formula is the same as finding the gradient.

$$\text{speed} = \frac{\text{distance}}{\text{time}} = \frac{12\text{ km}}{0.5\text{ hours}} = 24\text{ km/h}$$

d) **At 9:30 Henry turns round and cycles home at 24 km/h. Complete the graph to show this.**

You have to work out how long it will take Henry to cycle the 18 km home:

$$\text{time} = \frac{\text{distance}}{\text{speed}} = \frac{18\text{ km}}{24\text{ km/h}} = \underline{0.75\text{ hours}}$$

0.75×60 mins = <u>45 mins</u>

Decimal times are yuck, so convert it to <u>minutes</u>.

45 minutes after 9:30 is 10:15, so that's the time Henry gets home. Now you can complete the graph.

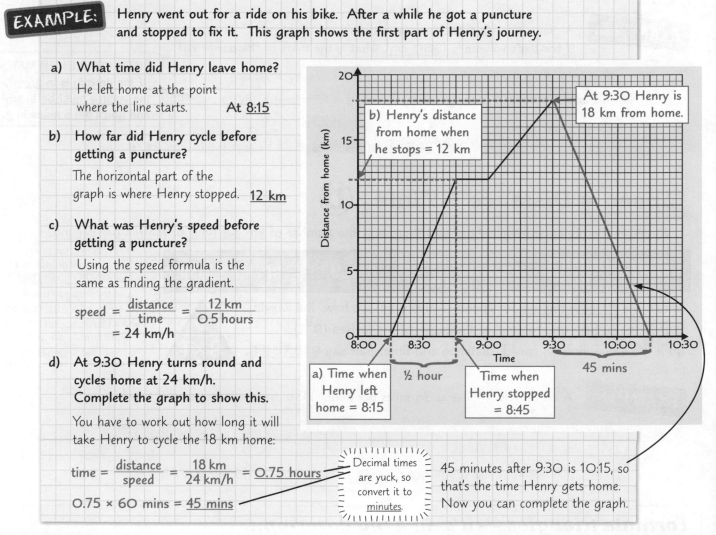

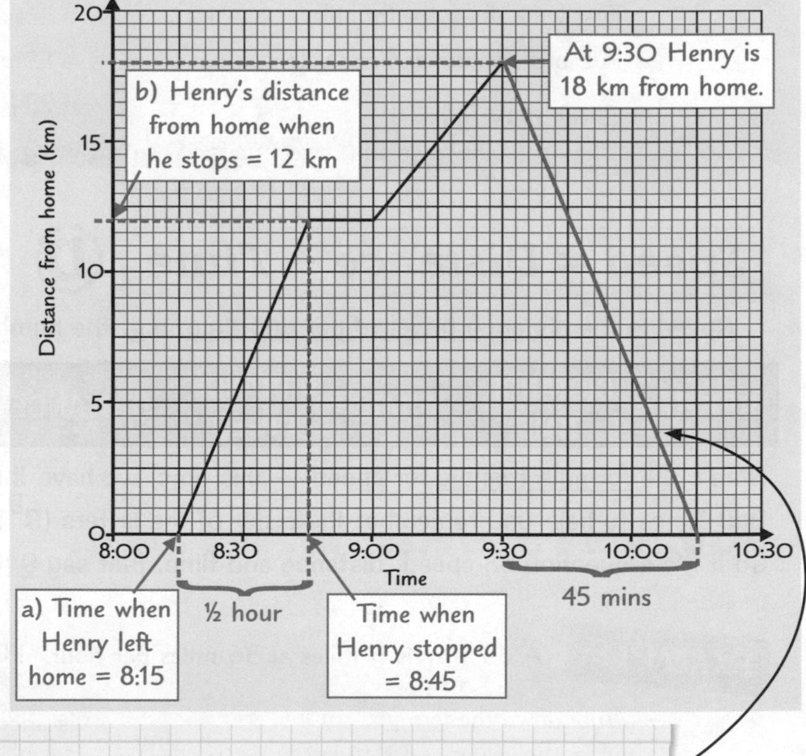

D-T Graphs — filled with highs and lows, an analogy of life...

The only way to get good at distance-time graphs is to practise, practise, practise...

Q1 a) Using the graph above, how long did Henry stop for? [1 mark] 

b) How far from home was Henry at 8:30? [1 mark] (D)

Unit Conversions

A nice easy page for a change — just some facts to learn. Hooray!

Metric Units

1) Length mm, cm, m, km
2) Area mm^2, cm^2, m^2, km^2,
3) Volume mm^3, cm^3, m^3, ml, litres
4) Weight g, kg, tonnes
5) Speed km/h, m/s

MEMORISE THESE KEY FACTS:

1 cm = 10 mm	1 tonne = 1000 kg
1 m = 100 cm	1 litre = 1000 ml
1 km = 1000 m	1 litre = 1000 cm^3
1 kg = 1000 g	1 cm^3 = 1 ml

Imperial Units

1) Length Inches, feet, yards, miles
2) Area Square inches, square feet, square miles
3) Volume Cubic inches, cubic feet, gallons, pints
4) Weight Ounces, pounds, stones, tons
5) Speed mph

IMPERIAL UNIT CONVERSIONS

1 Foot = 12 Inches
1 Yard = 3 Feet
1 Gallon = 8 Pints
1 Stone = 14 Pounds (lb)
1 Pound = 16 Ounces (oz)

Metric-Imperial Conversions

LEARN THESE — they don't promise to give you these in the exam and if they're feeling mean (as they often are), they won't.

To convert between units, multiply or divide by the conversion factor:

APPROXIMATE CONVERSIONS

2.2 pounds (lb) ≈ 1 kg
1 inch ≈ 2.5 cm
1 gallon ≈ 4.5 litres
5 miles ≈ 8 km

EXAMPLE:

a) Convert 10 pounds into kg.

1 kg ≈ 2.2 lb, so 10 lb ≈ 10 ÷ 2.2 ≈ 4.5 kg

b) Convert 12.25 gallons into litres.

1 gallon ≈ 4.5 litres, so 12.25 gallons ≈ 12.25 × 4.5 ≈ 55.1 litres

ALWAYS CHECK IT LOOKS SENSIBLE
1 kg ≈ 2.2 lb, so the no. of pounds should be about twice the no. of kg.
If it's not, then chances are you divided instead of multiplied, or vice versa.

Convert Speeds in Two Steps

Speeds are made up of two measures — a distance and a time. To convert from, say, miles per hour to metres per second, you have to convert the distance unit and the time unit separately.

EXAMPLE: A rabbit's top speed is 56 km/h. How fast is this in m/s?

First convert from km/h to m/h: 56 km/h = (56 × 1000) m/h = 56 000 m/h

Now convert from m/h to m/s: 56 000 m/h = (56 000 ÷ 3600) m/s
= 15.6 m/s (1 d.p.)

1 minute = 60 seconds
and 1 hour = 60 minutes
So 1 hour = 60 × 60
= 3600 seconds.

Imperial units — they're mint...

Learn all these conversions. Then turn over and write them down. Hmm, I don't know about you, but I quite fancy some conversion-based questions after all that. Which is convenient...

Q1 Convert 26.2 litres to gallons. Give your answer to 1 d.p. [2 marks] (E)

Q2 Convert 80 cm/s to miles per hour. [3 marks] (C)

Unit Conversions

Converting areas and volumes from one unit to another is an exam disaster that you have to know how to avoid. 1 m² definitely does NOT equal 100 cm². Remember this and read on for why.

Converting Area and Volume Measurements

$1\text{ m}^2 = 100\text{ cm} \times 100\text{ cm} = 10\,000\text{ cm}^2$
$1\text{ cm}^2 = 10\text{ mm} \times 10\text{ mm} = 100\text{ mm}^2$

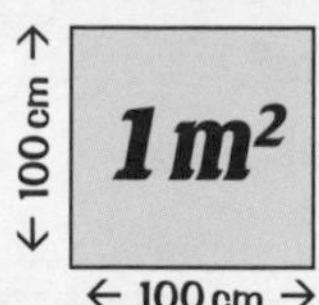

$1\text{ m}^3 = 100\text{ cm} \times 100\text{ cm} \times 100\text{ cm} = 1\,000\,000\text{ cm}^3$
$1\text{ cm}^3 = 10\text{ mm} \times 10\text{ mm} \times 10\text{ mm} = 1000\text{ mm}^3$

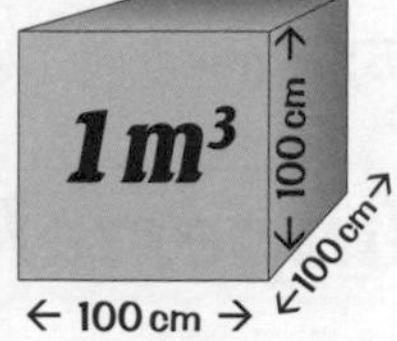

EXAMPLES:

1. Convert 9 m² to cm².
   To change area measurements from m² to cm² multiply by 100 twice.
   $9 \times 100 \times 100 = 90\,000\text{ cm}^2$

2. Convert 60 000 mm³ to cm³.
   To change volume measurements from mm³ to cm³ divide by 10 three times.
   $60\,000 \div (10 \times 10 \times 10) = 60\text{ cm}^3$

Conversion Graphs

Conversion graphs themselves are easy to use. That's why examiners often wrap them up in tricky questions.

METHOD FOR USING CONVERSION GRAPHS:

1) Draw a line from a value on one axis.
2) Keep going until you hit the LINE.
3) Then change direction and go straight to the other axis.
4) Read off the value from this axis. The two values are equivalent.

EXAMPLE: Sam went on holiday to Florida and paid $3.15 per gallon for petrol. How much is this in pounds (£) per litre?

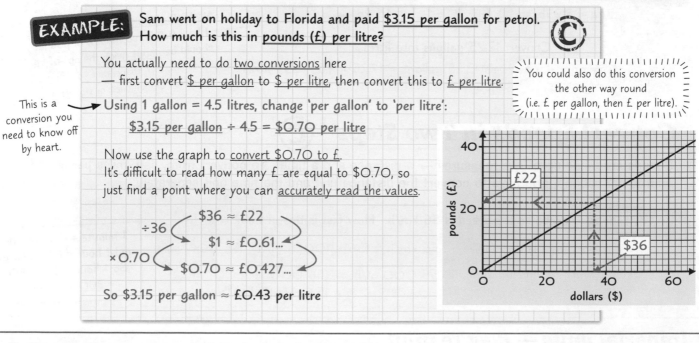

You actually need to do two conversions here — first convert \$ per gallon to \$ per litre, then convert this to £ per litre.

You could also do this conversion the other way round (i.e. £ per gallon, then £ per litre).

This is a conversion you need to know off by heart. → Using 1 gallon = 4.5 litres, change 'per gallon' to 'per litre':

\$3.15 per gallon ÷ 4.5 = \$0.70 per litre

Now use the graph to convert \$0.70 to £.
It's difficult to read how many £ are equal to \$0.70, so just find a point where you can accurately read the values.

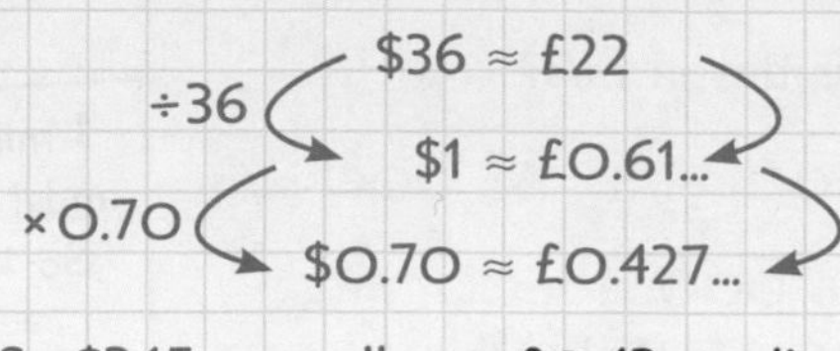

So \$3.15 per gallon ≈ £0.43 per litre

Graph: pounds (£) against dollars (\$), axes 0–40 (pounds) and 0–60 (dollars), with \$36 and £22 marked.

Learn how to convert areas into marks...

Conversion graphs in exams can be for all sorts of units. They all work the same way though.

Q1 Change 3 m³ to mm³. [2 marks] 

Q2 Use the conversion graph above to find the approximate value of £163 in dollars. [3 marks] 

Triangle Construction

How you construct a triangle depends on what info you're given about the triangle...

Three sides — use a Ruler and Compasses

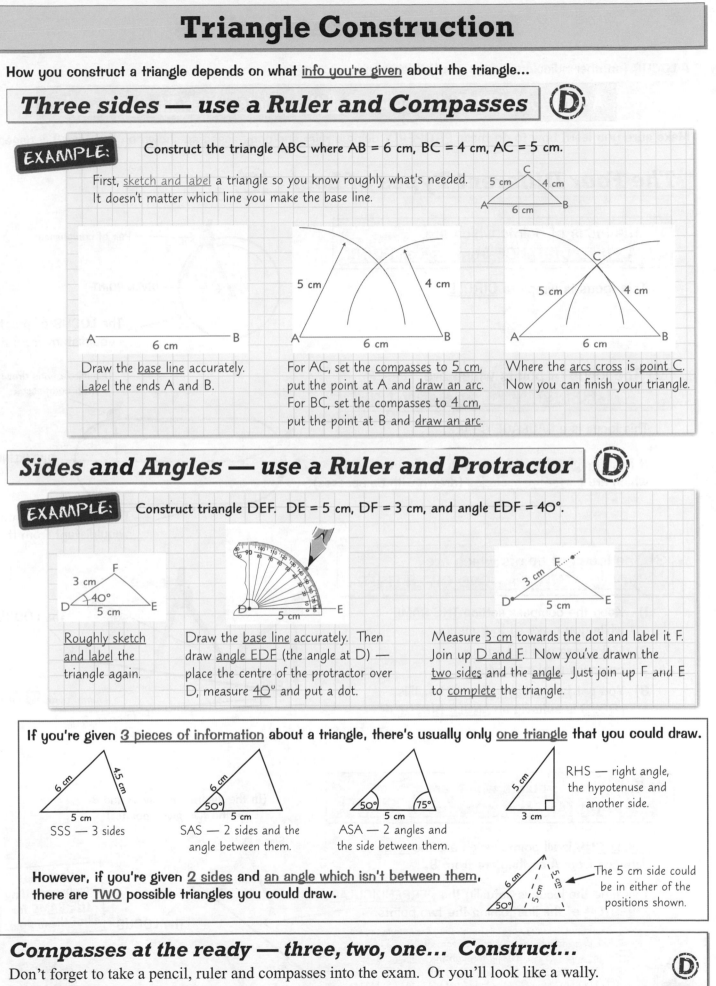

EXAMPLE: **Construct the triangle ABC where AB = 6 cm, BC = 4 cm, AC = 5 cm.**

First, sketch and label a triangle so you know roughly what's needed. It doesn't matter which line you make the base line.

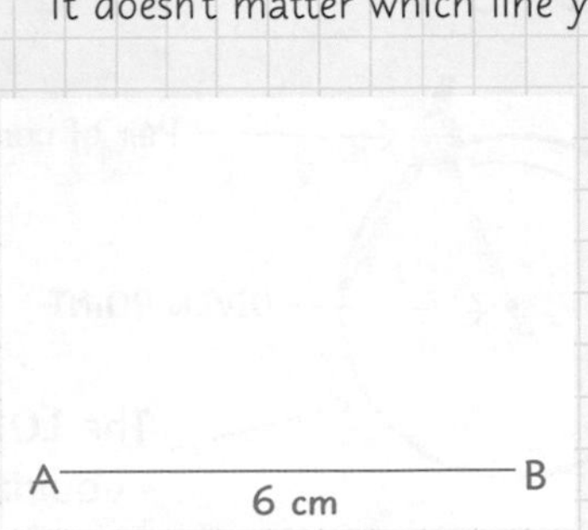

Draw the base line accurately. Label the ends A and B.

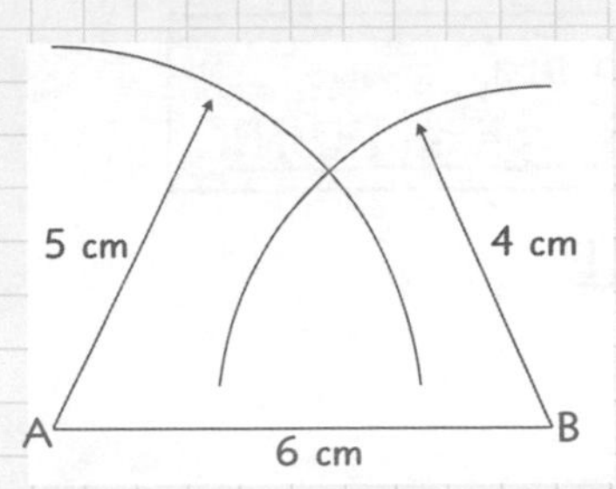

For AC, set the compasses to 5 cm, put the point at A and draw an arc. For BC, set the compasses to 4 cm, put the point at B and draw an arc.

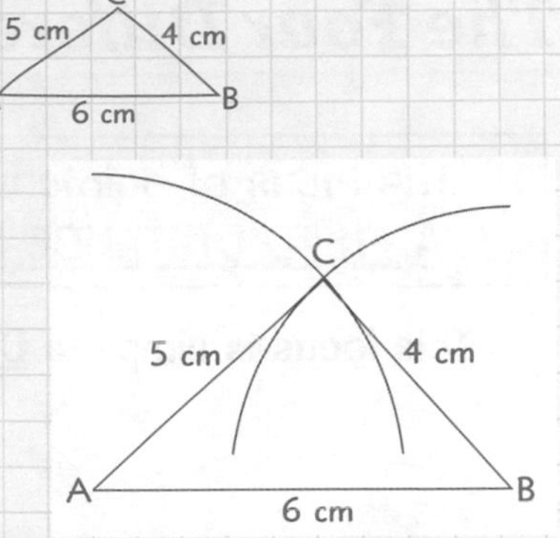

Where the arcs cross is point C. Now you can finish your triangle.

Sides and Angles — use a Ruler and Protractor

EXAMPLE: **Construct triangle DEF. DE = 5 cm, DF = 3 cm, and angle EDF = 40°.**

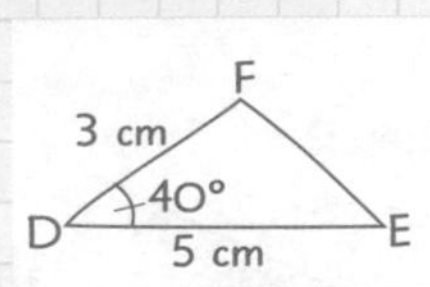

Roughly sketch and label the triangle again.

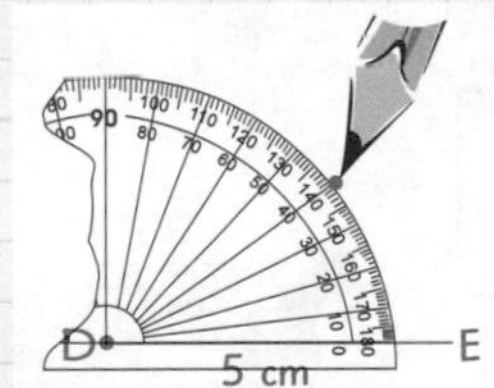

Draw the base line accurately. Then draw angle EDF (the angle at D) — place the centre of the protractor over D, measure 40° and put a dot.

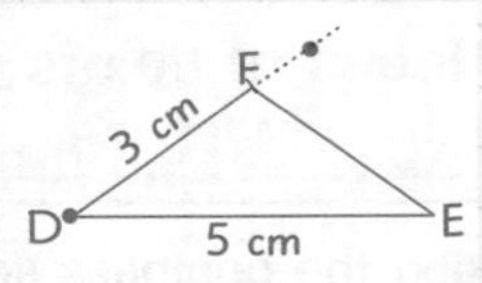

Measure 3 cm towards the dot and label it F. Join up D and F. Now you've drawn the two sides and the angle. Just join up F and E to complete the triangle.

If you're given 3 pieces of information about a triangle, there's usually only one triangle that you could draw.

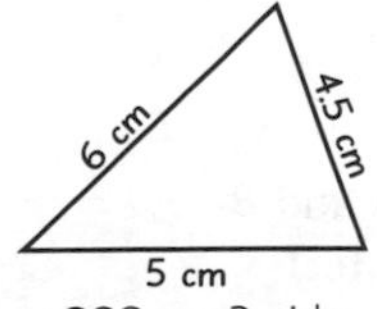

SSS — 3 sides

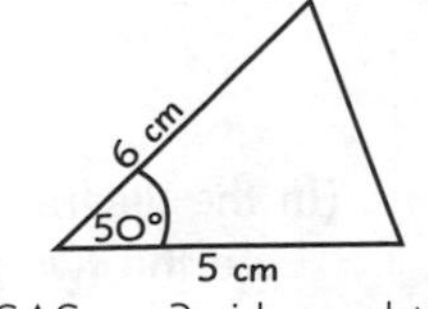

SAS — 2 sides and the angle between them.

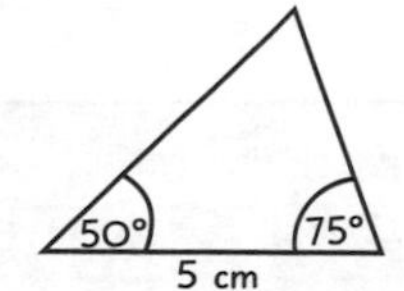

ASA — 2 angles and the side between them.

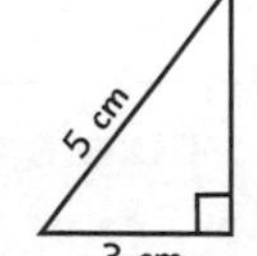

RHS — right angle, the hypotenuse and another side.

However, if you're given 2 sides and an angle which isn't between them, there are TWO possible triangles you could draw.

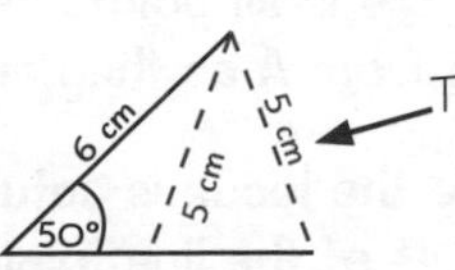

The 5 cm side could be in either of the positions shown.

Compasses at the ready — three, two, one... Construct...

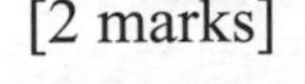

Don't forget to take a pencil, ruler and compasses into the exam. Or you'll look like a wally.

Q1 Construct an equilateral triangle with sides 5 cm. Leave your construction marks visible. [2 marks]

Q2 Construct and label triangle ABC. Angle ABC = 45°, angle BCA = 40°, side BC = 7.5 cm. [2 marks]

Loci and Constructions

A **LOCUS** (another ridiculous maths word) is simply:

A LINE or REGION that shows all the points which fit a given rule.

Make sure you learn how to do these **PROPERLY** using a ruler and compasses as shown on the next few pages.

The Four Different Types of Loci

Ⓒ

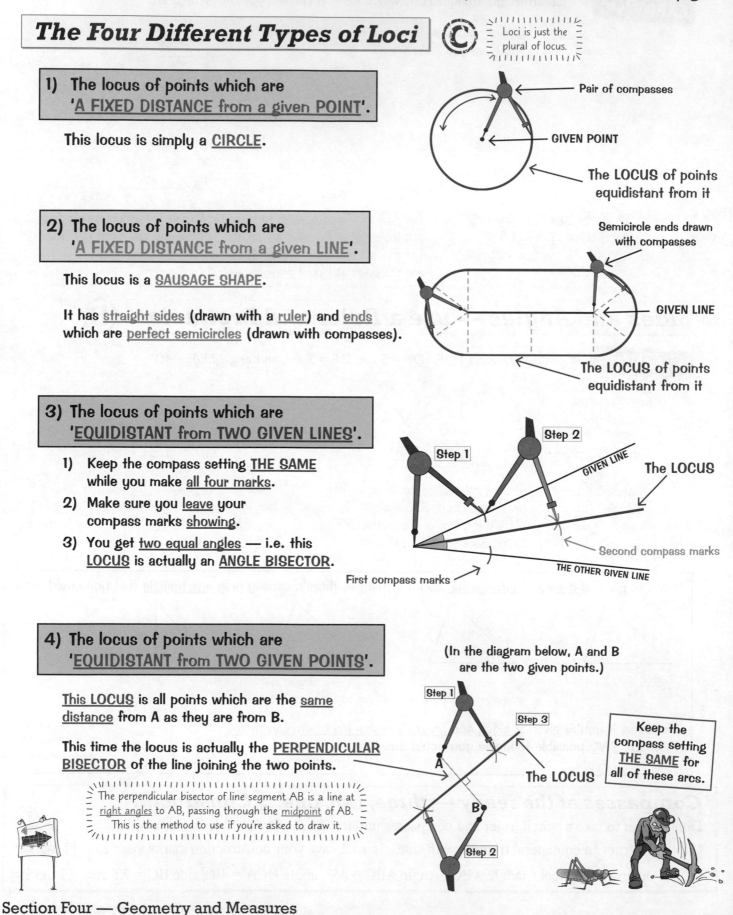

Loci is just the plural of locus.

1) The locus of points which are 'A FIXED DISTANCE from a given POINT'.

This locus is simply a **CIRCLE**.

Pair of compasses

GIVEN POINT

The LOCUS of points equidistant from it

2) The locus of points which are 'A FIXED DISTANCE from a given LINE'.

This locus is a **SAUSAGE SHAPE**.

It has straight sides (drawn with a ruler) and ends which are perfect semicircles (drawn with compasses).

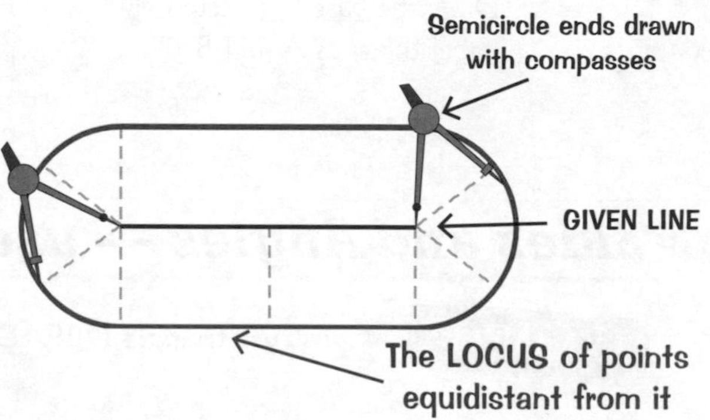

3) The locus of points which are 'EQUIDISTANT from TWO GIVEN LINES'.

1) Keep the compass setting **THE SAME** while you make all four marks.
2) Make sure you leave your compass marks showing.
3) You get two equal angles — i.e. this **LOCUS** is actually an **ANGLE BISECTOR**.

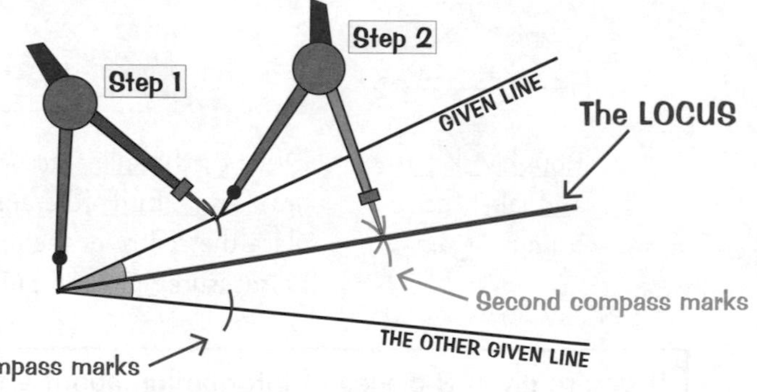

4) The locus of points which are 'EQUIDISTANT from TWO GIVEN POINTS'.

(In the diagram below, A and B are the two given points.)

This LOCUS is all points which are the same distance from A as they are from B.

This time the locus is actually the **PERPENDICULAR BISECTOR** of the line joining the two points.

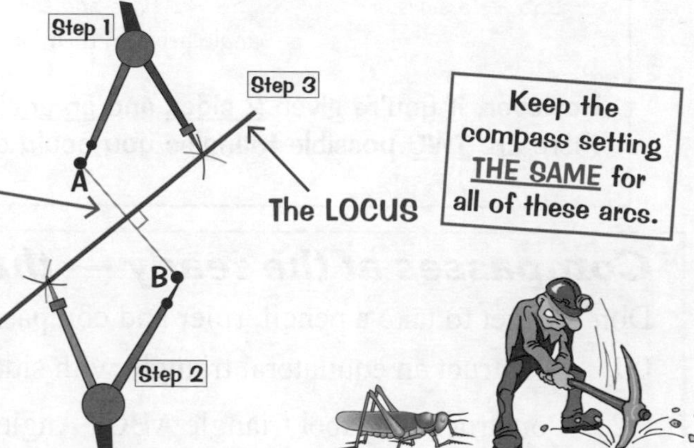

The perpendicular bisector of line segment AB is a line at right angles to AB, passing through the midpoint of AB. This is the method to use if you're asked to draw it.

Loci and Constructions

Don't just read the page through once and hope you'll remember it — get your ruler, compasses and pencil out and have a go. It's the only way of testing whether you really know this stuff.

Constructing Accurate 60° Angles

1) They may well ask you to draw an accurate 60° angle without a protractor.
2) One thing they're needed for is constructing equilateral triangles.
3) Follow the method shown in this diagram (make sure you leave the compass settings the same for each step).

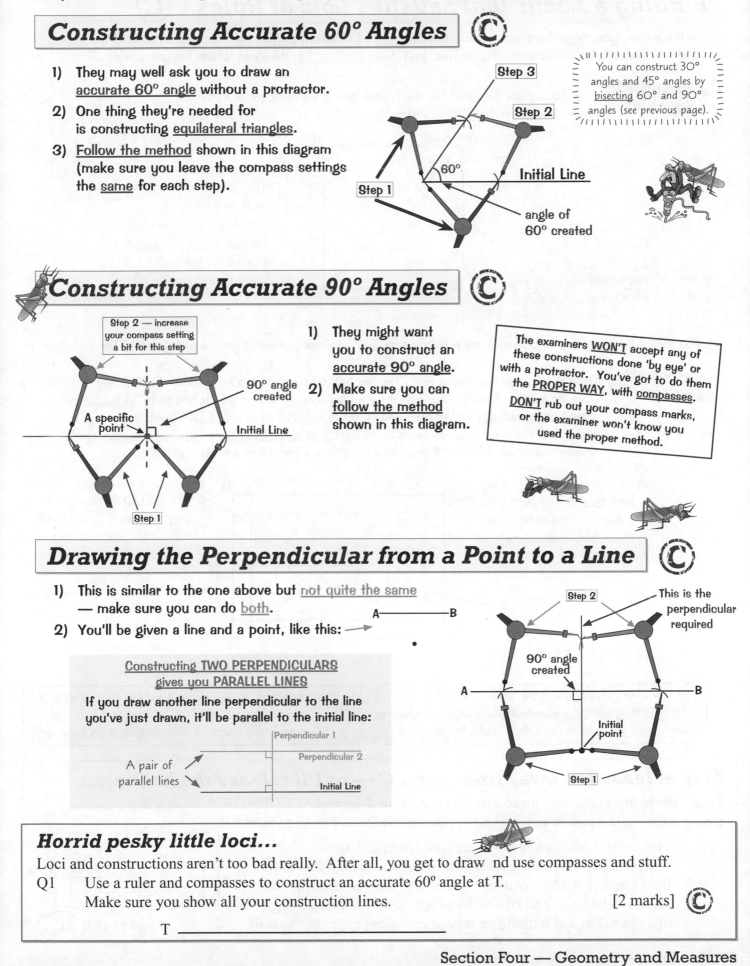

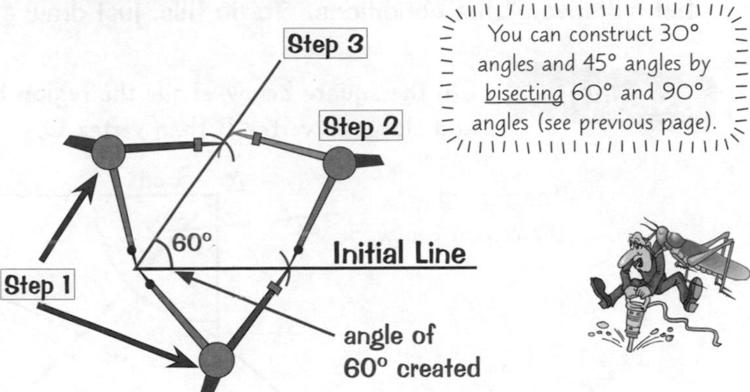

You can construct 30° angles and 45° angles by bisecting 60° and 90° angles (see previous page).

Constructing Accurate 90° Angles

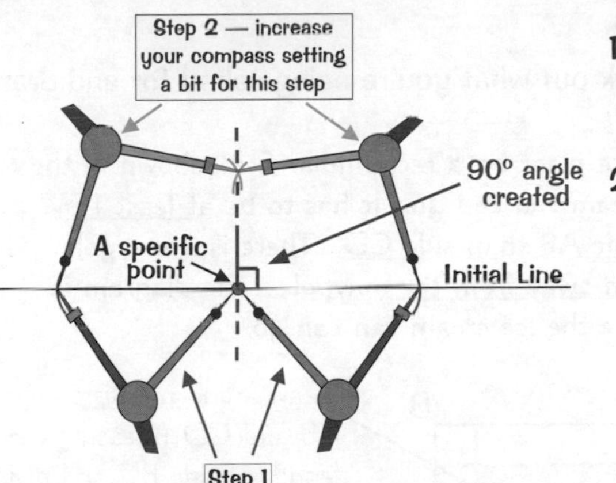

1) They might want you to construct an accurate 90° angle.
2) Make sure you can follow the method shown in this diagram.

The examiners WON'T accept any of these constructions done 'by eye' or with a protractor. You've got to do them the PROPER WAY, with compasses. DON'T rub out your compass marks, or the examiner won't know you used the proper method.

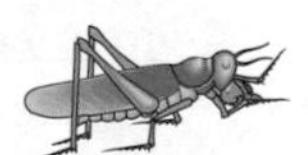

Drawing the Perpendicular from a Point to a Line

1) This is similar to the one above but not quite the same — make sure you can do both.
2) You'll be given a line and a point, like this:

A———B

•

Constructing TWO PERPENDICULARS gives you PARALLEL LINES

If you draw another line perpendicular to the line you've just drawn, it'll be parallel to the initial line:

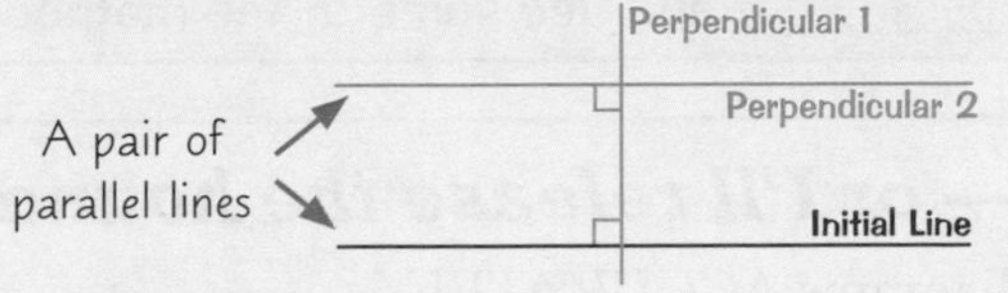

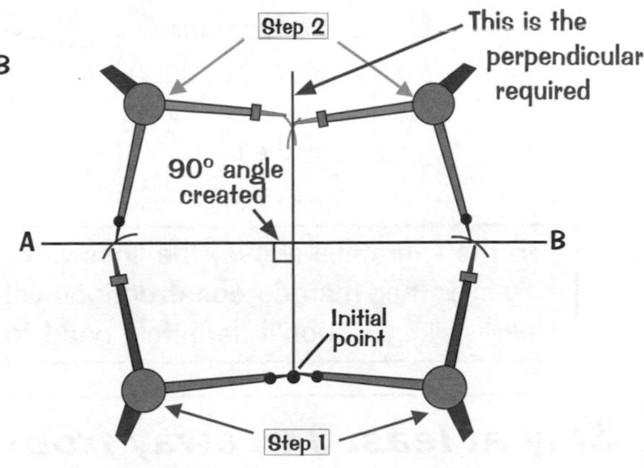

Horrid pesky little loci...

Loci and constructions aren't too bad really. After all, you get to draw nd use compasses and stuff.

Q1 Use a ruler and compasses to construct an accurate 60° angle at T.
Make sure you show all your construction lines. [2 marks]

T ____________________

Loci and Constructions — Worked Examples

Now you know what loci are, and how to do all the constructions you need, it's time to put them all together.

Finding a Locus that Satisfies Lots of Rules

In the exam, you might be given a situation with lots of different conditions, and asked to find the region that satisfies all the conditions. To do this, just draw each locus, then see which bit you want.

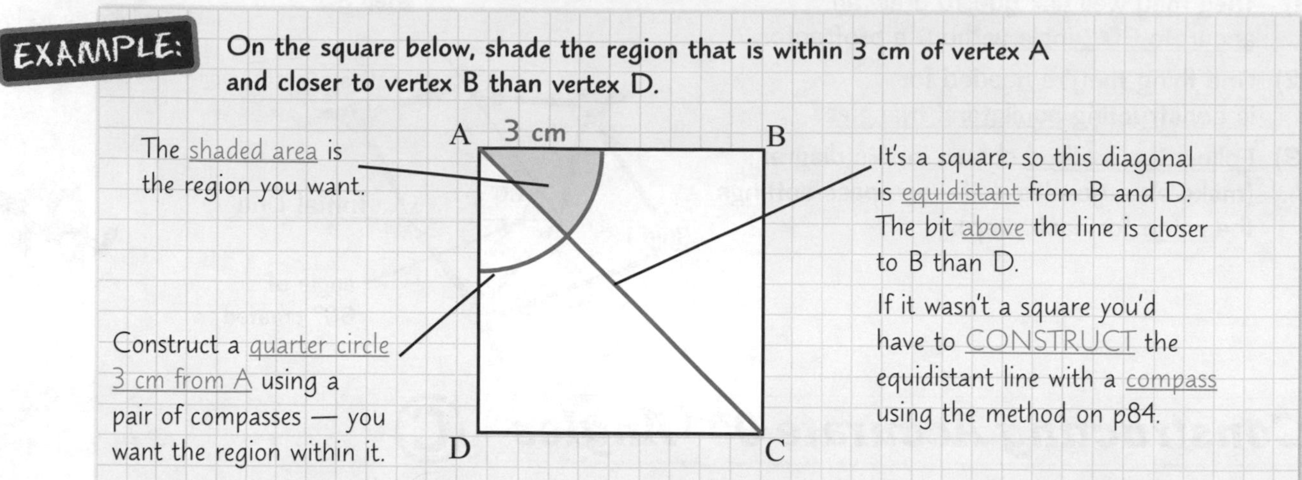

EXAMPLE:
On the square below, shade the region that is within 3 cm of vertex A and closer to vertex B than vertex D.
A
3 cm
B
C
D
The shaded area is the region you want.
It's a square, so this diagonal is equidistant from B and D. The bit above the line is closer to B than D.
If it wasn't a square you'd have to CONSTRUCT the equidistant line with a compass using the method on p84.
Construct a quarter circle 3 cm from A using a pair of compasses — you want the region within it.

You might be given the information as a wordy problem — work out what you're being asked for and draw it.

Tessa is organising a village fete. The fete will take place on a rectangular field, shown in the diagram below. Tessa is deciding where an ice cream van can go. It has to be at least 1 m away from each edge of the field, and closer to side AB than side CD. There is a maypole at M, and the ice cream van must be at least 2 m away from the maypole. The diagram is drawn to a scale of 1 cm = 1 m. Show on it where the ice cream van can go.

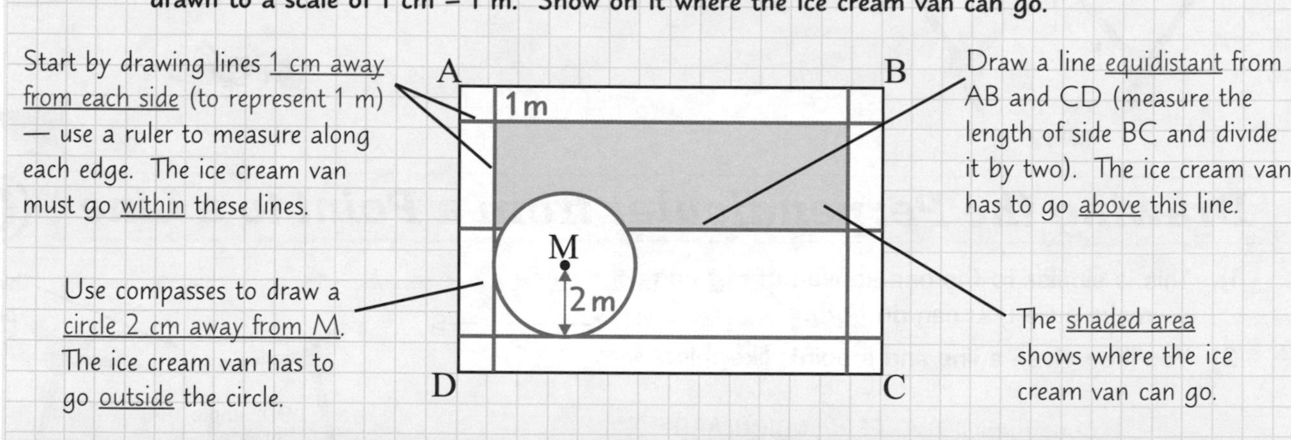

Start by drawing lines 1 cm away from each side (to represent 1 m) — use a ruler to measure along each edge. The ice cream van must go within these lines.
A
B
C
D
1 m
M
2 m
Draw a line equidistant from AB and CD (measure the length of side BC and divide it by two). The ice cream van has to go above this line.
Use compasses to draw a circle 2 cm away from M. The ice cream van has to go outside the circle.
The shaded area shows where the ice cream van can go.

In the examples above, the lines were all at right angles to each other, so you could just measure with a ruler rather than do constructions with compasses. If the question says "Leave your construction lines clearly visible", you'll definitely need to get your compasses out and use some of the methods on p84-85.

Stay at least 3 m away from point C — or I'll release the hounds...

I can't stress this enough — make sure you draw your diagrams ACCURATELY (using a ruler and a pair of compasses) — like in this Exam Practice Question:

Q1 The gardens of a stately home are shown on the diagram. The public can visit the gardens, but must stay at least 2 m away from the rectangular pond and at least 2 m away from each of the statues (labelled A and B). Make a copy of this diagram using a scale of 1 cm = 2 m and indicate on it the areas where the public can go.

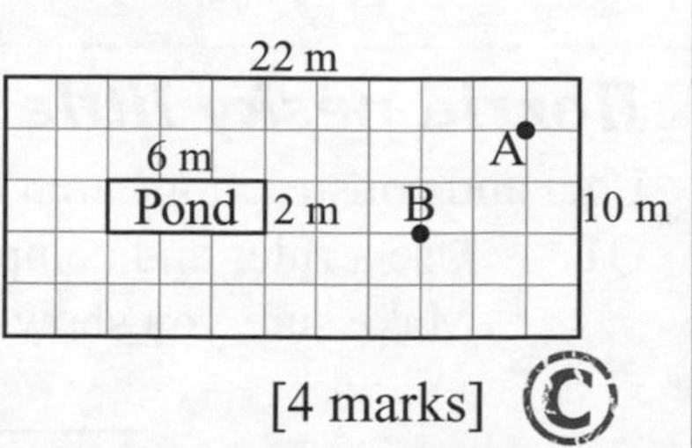

[4 marks] C

Bearings

Bearings. They'll be useful next time you're off sailing. Or in your Maths exam.

Bearings

To find or plot a bearing you must remember the three key words:

1) 'FROM' Find the word 'FROM' in the question, and put your pencil on the diagram at the point you are going 'from'.

2) NORTHLINE At the point you are going FROM, draw in a NORTHLINE. (There'll often be one drawn for you in exam questions.)

3) CLOCKWISE Now draw in the angle CLOCKWISE from the northline to the line joining the two points. This angle is the required bearing.

N The bearing of A from B

A

B

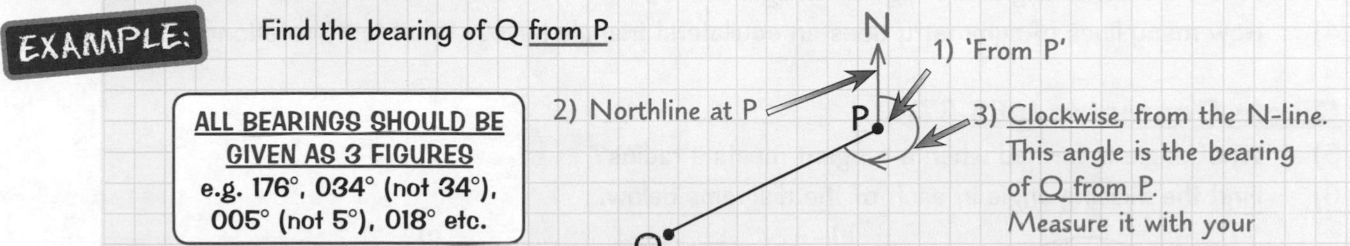

EXAMPLE:
Find the bearing of Q from P.
ALL BEARINGS SHOULD BE GIVEN AS 3 FIGURES
e.g. 176°, 034° (not 34°), 005° (not 5°), 018° etc.
N
1) 'From P'
2) Northline at P
P
3) Clockwise, from the N-line. This angle is the bearing of Q from P. Measure it with your protractor — 245°.
Q

EXAMPLE: The bearing of Z from Y is 110°.
Find the bearing of Y from Z.

See page 62 for interior angles.

First sketch a diagram so you can see what's going on.
Angles a and b are interior angles, so they add up to 180°.
Angle b = 180° – 110° = 70°
So bearing of Y from Z = 360° – 70° = 290°.

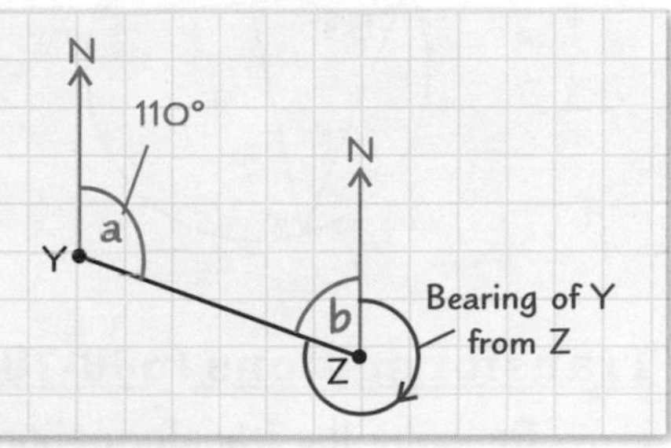

Bearings Questions and Scale Drawings

EXAMPLE: A hiker walks 2 km from point A, on a bearing of 036°.
If the scale of the map below is 2 cm to 1 km, how far is the hiker now from his car?

If you are asked to CALCULATE a distance or an angle, you'll need to use the cosine or sine rule (see p93).

First, draw a line at a bearing of 036° from point A. 1 km is 2 cm on the map and the hiker walks 2 km, so make the line from A 4 cm long.
You want the distance of the hiker from the car, so use a ruler to measure it on the map, then use the scale to work out the real distance it represents.

Distance to car on map = 3 cm. 2 cm = 1 km, so 1 cm = 0.5 km, therefore 3 cm = 1.5 km.

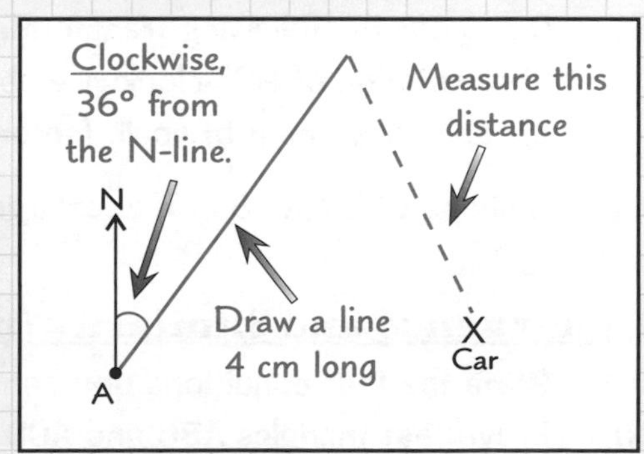

Please bear with me while I figure out where we are...

Learn the three key words above and scribble them out from memory. Now try these practice questions.

Q1 Measure the bearing of T from H. [1 mark] (D)

T N H

Q2 A ship sails 12 km on a bearing of 050°, then 20 km on a bearing of 100°.
It then sails directly back to its starting position. Calculate this distance to 1 d.p. [5 marks]

Revision Questions for Section Four

There are lots of opportunities to show off your artistic skills here (as long as you use them to answer the questions).

- Try these questions and tick off each one when you get it right.
- When you've done all the questions for a topic and are completely happy with it, tick off the topic.

Angles and Polygons (p61-65)

1) What do angles in a quadrilateral add up to?
2) Find the missing angles in the diagrams below.

a)

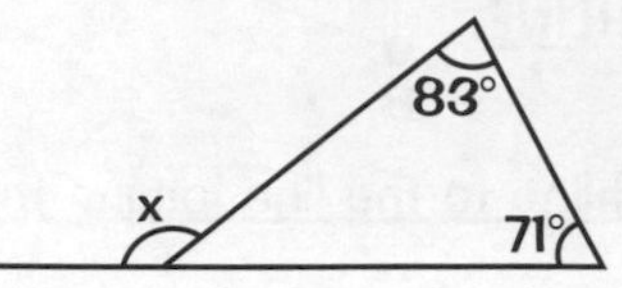

b)

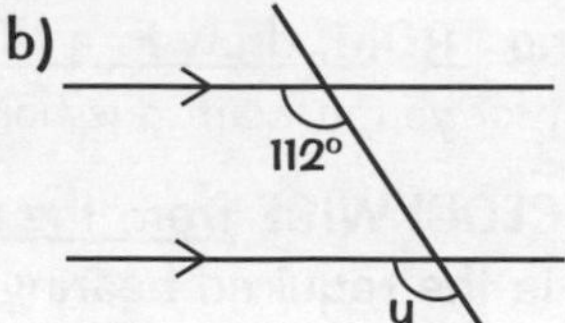

c)

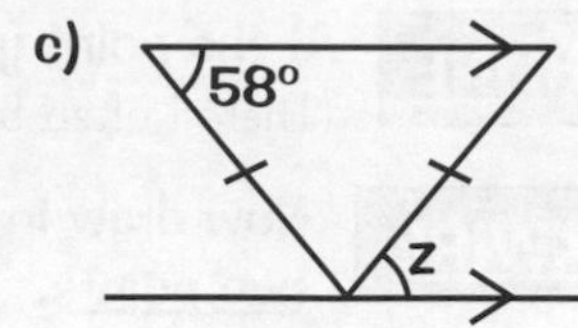

3) Find the exterior angle of a regular hexagon.
4) How many lines of symmetry does an equilateral triangle have? What is its rotational symmetry?

Circle Geometry (p66-67)

5) What angle is formed when a tangent meets a radius?
6) Find the missing angle in each of the diagrams below.

a)

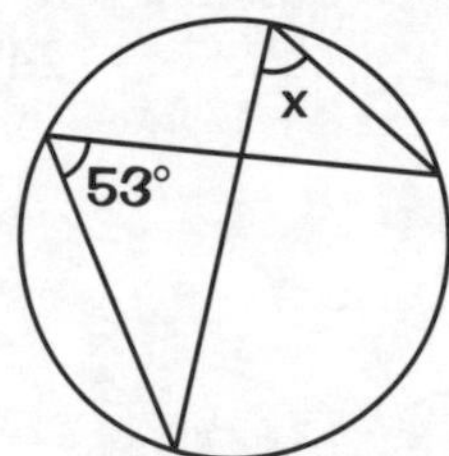

b)

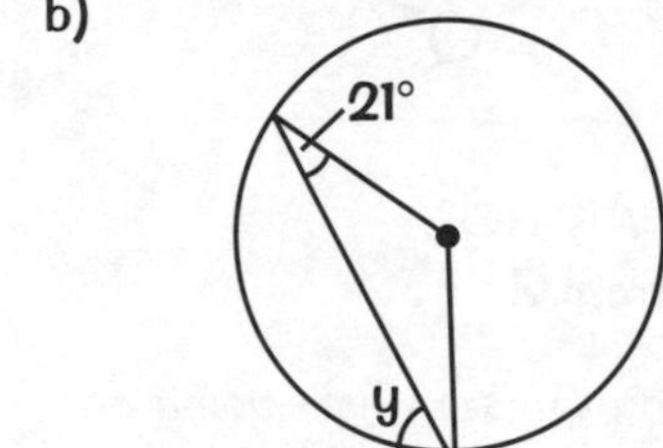

c)

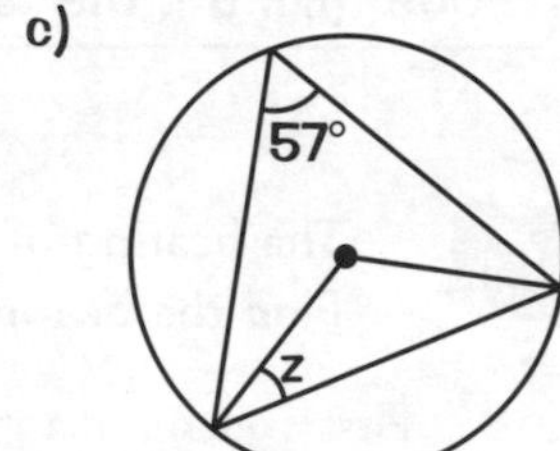

Transformations (p68-70)

7) Describe the transformation that maps:
   a) Shape A onto Shape B
   b) Shape A onto shape C

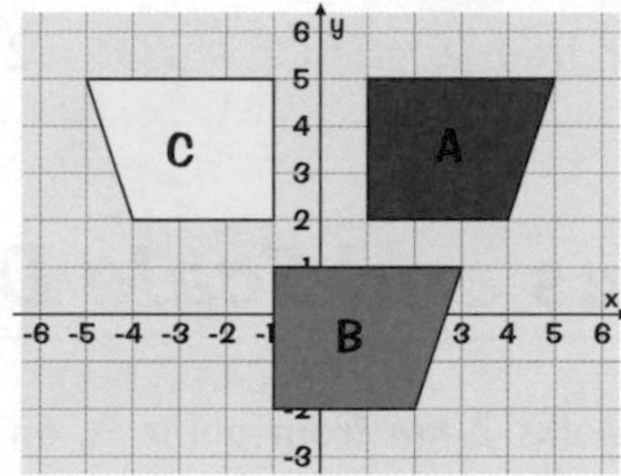

8) Carry out the following transformations on the triangle X, which has vertices (1, 1), (4, 1) and (2, 3):
   a) a rotation of 90° clockwise about (1, 1)   b) a translation by the vector $\begin{pmatrix}-3\\-4\end{pmatrix}$
   c) an enlargement of scale factor 2, centre (1, 1)
9) A shape with area 5 cm² is enlarged by a scale factor of 4. What is the area of the enlarged shape?

Congruence and Similarity (p71-72)

10) State the four conditions you can use to prove that two triangles are congruent.
11) Prove that triangles ABC and ACD on the right are congruent.

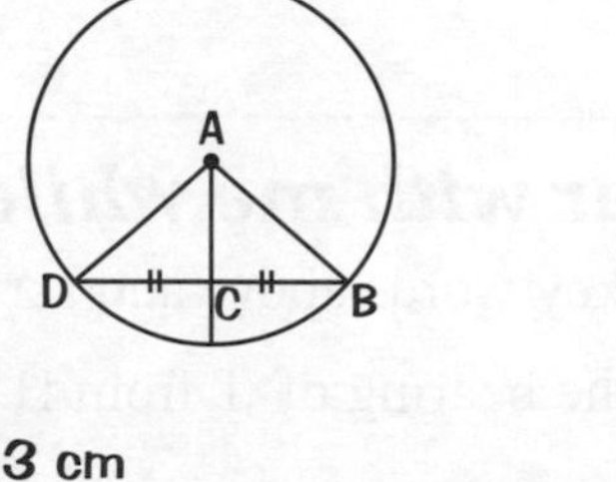

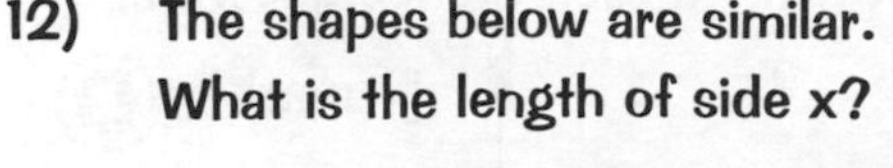

12) The shapes below are similar.
    What is the length of side x?

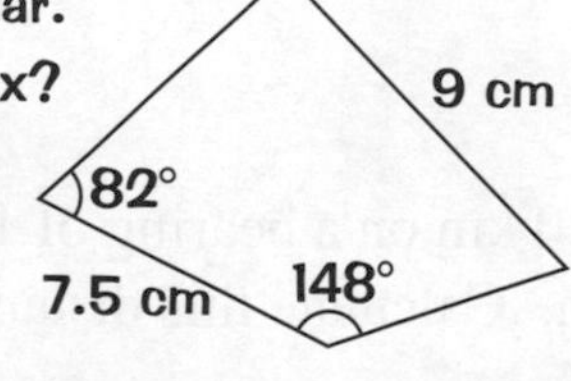

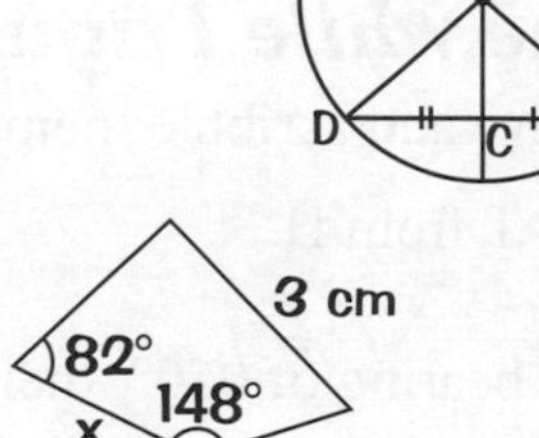

Revision Questions for Section Four

Projections (p73)

13) What is a plan view?
14) On squared paper, draw the front elevation of the shape on the right.

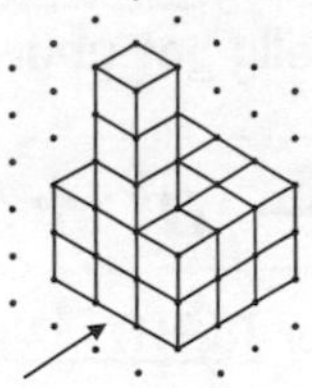

Area, Surface Area and Volume (p74-78)

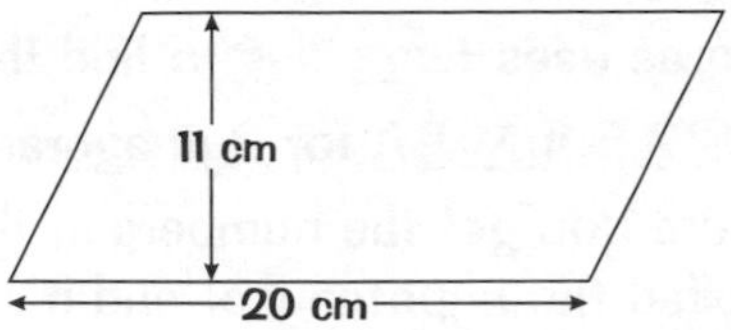

15) What is the formula for finding the area of a trapezium?
16) Find the area of the shape on the right.
17) A circle has diameter 16 cm. Find its exact circumference and area.
18) Find the area of the sector with radius 10 cm and angle 45° to 2 d.p.
19) What is the formula for finding the surface area of a sphere?
20) The shape on the right is made from a cylinder and a hemisphere. Find its exact surface area. 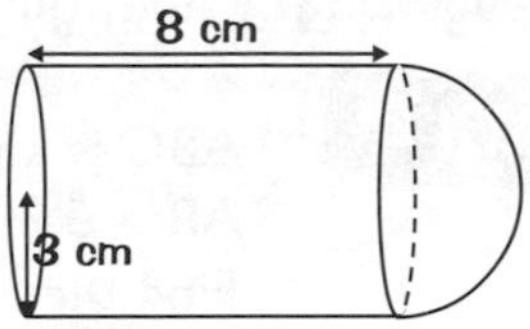
21) Find the volume of a hexagonal prism with a cross-sectional area 36 cm² and length 11 cm.
22) Find the volume of the solid on the right (to 2 d.p.):

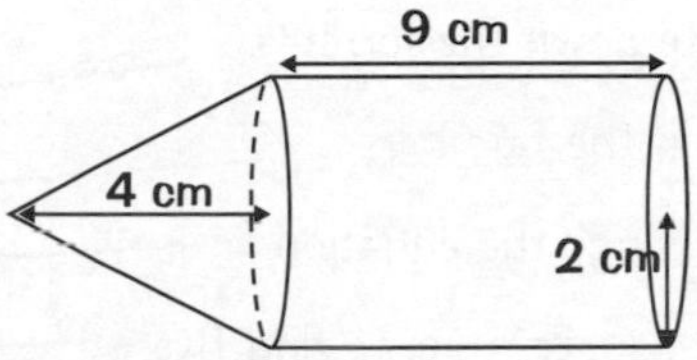

11 cm
20 cm
8 cm
3 cm
9 cm
4 cm
2 cm

Density, Speed and Distance-Time Graphs (p79-80)

23) Find the volume of a snowman if its density is 0.4 g/cm³ and its mass is 5 kg.
24) Find the average speed of a car if it travels 63 miles in an hour and a half.
25) What does a horizontal line mean on a distance-time graph?
26) The graph on the right shows Ben's car journey. What speed did he drive home at?

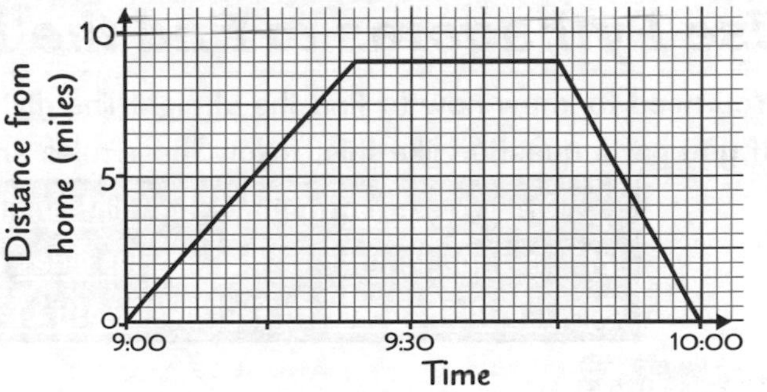

Distance from home (miles)
10
5
0
9:00
9:30
10:00
Time

Unit Conversions (p81-82)

27) Convert a) 5.6 litres to cm³, b) 8 pounds to kg, c) 3 m/s to km/h, d) 569 m² to cm².
28) Convert 12 km to miles. Construct a graph if you need to.

Constructions and Loci (p83-86)

29) Construct triangle XYZ, where XY = 5.6 cm, XZ = 7.2 cm and angle YXZ = 55°.
30) Construct two triangles, ABC, with angle A = 40°, AB = 6 cm, BC = 4.5 cm.
31) What shape does the locus of points that are a fixed distance from a given point make?
32) Construct an accurate 45° angle.
33) Draw a square with sides of length 6 cm and label it ABCD. Shade the region that is nearer to AB than CD and less than 4 cm from vertex A.

Bearings (p87)

34) Describe how to find a bearing from point A to point B.
35) A helicopter flies 25 km on a bearing of 210°, then 20 km on a bearing of 040°. Draw a scale diagram to show this. Use a scale of 1 cm = 5 km.

Pythagoras' Theorem

Pythagoras' theorem sounds hard but it's actually dead simple.
It's also dead important, so make sure you really get your teeth into it.

Pythagoras' Theorem — $a^2 + b^2 = c^2$

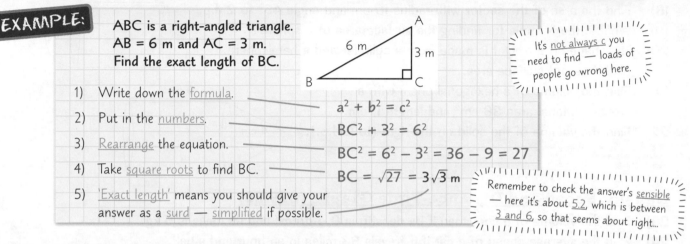

1) PYTHAGORAS' THEOREM only works for RIGHT-ANGLED TRIANGLES.
2) Pythagoras uses two sides to find the third side.
3) The BASIC FORMULA for Pythagoras is $a^2 + b^2 = c^2$
4) Make sure you get the numbers in the RIGHT PLACE. c is the longest side (called the hypotenuse) and it's always opposite the right angle.
5) Always CHECK that your answer is SENSIBLE.

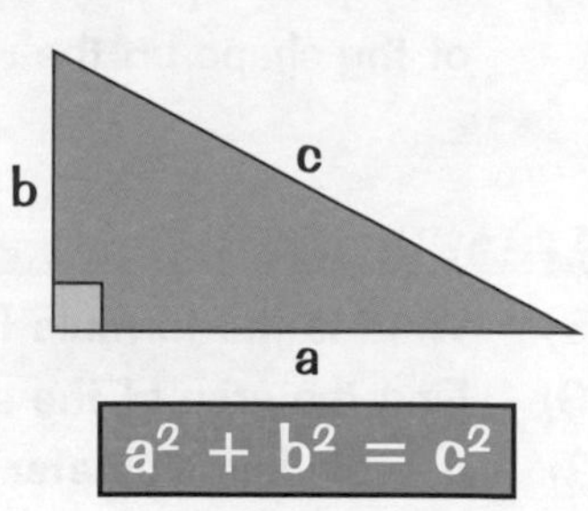

ABC is a right-angled triangle.
AB = 6 m and AC = 3 m.
Find the exact length of BC.

1) Write down the formula. — $a^2 + b^2 = c^2$
2) Put in the numbers. — $BC^2 + 3^2 = 6^2$
3) Rearrange the equation. — $BC^2 = 6^2 - 3^2 = 36 - 9 = 27$
4) Take square roots to find BC. — $BC = \sqrt{27} = 3\sqrt{3}$ m
5) 'Exact length' means you should give your answer as a surd — simplified if possible.

Remember to check the answer's sensible — here it's about 5.2, which is between 3 and 6, so that seems about right...

Use Pythagoras to find the Distance Between Points

You need to know how to find the straight-line distance between two points on a graph.
If you get a question like this, follow these rules and it'll all become breathtakingly simple:

1) Draw a sketch to show the right-angled triangle.
2) Find the lengths of the shorter sides of the triangle.
3) Use Pythagoras to find the length of the hypotenuse. (That's your answer.)

EXAMPLE: **Point P has coordinates (8, 3) and point Q has coordinates (–4, 8). Find the length of the line PQ.**

①

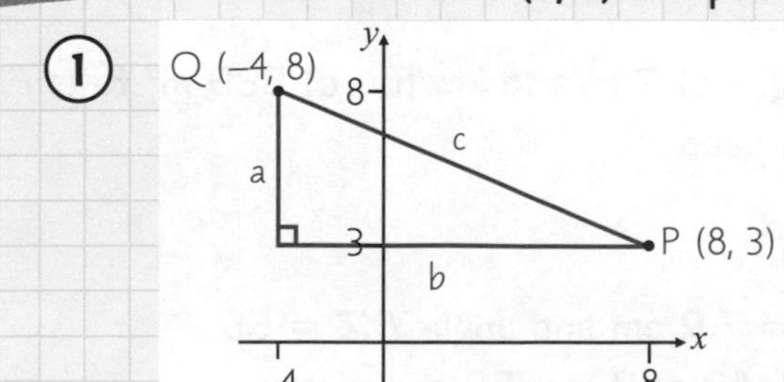

② Length of side a = 8 – 3 = 5
Length of side b = 8 – –4 = 12

③ Use Pythagoras to find side c:
$c^2 = a^2 + b^2 = 5^2 + 12^2 = 25 + 144 = 169$
So: $c = \sqrt{169} = 13$

Remember, if it's not a right angle, it's a wrong angle...

Once you've learned all the Pythagoras facts on this page, try these Exam Practice Questions.

Q1 Find the length of AC correct to 1 decimal place.

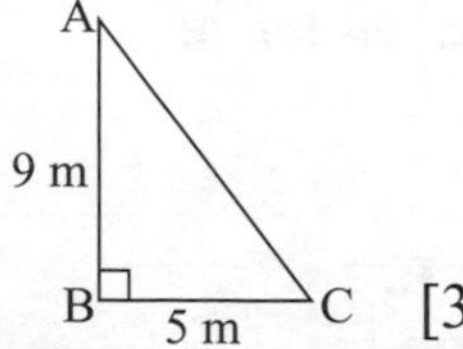

[3 marks]

Q2 Find the exact length of BC.

B
17 m
A
15 m
C

[3 marks]

Q3 Point A has coordinates (10, 15) and point B has coordinates (6, 12).
Find the length of the line AB. [4 marks]

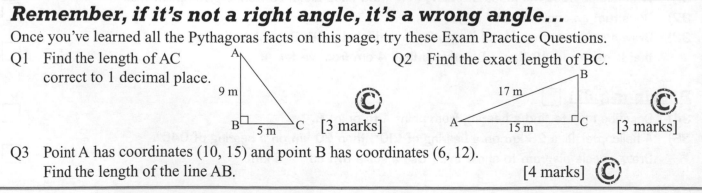

Trigonometry — Sin, Cos, Tan

Trigonometry — it's a big scary word. But it's not a big scary topic. An important topic, yes. An always cropping up topic, definitely. But scary? Pur-lease. Takes more than a triangle to scare me. Read on...

The 3 Trigonometry Formulas

There are three basic trig formulas — each one links two sides and an angle of a right-angled triangle.

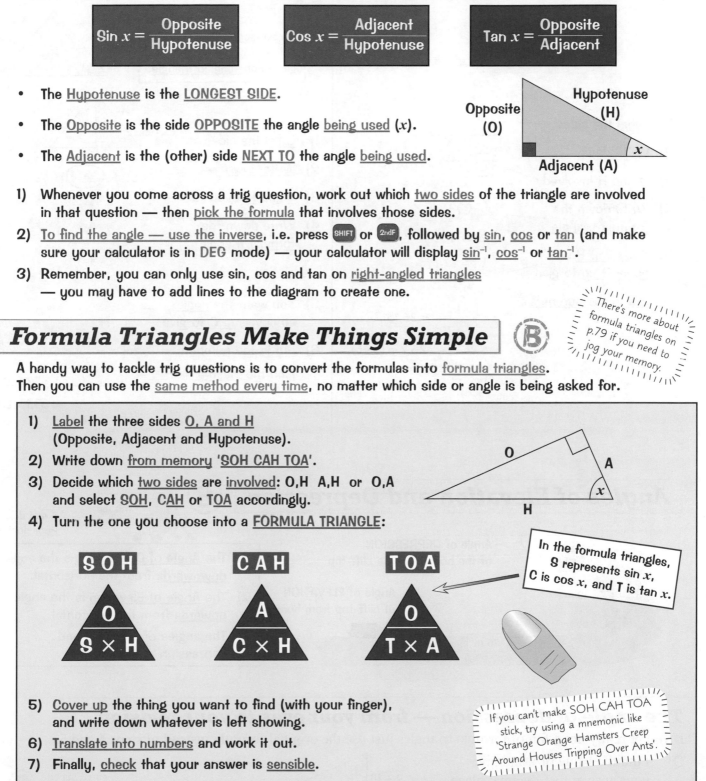

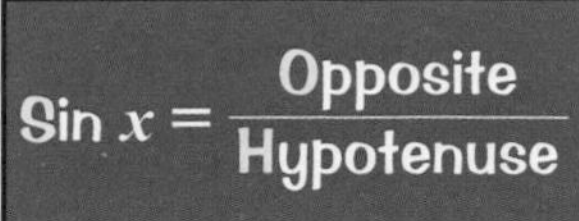

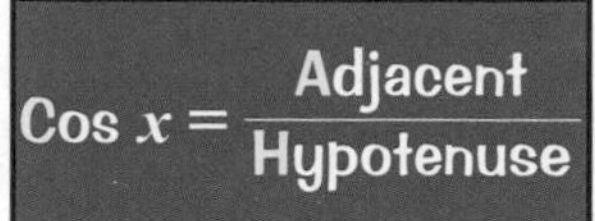

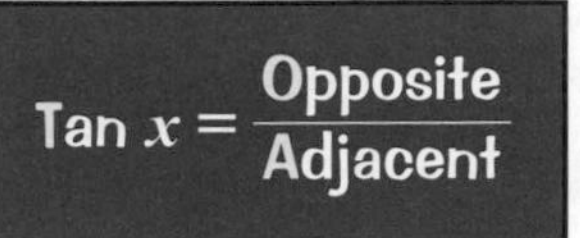

- The Hypotenuse is the LONGEST SIDE.
- The Opposite is the side OPPOSITE the angle being used (x).
- The Adjacent is the (other) side NEXT TO the angle being used.

Hypotenuse (H)
Opposite (O)
Adjacent (A)
x

1) Whenever you come across a trig question, work out which two sides of the triangle are involved in that question — then pick the formula that involves those sides.
2) To find the angle — use the inverse, i.e. press SHIFT or 2ndF, followed by sin, cos or tan (and make sure your calculator is in DEG mode) — your calculator will display $\sin^{-1}$, $\cos^{-1}$ or $\tan^{-1}$.
3) Remember, you can only use sin, cos and tan on right-angled triangles — you may have to add lines to the diagram to create one.

There's more about formula triangles on p.79 if you need to jog your memory.

Formula Triangles Make Things Simple

A handy way to tackle trig questions is to convert the formulas into formula triangles. Then you can use the same method every time, no matter which side or angle is being asked for.

1) Label the three sides O, A and H (Opposite, Adjacent and Hypotenuse).
2) Write down from memory 'SOH CAH TOA'.
3) Decide which two sides are involved: O,H A,H or O,A and select SOH, CAH or TOA accordingly.
4) Turn the one you choose into a FORMULA TRIANGLE:

O
A
H
x

SOH

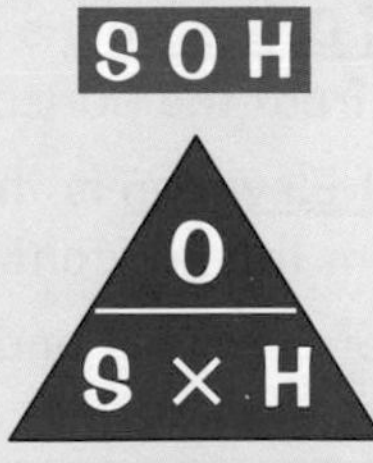

CAH

TOA

In the formula triangles, S represents sin x, C is cos x, and T is tan x.

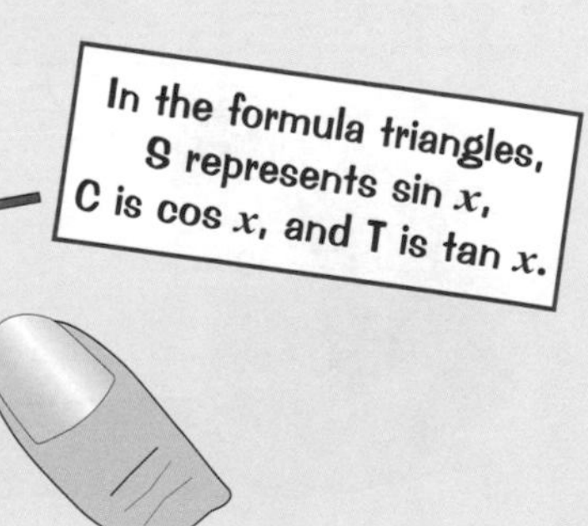

5) Cover up the thing you want to find (with your finger), and write down whatever is left showing.
6) Translate into numbers and work it out.
7) Finally, check that your answer is sensible.

If you can't make SOH CAH TOA stick, try using a mnemonic like 'Strange Orange Hamsters Creep Around Houses Tripping Over Ants'.

SOH CAH TOA — the not-so-secret formula for success...

You need to know this stuff off by heart — so go over this page a few times until you've got those formulas firmly lodged and all ready to reel off in the exam. All set? Time for some examples...

Trigonometry — Sin, Cos, Tan

Here are some lovely examples to help you through the trials of trig.

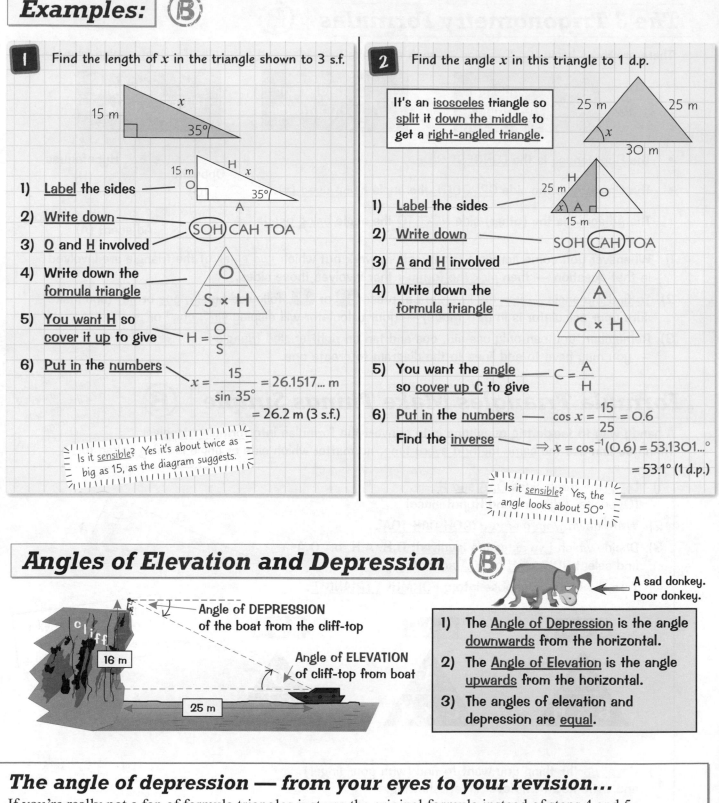

Examples: 

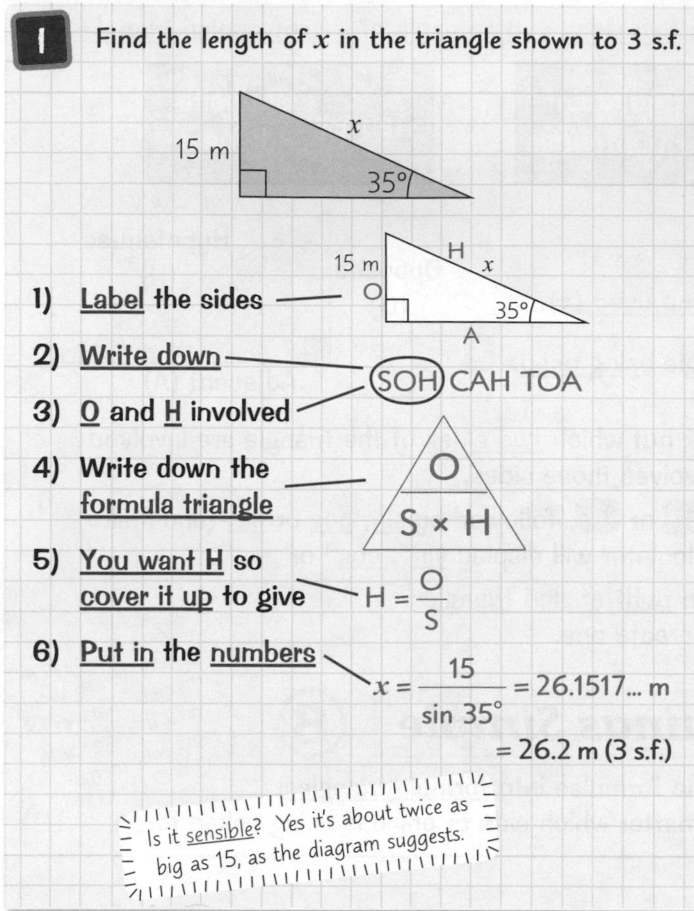

1 Find the length of x in the triangle shown to 3 s.f.
15 m
x
35°
15 m
H
x
O
35°
A
1) Label the sides
2) Write down
3) O and H involved
SOH CAH TOA
4) Write down the formula triangle
O
S × H
5) You want H so cover it up to give
H = O/S
6) Put in the numbers
x = 15 / sin 35° = 26.1517... m
= 26.2 m (3 s.f.)
Is it sensible? Yes it's about twice as big as 15, as the diagram suggests.

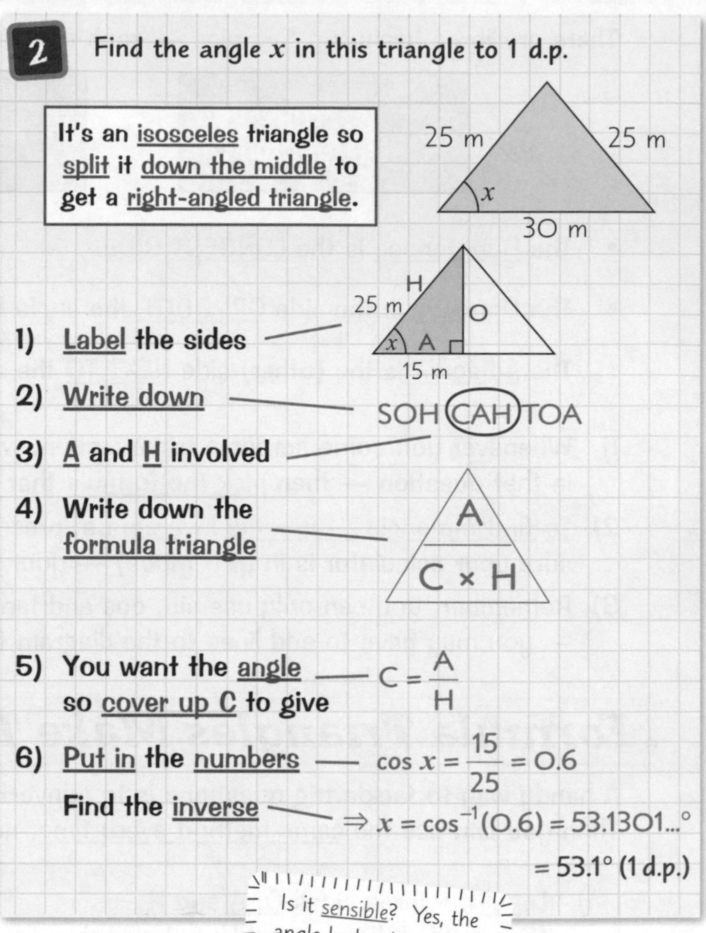

2 Find the angle x in this triangle to 1 d.p.
It's an isosceles triangle so split it down the middle to get a right-angled triangle.
25 m
25 m
x
30 m
H
25 m
O
x
A
15 m
1) Label the sides
2) Write down
SOH CAH TOA
3) A and H involved
4) Write down the formula triangle
A
C × H
5) You want the angle so cover up C to give
C = A/H
6) Put in the numbers
cos x = 15/25 = 0.6
Find the inverse
⇒ x = cos⁻¹(0.6) = 53.1301...°
= 53.1° (1 d.p.)
Is it sensible? Yes, the angle looks about 50°.

Angles of Elevation and Depression

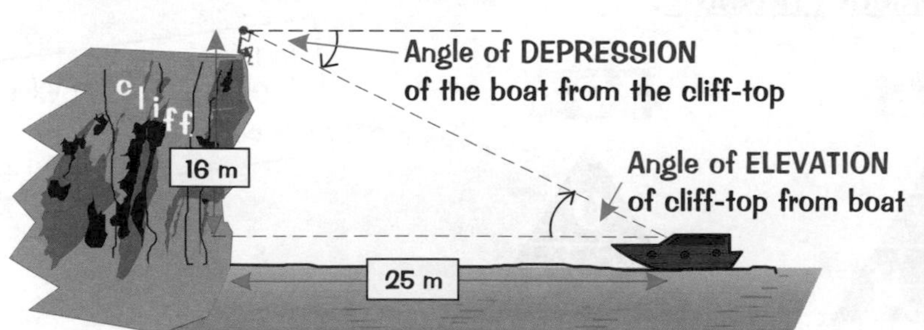

1) The Angle of Depression is the angle downwards from the horizontal.
2) The Angle of Elevation is the angle upwards from the horizontal.
3) The angles of elevation and depression are equal.

The angle of depression — from your eyes to your revision...

If you're really not a fan of formula triangles just use the original formula instead of steps 4 and 5.

Q1 Find the value of x and give your answer to 1 decimal place.

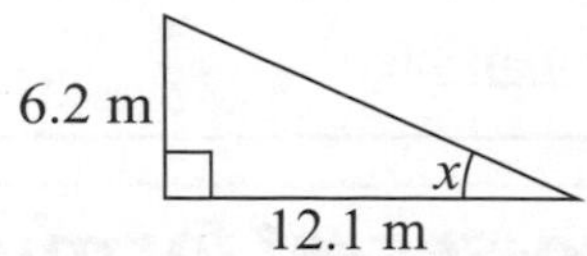

[3 marks] 

Q2 A 3.2 m ladder is leaning against a vertical wall. It is at an angle of 68° to the horizontal ground. How far does the ladder reach up the wall? Give your answer to 3 s.f. [3 marks] (B)

The Sine and Cosine Rules

Normal trigonometry using SOH CAH TOA etc. can only be applied to right-angled triangles. Which leaves us with the question of what to do with other-angled triangles. Step forward the Sine and Cosine Rules...

Labelling the Triangle

This is very important. You must label the sides and angles properly so that the letters for the sides and angles correspond with each other. Use lower case letters for the sides and capitals for the angles.

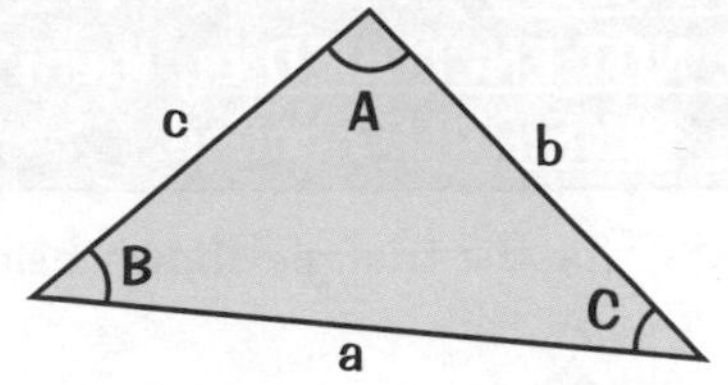

Remember, side 'a' is opposite angle A etc.

It doesn't matter which sides you decide to call a, b, and c, just as long as the angles are then labelled properly.

Three Formulas to Learn:

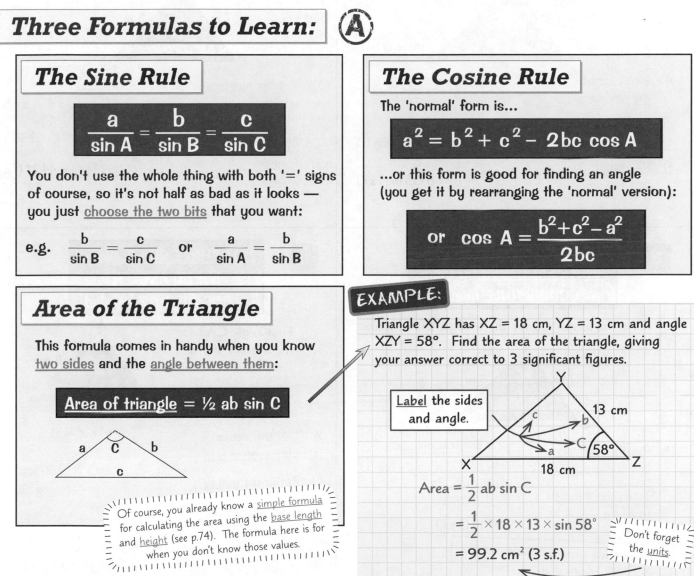

The Sine Rule

$$\frac{a}{\sin A} = \frac{b}{\sin B} = \frac{c}{\sin C}$$

You don't use the whole thing with both '=' signs of course, so it's not half as bad as it looks — you just choose the two bits that you want:

e.g. $\frac{b}{\sin B} = \frac{c}{\sin C}$ or $\frac{a}{\sin A} = \frac{b}{\sin B}$

The Cosine Rule

The 'normal' form is...

$$a^2 = b^2 + c^2 - 2bc \cos A$$

...or this form is good for finding an angle (you get it by rearranging the 'normal' version):

$$\text{or} \quad \cos A = \frac{b^2 + c^2 - a^2}{2bc}$$

Area of the Triangle

This formula comes in handy when you know two sides and the angle between them:

Area of triangle = ½ ab sin C

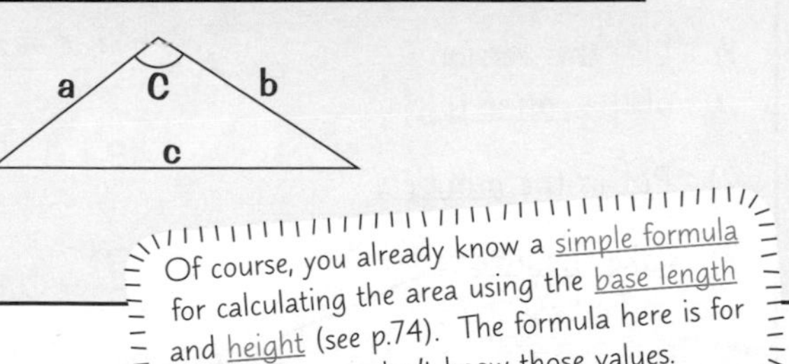

Of course, you already know a simple formula for calculating the area using the base length and height (see p.74). The formula here is for when you don't know those values.

EXAMPLE:

Triangle XYZ has XZ = 18 cm, YZ = 13 cm and angle XZY = 58°. Find the area of the triangle, giving your answer correct to 3 significant figures.

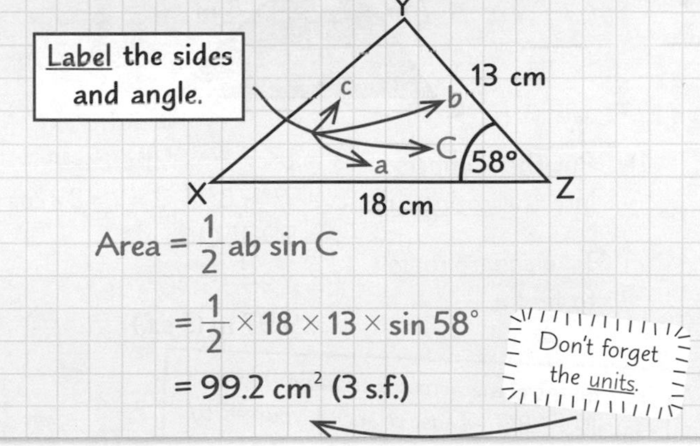

$\text{Area} = \frac{1}{2} ab \sin C$

$= \frac{1}{2} \times 18 \times 13 \times \sin 58°$

$= 99.2 \text{ cm}^2$ (3 s.f.)

Don't forget the units.

...and step back again. Hope you enjoyed a moment in the spotlight...

You need to learn all of these formulas off by heart and practise using them.
Here's an area question to have a go at, and fear not, you'll get your chance to tackle some sine and cosine rule problems on the next page...

Q1 Triangle FGH has FG = 9 cm, FH = 12 cm and angle GFH = 37°. Find its area, giving your answer correct to 3 significant figures. [2 marks]

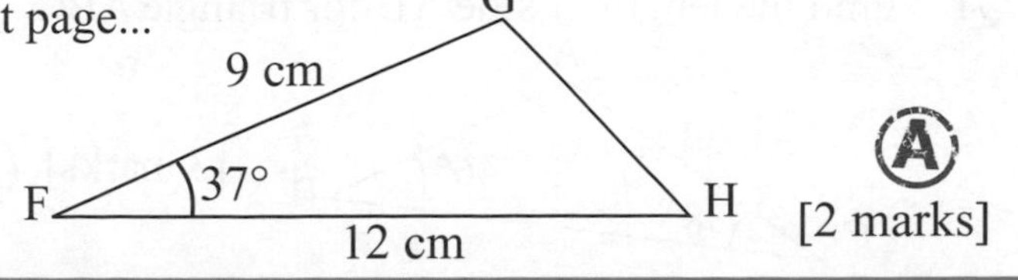

The Sine and Cosine Rules

Amazingly, there are only FOUR question types where the sine and cosine rules would be applied. So learn the exact details of these four examples and you'll be laughing. WARNING: if you laugh too much people will think you're crazy.

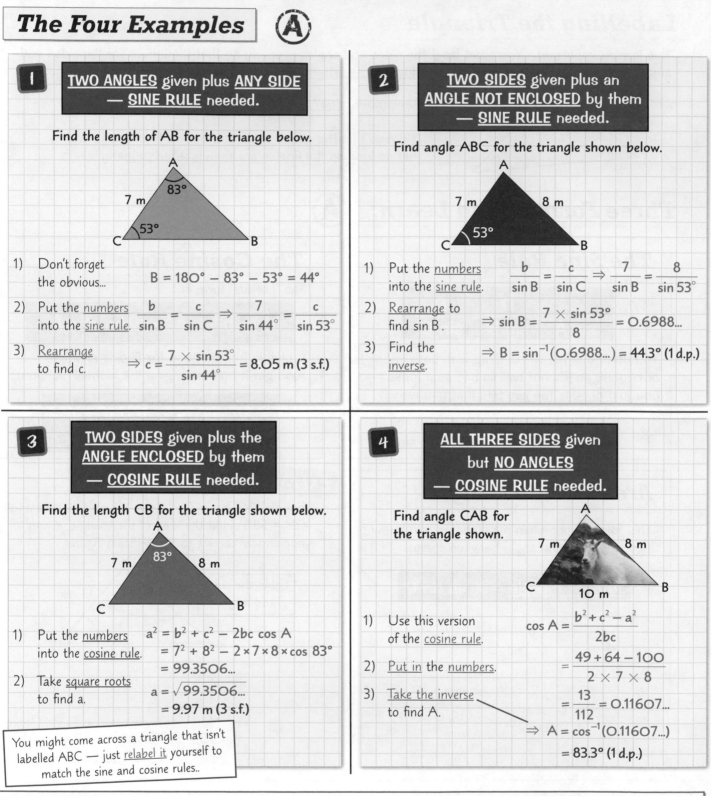

The Four Examples

1

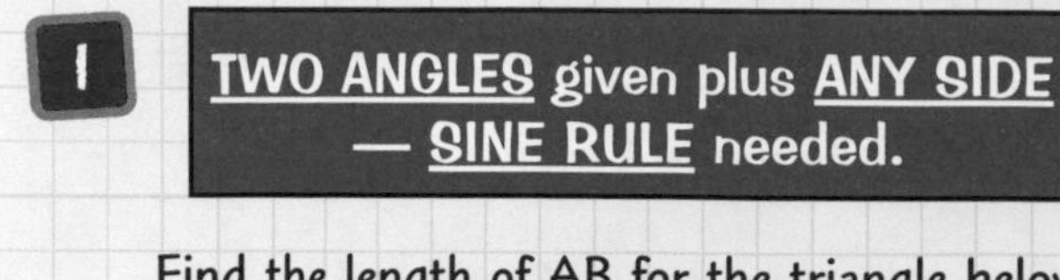

TWO ANGLES given plus ANY SIDE
— SINE RULE needed.

Find the length of AB for the triangle below.

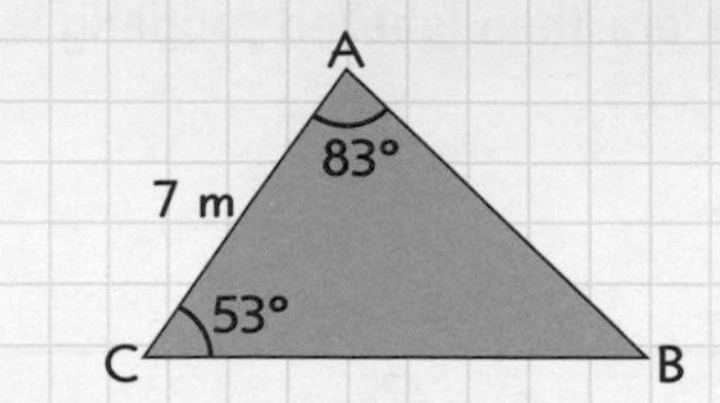

1) Don't forget the obvious... $B = 180° - 83° - 53° = 44°$
2) Put the numbers into the sine rule. $\frac{b}{\sin B} = \frac{c}{\sin C} \Rightarrow \frac{7}{\sin 44°} = \frac{c}{\sin 53°}$
3) Rearrange to find c. $\Rightarrow c = \frac{7 \times \sin 53°}{\sin 44°} = 8.05 \text{ m (3 s.f.)}$

2

TWO SIDES given plus an ANGLE NOT ENCLOSED by them — SINE RULE needed.

Find angle ABC for the triangle shown below.

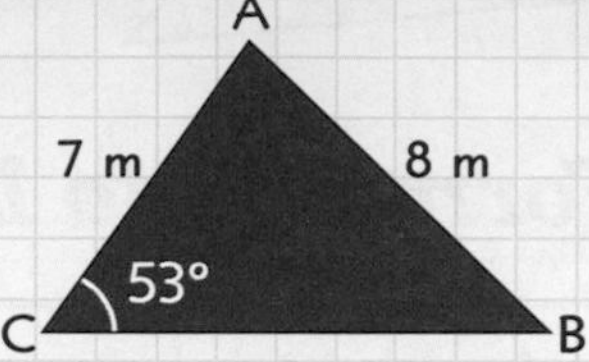

1) Put the numbers into the sine rule. $\frac{b}{\sin B} = \frac{c}{\sin C} \Rightarrow \frac{7}{\sin B} = \frac{8}{\sin 53°}$
2) Rearrange to find sin B. $\Rightarrow \sin B = \frac{7 \times \sin 53°}{8} = 0.6988...$
3) Find the inverse. $\Rightarrow B = \sin^{-1}(0.6988...) = 44.3° \text{ (1 d.p.)}$

3

TWO SIDES given plus the ANGLE ENCLOSED by them — COSINE RULE needed.

Find the length CB for the triangle shown below.

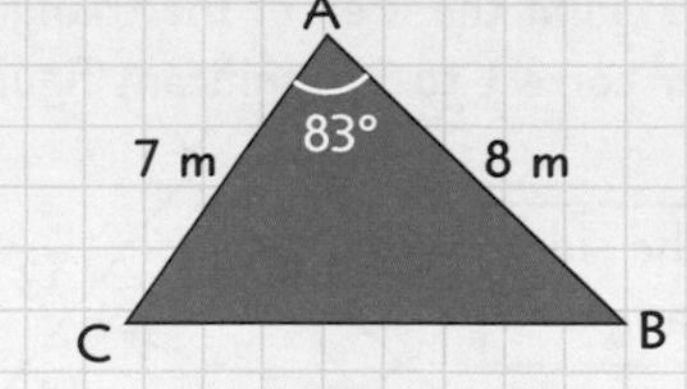

1) Put the numbers into the cosine rule.
$a^2 = b^2 + c^2 - 2bc \cos A$
$= 7^2 + 8^2 - 2 \times 7 \times 8 \times \cos 83°$
$= 99.3506...$
2) Take square roots to find a.
$a = \sqrt{99.3506...}$
$= 9.97 \text{ m (3 s.f.)}$

You might come across a triangle that isn't labelled ABC — just relabel it yourself to match the sine and cosine rules..

4

ALL THREE SIDES given but NO ANGLES — COSINE RULE needed.

Find angle CAB for the triangle shown.

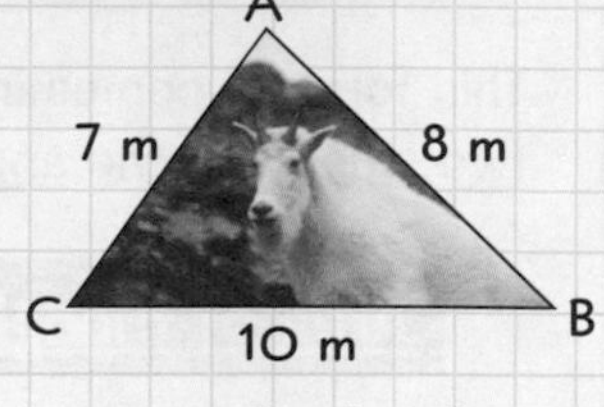

1) Use this version of the cosine rule. $\cos A = \frac{b^2 + c^2 - a^2}{2bc}$
2) Put in the numbers. $= \frac{49 + 64 - 100}{2 \times 7 \times 8}$
3) Take the inverse to find A. $= \frac{13}{112} = 0.11607...$
$\Rightarrow A = \cos^{-1}(0.11607...)$
$= 83.3° \text{ (1 d.p.)}$

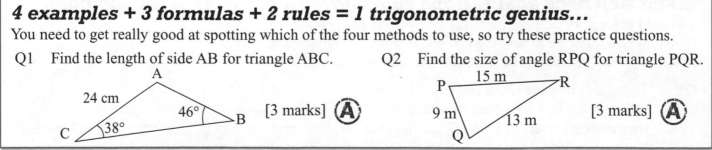

4 examples + 3 formulas + 2 rules = 1 trigonometric genius...

You need to get really good at spotting which of the four methods to use, so try these practice questions.

Q1 Find the length of side AB for triangle ABC.

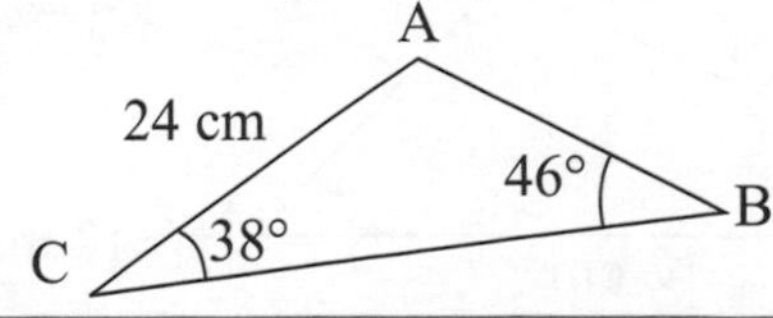

[3 marks] (A)

Q2 Find the size of angle RPQ for triangle PQR.

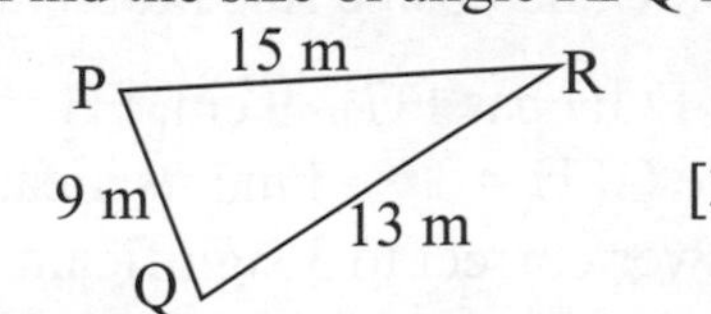

[3 marks] (A)

3D Pythagoras

This is a 3D version of the 2D Pythagoras theorem you saw on page 90.
There's just one simple formula — learn it and the world's your oyster...

3D Pythagoras for Cuboids — $a^2 + b^2 + c^2 = d^2$

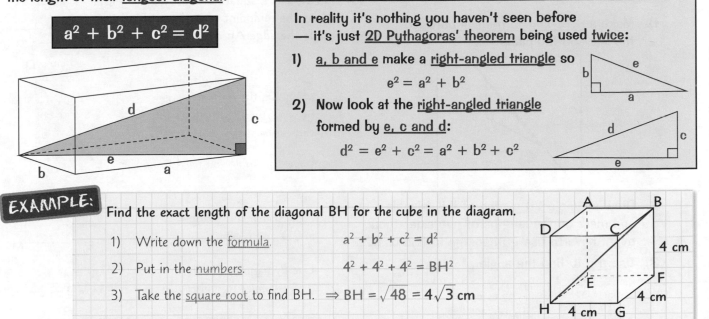

Cuboids have their own formula for calculating the length of their longest diagonal:

$$a^2 + b^2 + c^2 = d^2$$

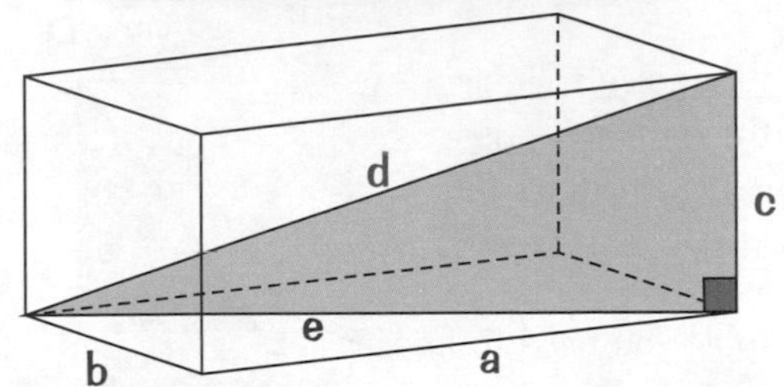

In reality it's nothing you haven't seen before — it's just 2D Pythagoras' theorem being used twice:

1) a, b and e make a right-angled triangle so

$$e^2 = a^2 + b^2$$

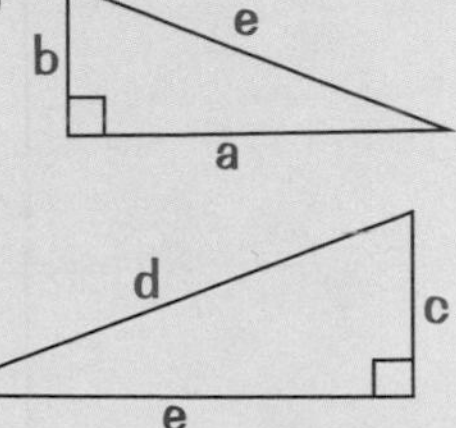

2) Now look at the right-angled triangle formed by e, c and d:

$$d^2 = e^2 + c^2 = a^2 + b^2 + c^2$$

EXAMPLE:

Find the exact length of the diagonal BH for the cube in the diagram.

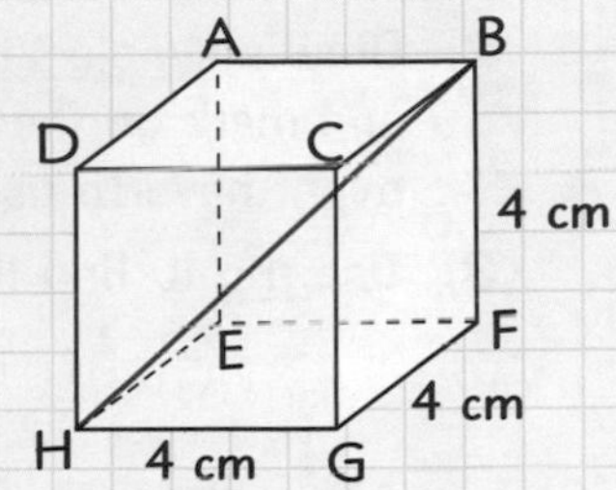

1) Write down the formula. $a^2 + b^2 + c^2 = d^2$

2) Put in the numbers. $4^2 + 4^2 + 4^2 = BH^2$

3) Take the square root to find BH. $\Rightarrow BH = \sqrt{48} = 4\sqrt{3}$ cm

The Cuboid Formula can be used in Other 3D Shapes

EXAMPLE:

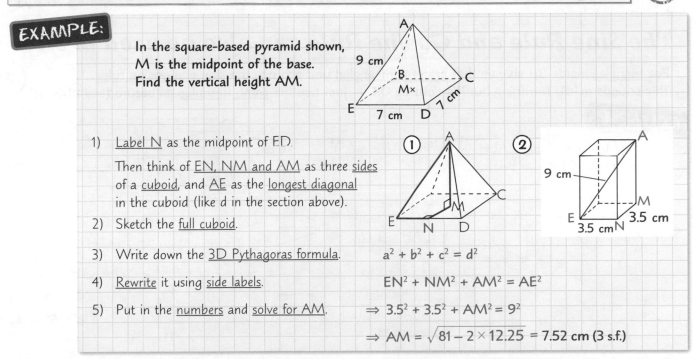

In the square-based pyramid shown, M is the midpoint of the base. Find the vertical height AM.

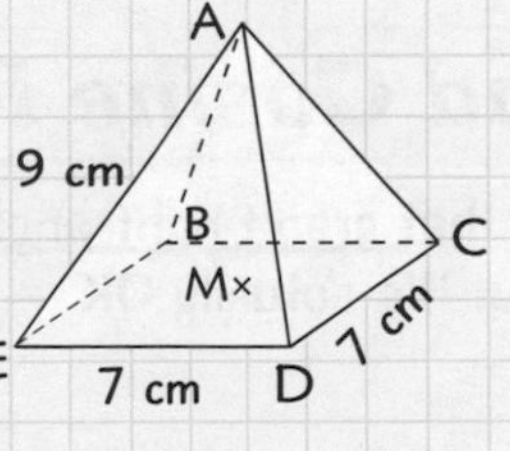

1) Label N as the midpoint of ED.
Then think of EN, NM and AM as three sides of a cuboid, and AE as the longest diagonal in the cuboid (like d in the section above).

2) Sketch the full cuboid.

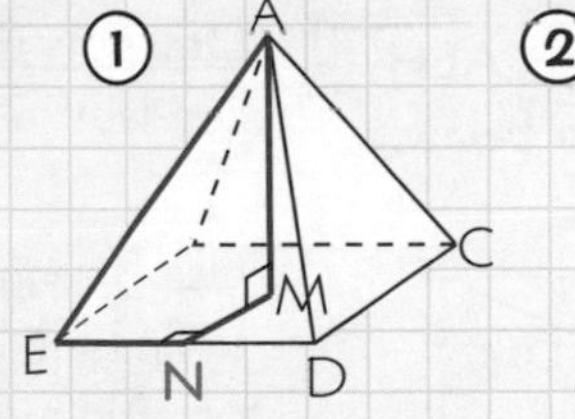

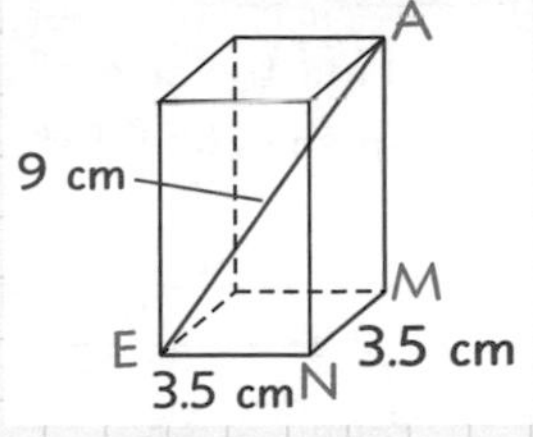
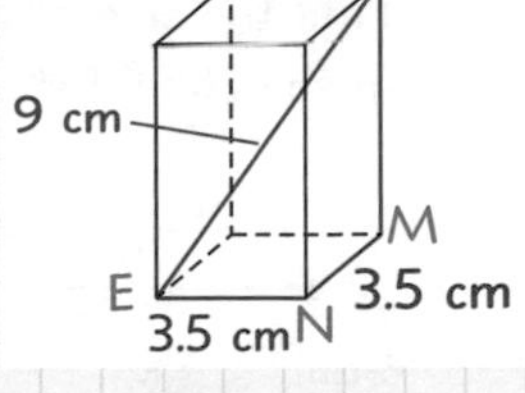

3) Write down the 3D Pythagoras formula. $a^2 + b^2 + c^2 = d^2$

4) Rewrite it using side labels. $EN^2 + NM^2 + AM^2 = AE^2$

5) Put in the numbers and solve for AM. $\Rightarrow 3.5^2 + 3.5^2 + AM^2 = 9^2$

$\Rightarrow AM = \sqrt{81 - 2 \times 12.25} = 7.52$ cm (3 s.f.)

Wow — just what can't right-angled triangles do?...

You need to be ready to tackle 3D questions in the exam, so have a go at this Exam Practice Question.

Q1 Find the length AH in the cuboid shown to 3 s.f.

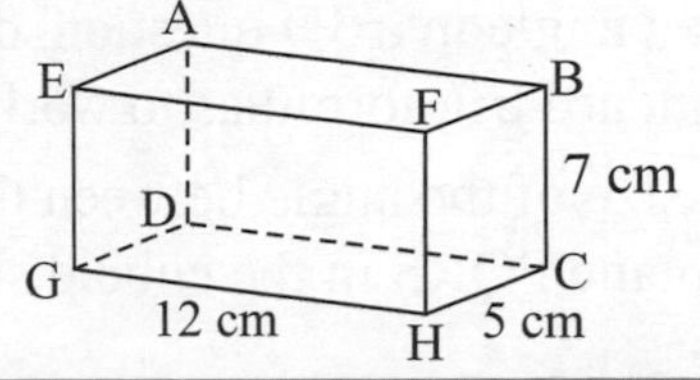

[3 marks]

3D Trigonometry

3D trig may sound tricky, and in many ways it is... but it's actually just using the same old rules.

Angle Between Line and Plane — Use a Diagram

Learn the 3-Step Method

1) **Make a right-angled triangle between the line and the plane.**

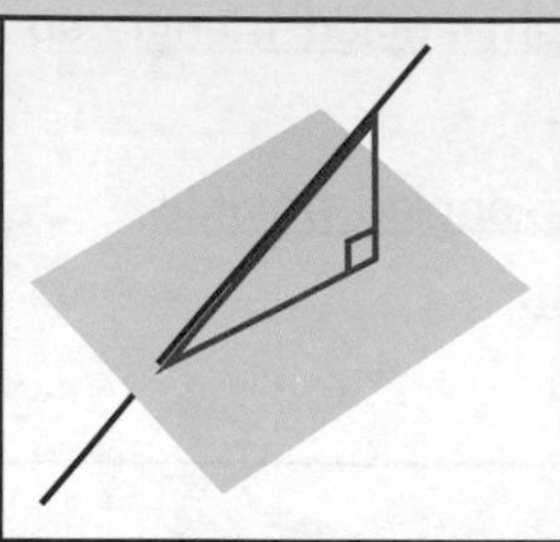

2) **Draw a simple 2D sketch of this triangle and mark on the lengths of two sides (you might have to use Pythagoras to find one).**
3) **Use trig to find the angle.**

Have a look at p.90-92 to jog your memory about Pythagoras and trig.

EXAMPLE:

ABCDE is a square-based pyramid with M as the midpoint of its base. Find the angle the edge AE makes with the base.

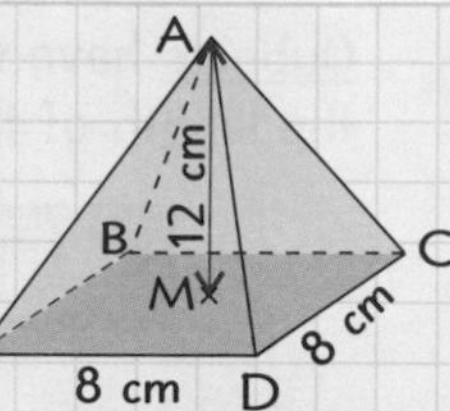

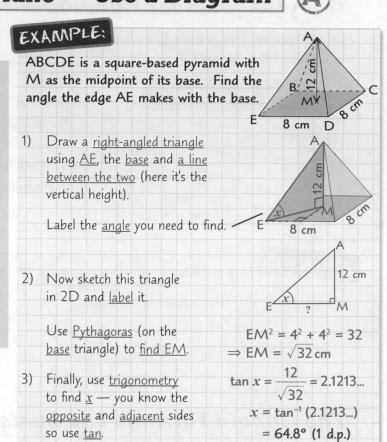

1) Draw a right-angled triangle using AE, the base and a line between the two (here it's the vertical height).

   Label the angle you need to find.

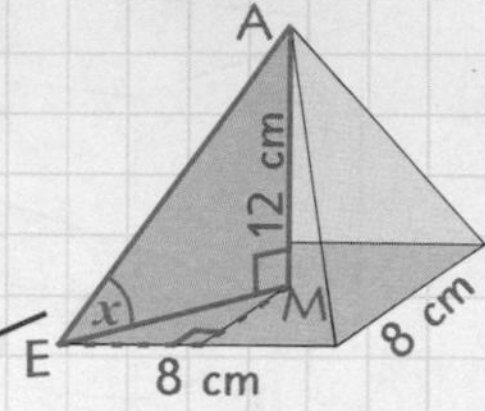

2) Now sketch this triangle in 2D and label it.

   (Triangle: A, E, M; 12 cm; x; ?)

   Use Pythagoras (on the base triangle) to find EM.

$$EM^2 = 4^2 + 4^2 = 32$$
$$\Rightarrow EM = \sqrt{32}\text{ cm}$$

3) Finally, use trigonometry to find x — you know the opposite and adjacent sides so use tan.

$$\tan x = \frac{12}{\sqrt{32}} = 2.1213...$$
$$x = \tan^{-1}(2.1213...)$$
$$= 64.8° \text{ (1 d.p.)}$$

The Sine Rule and Cosine Rule can also be used in 3D

For triangles inside 3D shapes that aren't right-angled you can use the sine and cosine rules.
This sounds mildly terrifying but it's actually OK — just use the same formulas as before (see p93-4).

EXAMPLE:

Find the size of angle AEH in the cuboid shown below.

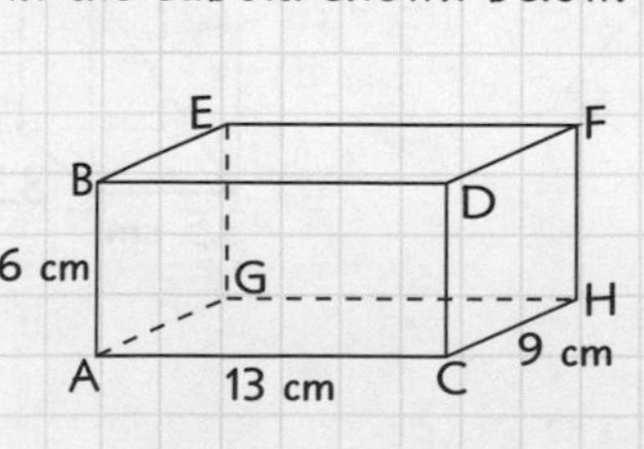

1) Draw the triangle AEH and label angle AEH as x.

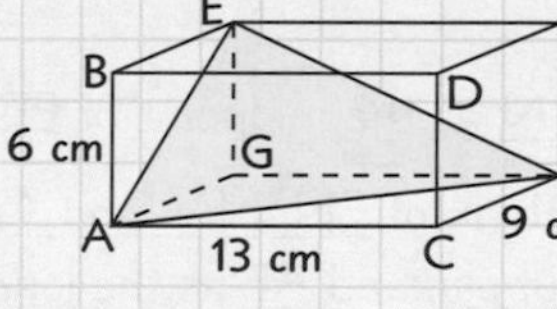

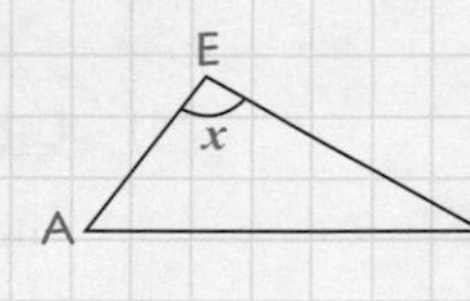

2) Use Pythagoras' theorem to find the lengths of AE, AH and EH.

$$AH^2 = 13^2 + 9^2 = 250 \Rightarrow AH = \sqrt{250}$$
$$AE^2 = 6^2 + 9^2 = 117 \Rightarrow AE = \sqrt{117}$$
$$EH^2 = 6^2 + 13^2 = 205 \Rightarrow EH = \sqrt{205}$$

3) Find x using the cosine rule: $AH^2 = AE^2 + EH^2 - 2 \times AE \times EH \times \cos x$

   Put in the numbers. $250 = 117 + 205 - 2\sqrt{117}\sqrt{205}\cos x$

   Rearrange and take the inverse to find x. $x = \cos^{-1}\left(\frac{117 + 205 - 250}{2\sqrt{117 \times 205}}\right) = 76.6° \text{ (1 d.p.)}$

The Return of the Cosine Rule — out now in 3D...

If you need to find an angle in a 3D question, don't panic — just put those standard trig formulas to work.

Q1 Find the size of the angle between the line PV and the plane PQRS in the cuboid shown. [4 marks]

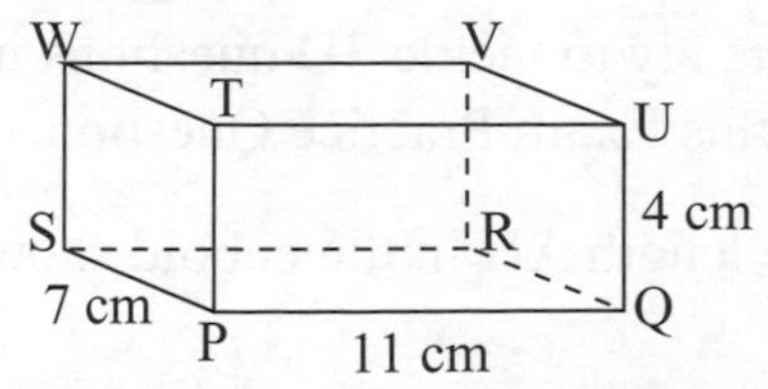

Vectors

Vectors represent a movement of a certain size in a certain direction.
They might seem a bit weird at first, but there are really just a few facts to get to grips with...

The Vector Notations

There are several ways to write vectors...

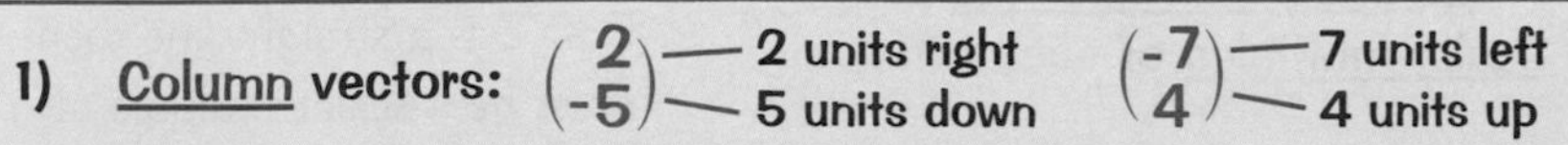

1) Column vectors: $\begin{pmatrix} 2 \\ -5 \end{pmatrix}$ — 2 units right, 5 units down; $\begin{pmatrix} -7 \\ 4 \end{pmatrix}$ — 7 units left, 4 units up

2) **a** —— exam questions use bold like this

3) $\underset{\sim}{a}$ or $\underline{a}$ — you should always underline them (unless you have a magical bolding pen...)

4) $\overrightarrow{AB}$ this means the vector from point A to point B

They're represented on a diagram by an arrow.

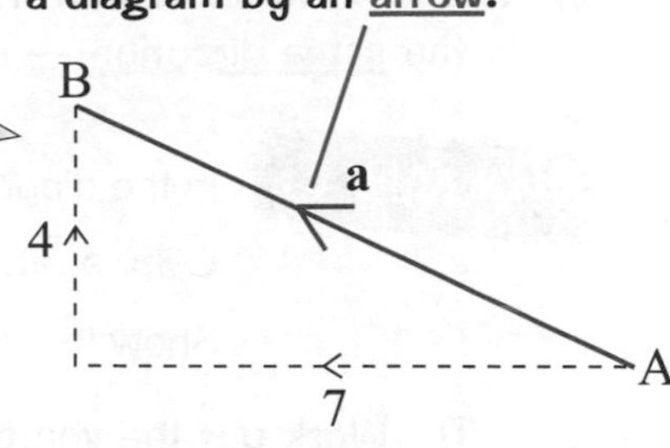

Multiplying a Vector by a Scalar

Multiplying a vector by a positive number changes the vector's size but not its direction — it scales the vector. If the number's negative then the direction gets switched.

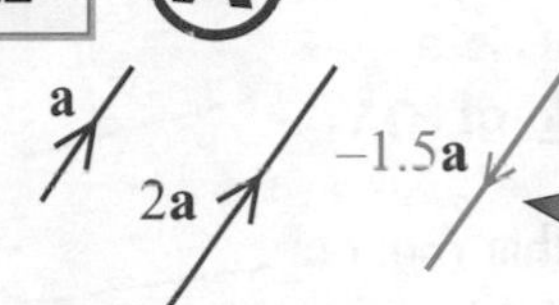

Vectors that are scalar multiples of each other are parallel.

Adding and Subtracting Vectors

You can describe movements between points by adding and subtracting known vectors. All vector exam questions are based around this.

"$\underset{\sim}{a} + \underset{\sim}{b}$" means 'go along $\underset{\sim}{a}$ then $\underset{\sim}{b}$'.

"$\underset{\sim}{c} - \underset{\sim}{d}$" means 'go along $\underset{\sim}{c}$ then backwards along $\underset{\sim}{d}$' (the minus sign means go the opposite way).

In the diagrams, $\overrightarrow{PR} = \underset{\sim}{a} + \underset{\sim}{b}$ and $\overrightarrow{XZ} = \underset{\sim}{c} - \underset{\sim}{d}$.

EXAMPLE:

In the diagram below, M is the midpoint of BC.
Find vectors $\overrightarrow{AM}$, $\overrightarrow{OC}$ and $\overrightarrow{AC}$ in terms of **a**, **b** and **m**.

To obtain the unknown vector just 'get there' by any route made up of known vectors.

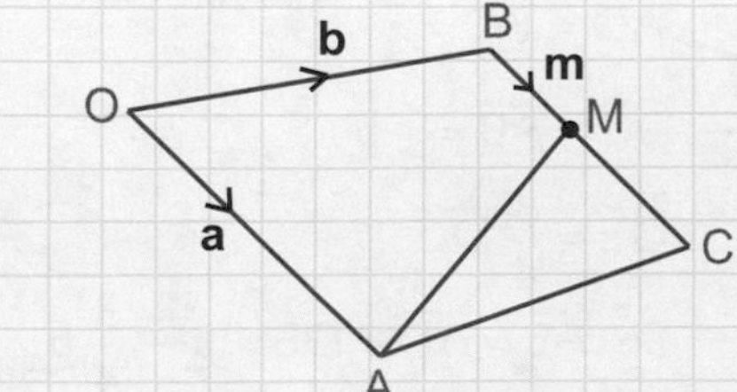

$\overrightarrow{AM} = -\underset{\sim}{a} + \underset{\sim}{b} + \underset{\sim}{m}$ — A to M via O and B

$\overrightarrow{OC} = \underset{\sim}{b} + 2\underset{\sim}{m}$ — O to C via B and M — M's half-way between B and C so $\overrightarrow{BC} = 2\underset{\sim}{m}$

$\overrightarrow{AC} = -\underset{\sim}{a} + \underset{\sim}{b} + 2\underset{\sim}{m}$ — A to C via O, B and M

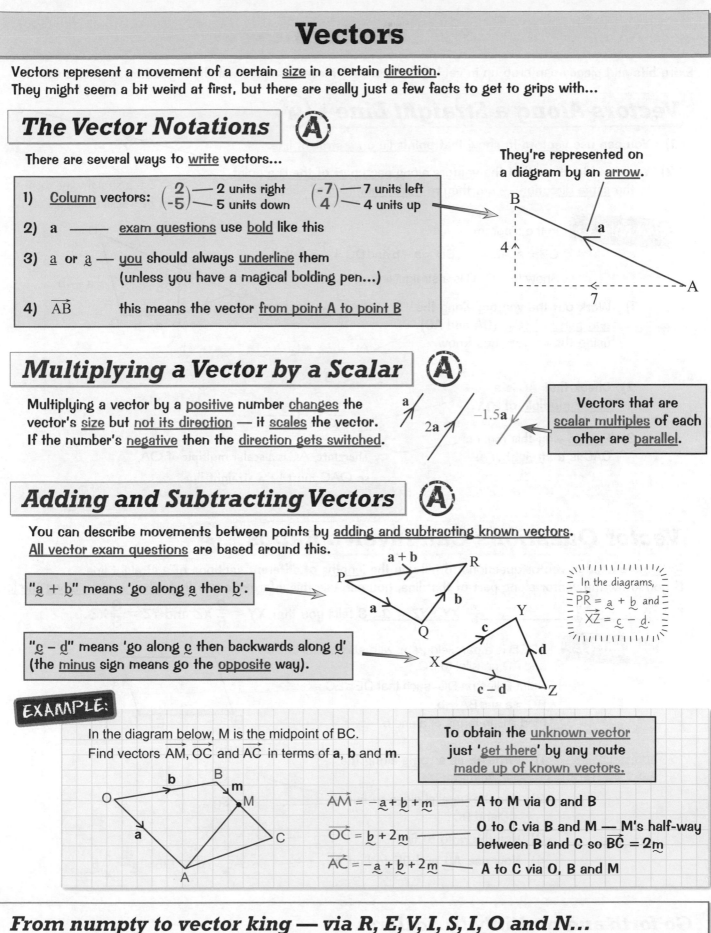

From numpty to vector king — via R, E, V, I, S, I, O and N...

You need to get to grips with questions like the one above, so here's one to have a go at...

Q1 In triangle ABC, M is the midpoint of BC and N is the midpoint of AB. $\overrightarrow{AC}$ = **p** and $\overrightarrow{BM}$ = **q**. Find $\overrightarrow{AB}$ and $\overrightarrow{NA}$ in terms of **p** and **q**. [4 marks]

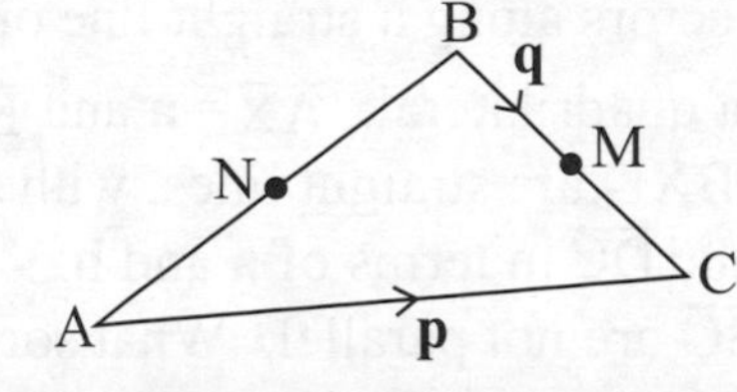

Vectors

Extra bits and pieces can crop up in vector questions — these examples will show you how to tackle them...

Vectors Along a Straight Line

1) You can use vectors to show that points lie on a straight line.
2) You need to show that the vectors along each part of the line point in the same direction — i.e. they're scalar multiples of each other.

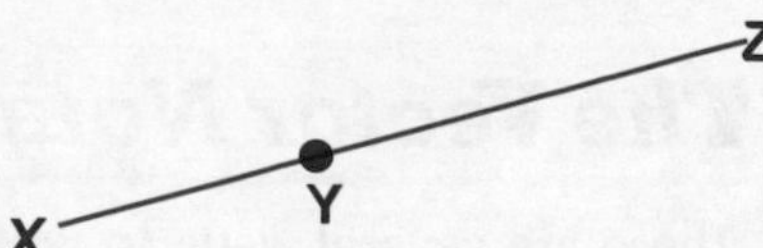

If XYZ is a straight line then $\overrightarrow{XY}$ must be a scalar multiple of $\overrightarrow{YZ}$.

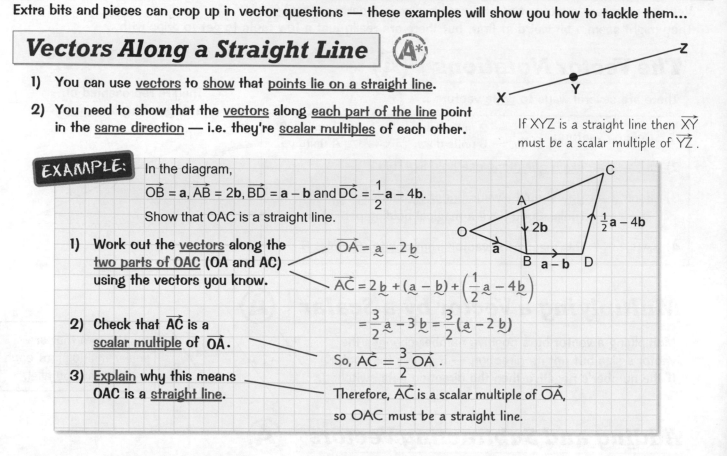

In the diagram,
$\overrightarrow{OB} = \mathbf{a}$, $\overrightarrow{AB} = 2\mathbf{b}$, $\overrightarrow{BD} = \mathbf{a} - \mathbf{b}$ and $\overrightarrow{DC} = \frac{1}{2}\mathbf{a} - 4\mathbf{b}$.
Show that OAC is a straight line.

1) **Work out the vectors along the two parts of OAC (OA and AC) using the vectors you know.**

$\overrightarrow{OA} = \underset{\sim}{a} - 2\underset{\sim}{b}$

$\overrightarrow{AC} = 2\underset{\sim}{b} + (\underset{\sim}{a} - \underset{\sim}{b}) + \left(\frac{1}{2}\underset{\sim}{a} - 4\underset{\sim}{b}\right)$

$= \frac{3}{2}\underset{\sim}{a} - 3\underset{\sim}{b} = \frac{3}{2}(\underset{\sim}{a} - 2\underset{\sim}{b})$

2) **Check that $\overrightarrow{AC}$ is a scalar multiple of $\overrightarrow{OA}$.**

So, $\overrightarrow{AC} = \frac{3}{2}\overrightarrow{OA}$.

3) **Explain why this means OAC is a straight line.**

Therefore, $\overrightarrow{AC}$ is a scalar multiple of $\overrightarrow{OA}$, so OAC must be a straight line.

Vector Questions Can Involve Ratios

Ratios are used in vector questions to tell you the lengths of different sections of a straight line. If you know the vector along part of that line, you can use this information to find other vectors along the line.

E.g. X ——— Y ———— Z **XY : YZ = 2 : 3 tells you that $\overrightarrow{XY} = \frac{2}{5}\overrightarrow{XZ}$ and $\overrightarrow{YZ} = \frac{3}{5}\overrightarrow{XZ}$.**

EXAMPLE:

ABCD is a parallelogram, with AB parallel to DC and AD parallel to BC.
Point E lies on DC, such that DE : EC = 3 : 1.
$\overrightarrow{BC} = \mathbf{a}$ and $\overrightarrow{BA} = \mathbf{b}$.
Find $\overrightarrow{AE}$ in terms of **a** and **b**.

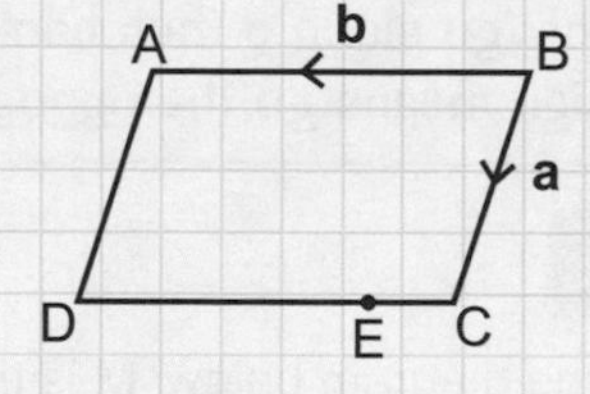

1) **Write $\overrightarrow{AE}$ as a route along the parallelogram.** — $\overrightarrow{AE} = \overrightarrow{AD} + \overrightarrow{DE}$
2) **Use the parallel sides to find $\overrightarrow{AD}$ and $\overrightarrow{DC}$.** — $\overrightarrow{AD} = \overrightarrow{BC} = \underset{\sim}{a}$, $\overrightarrow{DC} = \overrightarrow{AB} = -\underset{\sim}{b}$
3) **Use the ratio to find $\overrightarrow{DE}$.** — $\overrightarrow{DE} = \frac{3}{4}\overrightarrow{DC}$
4) **Now use $\overrightarrow{AD}$ and $\overrightarrow{DE}$ to find $\overrightarrow{AE}$.** — So $\overrightarrow{AE} = \overrightarrow{AD} + \overrightarrow{DE} = \underset{\sim}{a} - \frac{3}{4}\underset{\sim}{b}$

Go forth and multiply by scalars...

So remember — vectors along a straight line or on parallel lines are just scalar multiples of each other.

Q1 ABCD is a quadrilateral. $\overrightarrow{AX} = \mathbf{a}$ and $\overrightarrow{BX} = \mathbf{b}$.
AXC and BXD are straight lines, with AX : XC = BX : XD = 2 : 3.
Find $\overrightarrow{AB}$ and $\overrightarrow{DC}$ in terms of **a** and **b**.
$\overrightarrow{AD}$ and $\overrightarrow{BC}$ are not parallel. What sort of quadrilateral is ABCD? [6 marks]

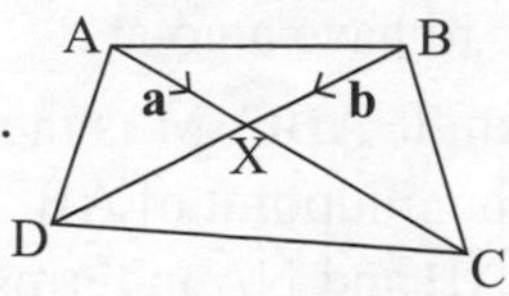

Revision Questions for Section Five

There are a good few facts and formulas in this section, so use this page to check you've got them all sorted.

- Try these questions and tick off each one when you get it right.
- When you've done all the questions for a topic and are completely happy with it, tick off the topic.

Pythagoras' Theorem (p90)

1) What is the formula for Pythagoras' theorem? What do you use it for?
2) A museum has a flight of stairs up to its front door (see diagram). A ramp is to be put over the top of the steps for wheelchair users. Calculate the length that the ramp would need to be to 3 s.f.

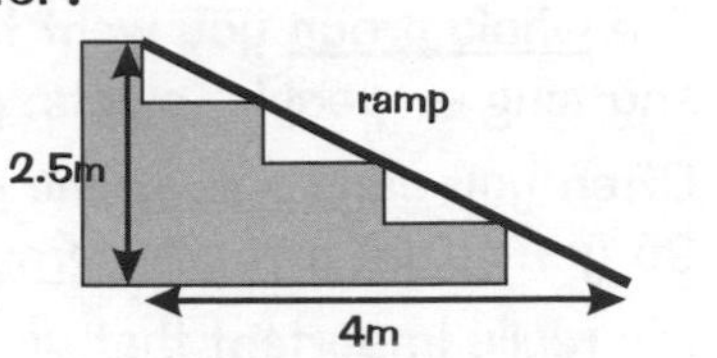

3) Point P has coordinates (-3, -2) and point Q has coordinates (2, 4). Calculate the length of the line PQ to 1 d.p.

Trigonometry — Sin, Cos, Tan (p91-92)

4) Write down the three trigonometry formula triangles.
5) Find the size of angle x in triangle ABC to 1 d.p.

9.1 cm, B, A, x, 7.6 cm, C

6) Find the length of side XZ of triangle XYZ to 3 s.f.

Z, 12.3 cm, 28°, X, Y

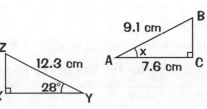

7) A seagull is sitting on top of a 2.8 m high lamp-post. It sees a bag of chips on the ground, 7.1 m away from the base of the lamp-post. Calculate the angle of depression of the chips from the top of the lamp-post, correct to 1 d.p.

The Sine and Cosine Rules (p93-94)

8) Write down the sine and cosine rules and the formula (involving sin) for the area of any triangle.
9) List the 4 different types of sine/cosine rule questions and which rule you need for each.
10) Triangle JKL has side JK = 7 cm, side JL = 11 cm and angle JLK = 32°. Find angle JKL.
11) In triangle FGH side FH = 8 cm, side GH = 9 cm and angle FHG = 47°. Find the length of side FG.
12) Triangle PQR has side PQ = 12 cm, side QR = 9 cm and angle PQR = 63°. Find its area.

3D Pythagoras (p95)

13) What is the formula for finding the length of the longest diagonal in a cuboid?
14) Find the length of the longest diagonal in the cuboid measuring 5 m × 6 m × 9 m.

3D Trigonometry (p96)

15) Find the angle between the line BH and the plane ABCD in this cuboid.

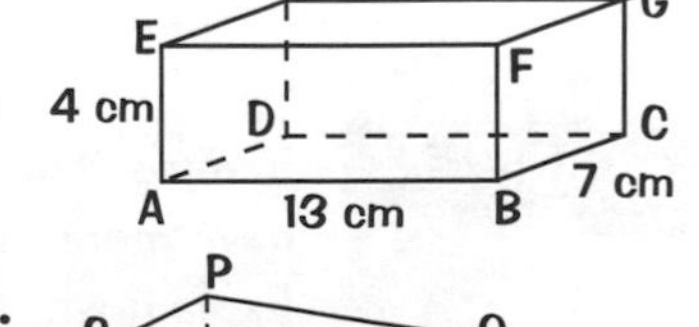

16) Find the size of angle WPU in the cuboid shown to the nearest degree.

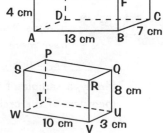

P, S, Q, R, 8 cm, T, W, U, 10 cm, V, 3 cm

Vectors (p97-98)

17) What is the effect of multiplying a vector by a scalar?
18) ABCD is a quadrilateral.
AXC is a straight line with AX : XC = 1 : 3.
  a) Find $\overrightarrow{AX}$.
  b) Find $\overrightarrow{DX}$ and $\overrightarrow{XB}$.
  c) Is DXB a straight line? Explain your answer.

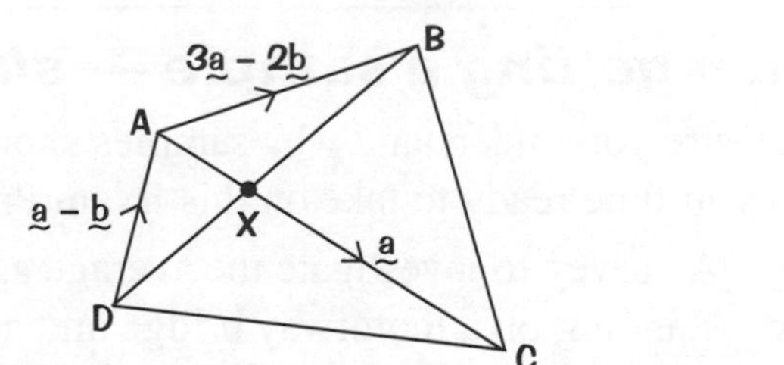

Sampling and Bias

To carry out any statistical investigation, you need to collect data. Ideally, you'd get data from every single member of the 'population' you're interested in. Alas, in reality you usually have to make do with a sample of them instead. But choosing a sample is fraught with danger, so you'd better read on...

Be Careful — Sample Data Must be Representative

1) The whole group you want to find out about is called the **POPULATION**. It can be a group of anything — people, plants, penguins, you name it.
2) Often you can't survey the whole population, e.g. because it's too big. So you select a smaller group from the population, called a **SAMPLE**, instead.
3) It's really important that your sample fairly represents the **WHOLE population**. This allows you to apply any conclusions from your survey to the whole population.

For a sample to be representative, it needs to be:

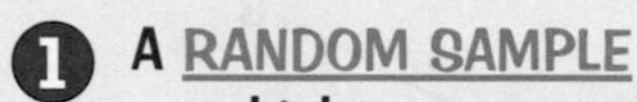

❶ A **RANDOM SAMPLE** — which means every member of the population has an equal chance of being in it.

❷ **BIG ENOUGH** for the size of the population.

See the next page for random sampling methods.

You Need to Spot Problems with Sampling Methods

A **BIASED sample** (or survey) is one that doesn't properly represent the whole population.

To **SPOT BIAS**, you need to think about:

1) **WHEN**, **WHERE** and **HOW** the sample is taken.
2) **HOW MANY** members are in it.

If certain groups are excluded, the **SAMPLE ISN'T RANDOM**. And that can lead to **BIAS** from things like age, gender, different interests, etc. If the sample is too small, it's also likely to be biased.

Tina wants to find out how often people travel by train. She decides to ask the people waiting for trains at her local train station one morning. Give one reason why this might not be a suitable sample to choose.

Think about when, where and how Tina selects her sample:

The sample is biased because it excludes people who never use the train and is likely to include a lot of people who use the train regularly.

You could also say that the sample is only taken at one particular place and time, so won't represent the whole population.

Samir's school has 800 pupils. Samir is interested in whether these pupils would like to have more music lessons. For his sample he selects 10 members of the school orchestra. Explain why Samir's sample is likely to be biased.

Firstly, a sample of 10 is too small to represent the whole school. The sample isn't random — only members of the orchestra are included, so it's likely to be biased in favour of more music lessons.

When getting a sample — size matters...

Make sure you understand why samples should be representative and how to spot when they're not. Then you'll be ready to take on this Exam Practice Question.

Q1 A survey to investigate the average age of cars on Britain's roads was done by standing on a motorway bridge and noting the registration of the first 100 cars. Give two reasons why this is a poor sampling technique. [2 marks]

Sampling Methods

To get a representative sample, you need to use random sampling. Here are two methods you can use.

Simple Random Sampling — choosing a Random Sample

One way to get a random sample is to use 'simple random sampling'. (C)

To SELECT a SIMPLE RANDOM SAMPLE...

1. Assign a number to every member of the population.
2. Create a list of random numbers, e.g. by using a computer, calculator or picking numbers out of a bag.
3. Match the random numbers to members of the population.

Albert wasn't so keen on the random thought generator

Use Stratified Sampling if there are different Groups

Sometimes the population can be split into groups where the members have something in common, e.g. age groups or gender. In these cases you can use STRATIFIED SAMPLING.
With this method, each group's share of the sample is calculated based on its share of the population — so bigger groups get more representation, and smaller groups get less.

To calculate the NUMBER of SAMPLE MEMBERS from EACH GROUP...

1. Find the proportion of the population contained in the group. → $\frac{\text{Number in group}}{\text{Total population}}$
2. Multiply by the sample size.

Once you've calculated the numbers, use simple random sampling within each group to create the sample.

EXAMPLE: The table on the right shows information about some of the students at Eastfield Secondary School.

	Year 9	Year 10	Year 11
Boys	206	219	120
Girls	194	181	80

a) A sample of 50 students is taken, stratified by year group and gender. Calculate the number of Year 10 girls in the sample.

1. Find the proportion of students that are Year 10 girls: $\frac{181}{206 + 219 + 120 + 194 + 181 + 80} = \frac{181}{1000}$ or 0.181 (Total population)
2. Multiply by the sample size: $0.181 \times 50 = 9.05$

Round to the nearest whole number. → There are 9 Year 10 girls.

b) A second sample of 100 students is taken, stratified by Year group. How many Year 11 students are in this sample?

1. Now you want the proportion of students in Year 11: $\frac{120 + 80}{1000} = \frac{2}{10}$ or 0.2
2. Multiply by the sample size: $0.2 \times 100 = 20$ Year 11 students

Another very stratifying page...

Cover the page and jot down how you would select a simple random sample and a stratified sample. Then have a go at this Exam Practice Question.

Q1 A company has 10 senior managers, 22 middle managers and 68 shop-floor workers.
A sample of 10 employees is to be taken, stratified by job title.
How many middle managers should be in the sample? [2 marks]

Collecting Data

Data is often collected to test a hypothesis (or theory). One way to do this is by using questionnaires. But your questions must be crystal clear, and you need to be able to record answers easily and accurately.

You can Organise your Data into Classes

1) Data can be qualitative (words), or quantitative (numbers). Quantitative data is either discrete — can only take certain exact values, or continuous — can take any value in a range.
2) To collect quantitative data, you often need to group it into classes. Discrete data classes should have 'gaps' between them, e.g. '0-1 goals', '2-3 goals' (jump from 1 to 2 — there are no values in between). Continuous data classes should have no 'gaps', so are often written using inequalities (see p107).
3) Whatever the data you have, make sure none of the classes overlap and they cover all the possible values.

EXAMPLE: Jonty wants to find out about the ages (in whole years) of people who use his local library. Design a data-collection sheet he could use to collect his data.

Include columns for: the data values, 'Tally' to record the answers and 'Frequency' to show the totals.

Use non-overlapping classes — with gaps because the data's discrete.

You can have gaps here, e.g. between 39 and 40, because the ages are all whole numbers, so you don't have to fit 39.5 in.

Age (whole years)	Tally	Frequency
0-19		
20-39		
40-59		
60-79		
80 or over		

Include classes like '...or over', '...or less' or 'other' to cover all options in a sensible number of classes.

Design your Questionnaire Carefully

Ⓒ

You need to be able to say what's wrong with questionnaire questions and write your own good questions. A GOOD question is:

1 CLEAR and EASY TO UNDERSTAND ✓

WATCH OUT FOR:
confusing wording or no time frame ✗

How much do you spend on food? ☐ a little ☐ average amount ☐ a lot

BAD: Wording is vague and no time frame is specified (e.g. each week or month).

BAD: Response boxes might be interpreted differently by different people.

2 EASY TO ANSWER ✓

WATCH OUT FOR:
response boxes that overlap, or don't allow for all possible answers ✗

How many pieces of fruit do you eat a day on average? ☐ 1-2 ☐ 2-3 ☐ 3-4 ☐ 4-5 ☐ > 5

BAD: Response boxes overlap and don't allow an answer of zero.

3 FAIR — NOT LEADING or BIASED ✓

WATCH OUT FOR:
wording that suggests an answer ✗

Do you agree that potatoes taste better than cabbage? ☐ Yes ☐ No

BAD: This is a leading question — you're more likely to say 'Yes'.

4 EASY TO ANALYSE afterwards ✓

WATCH OUT FOR:
open-ended questions, with no limit on the possible answers ✗

What is your favourite food? ....

BAD: Every answer could be different — it would be better to include response boxes to choose from.

Who wants to collect a questionnaire — the (not so exciting) quiz spin-off...

Make sure you learn the 4 key points for writing good questions, then have a go at this Practice Question.

Q1 The four questions on the page above are to be included on a questionnaire about food. Design a better version of each question to go on the questionnaire. [4 marks] 

Mean, Median, Mode and Range

Mean, median, mode and range pop up all the time in statistics questions — make sure you know what they are.

The Four Definitions

MODE = MOST common
MEDIAN = MIDDLE value (when values are in order of size)
MEAN = TOTAL of items ÷ NUMBER of items
RANGE = Difference between highest and lowest

REMEMBER:
Mode = most (emphasise the 'mo' in each when you say them)
Median = mid (emphasise the m*d in each when you say them)
Mean is just the average, but it's mean 'cos you have to work it out.

The Golden Rule

There's one vital step for finding the median that lots of people forget:

Always REARRANGE the data in ASCENDING ORDER (and check you have the same number of entries!)

You absolutely must do this when finding the median, but it's also really useful for working out the mode too.

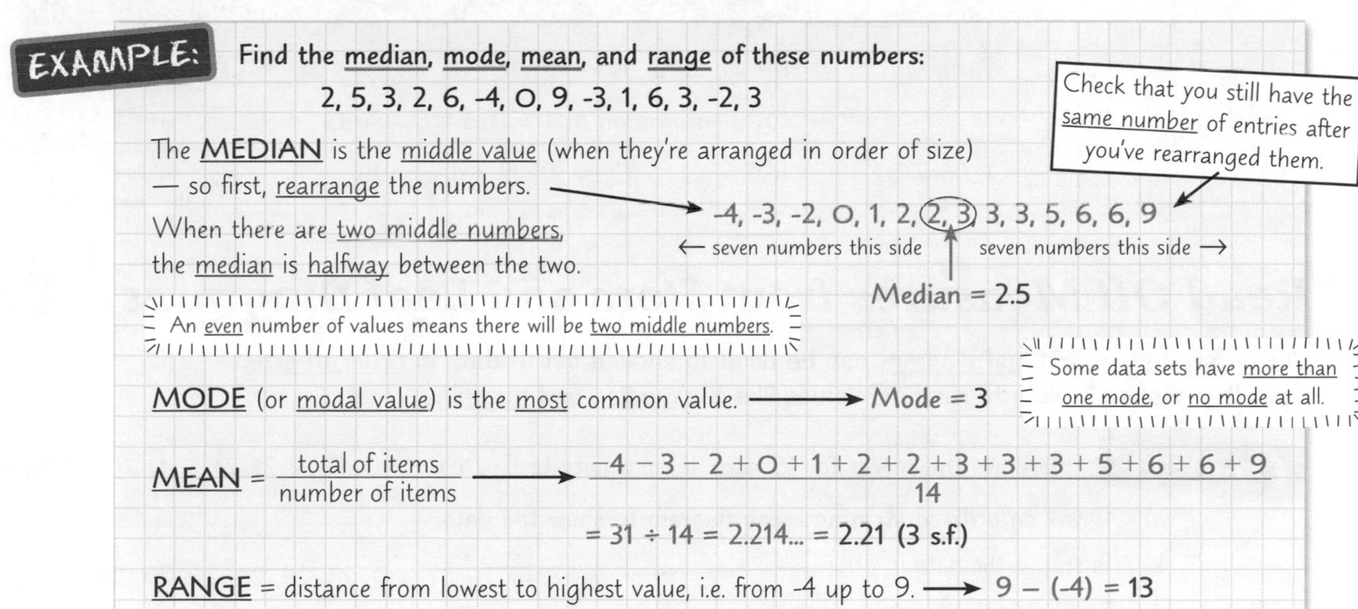

EXAMPLE: Find the median, mode, mean, and range of these numbers:
2, 5, 3, 2, 6, -4, 0, 9, -3, 1, 6, 3, -2, 3

Check that you still have the same number of entries after you've rearranged them.

The MEDIAN is the middle value (when they're arranged in order of size) — so first, rearrange the numbers. → -4, -3, -2, 0, 1, 2, (2, 3) 3, 3, 5, 6, 6, 9

← seven numbers this side | seven numbers this side →

When there are two middle numbers, the median is halfway between the two. Median = 2.5

An even number of values means there will be two middle numbers.

MODE (or modal value) is the most common value. → Mode = 3

Some data sets have more than one mode, or no mode at all.

$$\text{MEAN} = \frac{\text{total of items}}{\text{number of items}} \rightarrow \frac{-4-3-2+0+1+2+2+3+3+3+5+6+6+9}{14}$$

$= 31 \div 14 = 2.214... = 2.21$ (3 s.f.)

RANGE = distance from lowest to highest value, i.e. from -4 up to 9. → $9 - (-4) = 13$

Choose the Best Average

The mean, median and mode all have their advantages and disadvantages — LEARN THEM:

	Advantages	Disadvantages
Mean	Uses all the data. Usually most representative.	Isn't always a data value. May be distorted by extreme data values.
Median	Easy to find in ordered data. Not distorted by extreme data values.	Isn't always a data value. Not always a good representation of the data.
Mode	Easy to find in tallied data. Always a data value (when it exists).	Doesn't always exist or sometimes more than one. Not always a good representation of the data.

Strike a pose, there's nothing to it — mode...

Learn the four definitions and the extra step you have to do to find the median, then give this a go...

Q1 Find the mean, median, mode and range for the set of data below:
1, 3, 14, –5, 6, –12, 18, 7, 23, 10, –5, –14, 0, 25, 8. [4 marks] 

Averages and Spread

Measures of spread tell you how spread out data is. The range (see the previous page) is a measure of spread over all the data values. The interquartile range tells you the spread of the middle 50% of values.

Quartiles Divide the Data into Four Equal Groups

1) The quartiles are the lower quartile Q_1, the median Q_2 and the upper quartile Q_3.
2) If you put the data in ascending order, the quartiles are 25% (¼), 50% (½) and 75% (¾) of the way through the list. So if a data set has n values, you work out the positions of the quartiles using these formulas:

Q_1 position number = $(n + 1)/4$
Q_2 position number = $2(n + 1)/4$
Q_3 position number = $3(n + 1)/4$

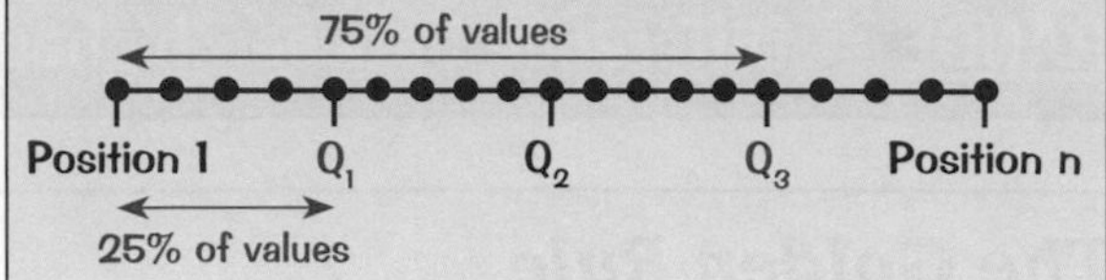

3) The INTERQUARTILE RANGE (IQR) is the difference between the upper quartile and the lower quartile and contains the middle 50% of values.

EXAMPLE: Here are the ages, in months, of a number of fine cheeses: 7, 12, 5, 4, 3, 9, 5, 11, 6, 5, 7
Find the interquartile range of the ages.

1. Put the data in order of size. → 3, 4, 5, 5, 5, 6, 7, 7, 9, 11, 12 ← Check you've got the right number of values — 11 ✓
2. Find Q_1 — n = 11, so Q_1 is in position (11 + 1)/4 = 3. So Q_1 = 5.
3. Find Q_3 — Q_3 is in position 3(11 + 1)/4 = 9. So Q_3 = 9.
4. Subtract Q_1 from Q_3 — IQR = $Q_3 - Q_1$ = 9 – 5 = 4 months

Careful — the formulas tell you the position of the quartile, not its value.

Read Off Measures from Stem and Leaf Diagrams

An ordered stem and leaf diagram can be used to show a set of data in order of size.
And that makes it easy to read off values like the median and quartiles.

EXAMPLE: Here are the scores for 15 dogs in an agility test: 26, 16, 29, 7, 12, 32, 29, 24, 13, 17, 20, 23, 24, 31, 34

a) Draw an ordered stem and leaf diagram to show the data.
1. First, put the data in order. → 7, 12, 13, 16, 17, 20, 23, 24, 24, 26, 29, 29, 31, 32, 34
2. Group the data into rows — you can group these values by 'number of tens'.
3. Remember to include a key.

Stem	Leaves
0	7
1	2 3 6 7
2	0 3 4 4 6 9 9
3	1 2 4

Key: 2|3 = 23

Write the first digit (number of tens) here...
...write the second digit for each value (number of units) here.

b) Find the median score and range of scores.
1. Find the position of the middle value and read off from the diagram. — Median is in position 2(n + 1)/4 = 2(15 + 1)/4 = 8. So median score = 24
2. The range is just the highest minus the lowest. — Range = 34 – 7 = 27

Chocolate-peanut-banana butter — not your average spread...

Learn how to find quartiles and to draw and read stem and leaf diagrams. Then do this Practice Question.

Q1 a) Draw an ordered stem and leaf diagram to show: 3.0, 1.6, 1.4, 2.2, 0.7, 1.1, 2.6 [3 marks] D
b) Use your diagram to find the interquartile range of the data. [3 marks]

Averages and Spread

The humble box plot might not look very fancy, but it tells you lots about the spread of data.

Box Plots show the Interquartile Range as a Box

1) Box plots give a good summary of the spread of a data set — they show a few key measures, rather than all the individual data values.
2) Make sure you can draw and interpret them.

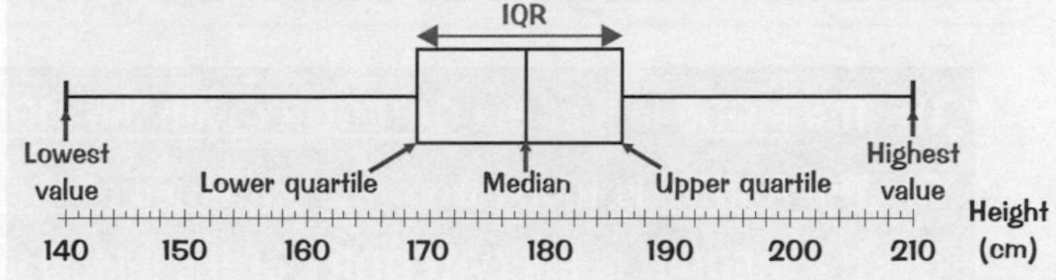

EXAMPLE: This table gives information about the numbers of rainy days last year in some cities. On the grid below, draw a box plot to show the information.

Minimum number	90
Maximum number	195
Lower quartile	130
Median	150
Upper quartile	175

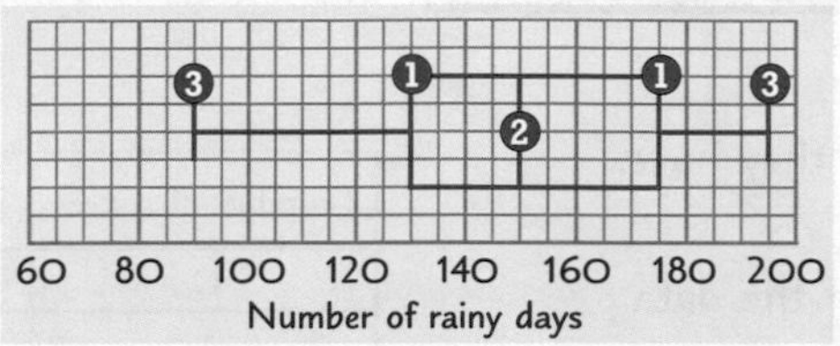

1. Mark on the quartiles and draw the box.
2. Draw a line at the median.
3. Mark on the minimum and maximum points and join them to the box with horizontal lines.

Compare Data using Averages and Spread

To compare two sets of data, you should look at:

You could also compare other key values like quartiles or min/max values.

1. AVERAGES — MEAN, MEDIAN or MODE
   Say which data set has the higher/lower value and what that means in the context of the data.
2. SPREAD — RANGE or INTERQUARTILE RANGE
   Say which data set has the larger/smaller value. A larger spread means the values are less consistent or there is more variation in the data.

If the data contains extreme values, it's better to use the IQR than the range.

EXAMPLE: An animal park is holding a 'guess the weight of the baby hippo' competition. These box plots show information about the weights guessed by a group of school children.

Compare the weights guessed by the boys and the girls.

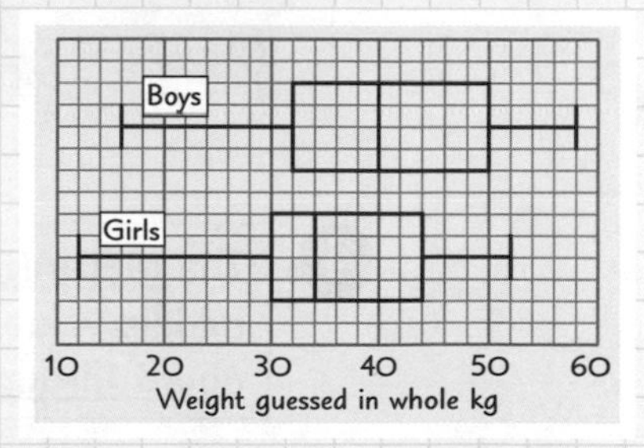

1. Compare averages by looking at the median values.
   The median for the boys is higher than the median for the girls.
   So the boys generally guessed heavier weights.
2. Compare the spreads by working out the range and IQR for each data set:
   Boys' range = 58 – 16 = 42 and IQR = 50 – 32 = 18.
   Girls' range = 52 – 12 = 40 and IQR = 44 – 30 = 14.
   Both the range and the IQR are smaller for the girls' guesses, so there is less variation in the weights guessed by the girls.

With my cunning plot, I'll soon control all the world's boxes...

Mwahaha... Make sure you get what's going on in the examples above, then do this Exam Practice Question.

Q1 Draw a box plot to represent this data: 5, 7, 12, 14, 10, 17, 22, 17, 9, 18, 12 [3 marks]

Frequency Tables — Finding Averages

The word FREQUENCY means HOW MANY, so a frequency table is just a 'How many in each category' table. You saw how to find averages and range on p.103 — it's the same ideas here, but with the data in a table.

Find Averages from Frequency Tables

1) The MODE is just the CATEGORY with the MOST ENTRIES.
2) The RANGE is found from the extremes of the first column.
3) The MEDIAN is the CATEGORY of the middle value in the second column.
4) To find the MEAN, you have to WORK OUT A THIRD COLUMN yourself.

The MEAN is then: 3rd Column Total ÷ 2nd Column Total

Categories — How many

Number of cats	Frequency	
0	17	
1	22	
2	15	
3	7	

Mysterious 3rd column...

EXAMPLE: Some people were asked how many sisters they have. The table opposite shows the results.

Find the mode, the range, the mean and the median of the data.

Number of sisters	Frequency
0	7
1	15
2	12
3	8
4	4
5	0

1. The MODE is the category with the most entries — i.e. the one with the highest frequency:

The highest frequency is 15 for '1 sister', so MODE = 1

2. The RANGE is the difference between the highest and lowest numbers of sisters — that's 4 sisters (no one has 5 sisters) and no sisters, so:

RANGE = 4 – 0 = 4

You can label the first column x and the frequency column f, then the third column is f × x.

3. To find the MEAN, add a 3rd column to the table showing 'number of sisters × frequency'. Add up these values to find the total number of sisters of all the people asked.

Number of sisters (x)	Frequency (f)	No. of sisters × Frequency (f × x)
0	7	0
1	15	15
2	12	24
3	8	24
4	4	16
5	0	0
Total	46	79

$$\text{MEAN} = \frac{\text{total number of sisters}}{\text{total number of people asked}} = \frac{79}{46} = 1.72 \text{ (3 s.f.)}$$

(79 = 3rd column total; 46 = 2nd column total)

4. The MEDIAN is the category of the middle value. Work out its position, then count through the 2nd column to find it.

It helps to imagine the data set out in an ordered list:
0000000111111111111111222222222222333333334444
(median)

There are 46 values, so the middle value is halfway between the 23rd and 24th values. There are a total of (7 + 15) = 22 values in the first two categories, and another 12 in the third category takes you to 34. So the 23rd and 24th values must both be in the category '2 sisters', which means the MEDIAN is 2.

My table has 5 columns, 6 rows and 4 legs...

Learn the four key points about averages, then try this fella.

Q1 50 people were asked how many times a week they play sport. The table opposite shows the results.

a) Find the median. [2 marks]

b) Calculate the mean. [3 marks] (D)

No. of times sport played	Frequency
0	8
1	15
2	17
3	6
4	4
5 or more	0

Grouped Frequency Tables

<u>Grouped frequency tables</u> group together the data into <u>classes</u>. They look like ordinary frequency tables, but they're a <u>slightly trickier</u> kettle of fish...

See p102 for grouped discrete data.

NON-OVERLAPPING CLASSES

- Use <u>inequality symbols</u> to cover all possible values.
- Here, <u>10</u> would go in the <u>1st</u> class, but <u>10.1</u> would go in the <u>2nd</u> class.

Height (h millimetres)	Frequency
$5 < h \le 10$	12
$10 < h \le 15$	15

To find MID-INTERVAL VALUES:

- Add together the <u>end values</u> of the <u>class</u> and <u>divide by 2</u>.
- E.g. $\frac{5+10}{2} = 7.5$

Find Averages from Grouped Frequency Tables

Unlike with ordinary frequency tables, you <u>don't know the actual data values</u>, only the <u>classes</u> they're in. So you have to <u>ESTIMATE THE MEAN</u>, rather than calculate it exactly. Again, you do this by <u>adding columns</u>:

1) Add a <u>3RD COLUMN</u> and enter the <u>MID-INTERVAL VALUE</u> for each class.
2) Add a <u>4TH COLUMN</u> to show '<u>FREQUENCY × MID-INTERVAL VALUE</u>' for each class.

And you'll be asked to find the <u>MODAL CLASS</u> and the <u>CLASS CONTAINING THE MEDIAN</u>, not exact values.

EXAMPLE: This table shows information about the weights, in kilograms, of 60 school children.

a) Write down the <u>modal class</u>.
b) Write down the <u>class containing the median</u>.
c) Calculate an <u>estimate for the mean weight</u>.

Weight (w kg)	Frequency
$30 < w \le 40$	8
$40 < w \le 50$	16
$50 < w \le 60$	18
$60 < w \le 70$	12
$70 < w \le 80$	6

a) **The <u>modal class</u> is the one with the <u>highest frequency</u>.**

Modal class is $50 < w \le 60$

b) **Work out the <u>position</u> of the <u>median</u>, then <u>count through</u> the <u>2nd column</u>.**

There are 60 values, so the median is halfway between the 30th and 31st values. Both these values are in the third class, so the class containing the median is $50 < w \le 60$.

c) **Add extra columns for '<u>mid-interval value</u>' and '<u>frequency × mid-interval value</u>'. Add up the values in the 4th column to estimate the <u>total weight</u> of the 60 children.**

Weight (w kg)	Frequency (f)	Mid-interval value (x)	fx
$30 < w \le 40$	8	35	280
$40 < w \le 50$	16	45	720
$50 < w \le 60$	18	55	990
$60 < w \le 70$	12	65	780
$70 < w \le 80$	6	75	450
Total	60	—	3220

Don't add up the mid-interval values.

$$\text{Mean} \approx \frac{\text{total weight}}{\text{number of children}}$$

← <u>4th column total</u>
← <u>2nd column total</u>

$$= \frac{3220}{60}$$

$$= 53.7 \text{ kg (3 s.f.)}$$

Mid-interval value — cheap ice creams...

Learn all the details on the page, then give this Exam Practice Question a go.

Q1 Estimate the mean of this data. Give your answer to 3 significant figures. [4 marks]

Length (l cm)	$15.5 \le l < 16.5$	$16.5 \le l < 17.5$	$17.5 \le l < 18.5$	$18.5 \le l < 19.5$
Frequency	12	18	23	8

Cumulative Frequency

Cumulative frequency just means adding it up as you go along — i.e. the total frequency so far.
A cumulative frequency graph shows cumulative frequency up the side and the range of data values along the bottom. You need to be able to draw the graph, read it and make estimates from it.

EXAMPLE: The table below shows information about the heights of a group of people.
a) Draw a cumulative frequency graph for the data.
b) Use your graph to estimate the median and interquartile range of the heights.

Height (h cm)	Frequency	Cumulative Frequency
140 < h ≤ 150	4	4
150 < h ≤ 160	9	4 + 9 = 13
160 < h ≤ 170	20	13 + 20 = 33
170 < h ≤ 180	33	33 + 33 = 66
180 < h ≤ 190	36	66 + 36 = 102
190 < h ≤ 200	15	102 + 15 = 117
200 < h ≤ 210	3	117 + 3 = 120

Total number of people surveyed

Cumulative frequency

Height (cm)

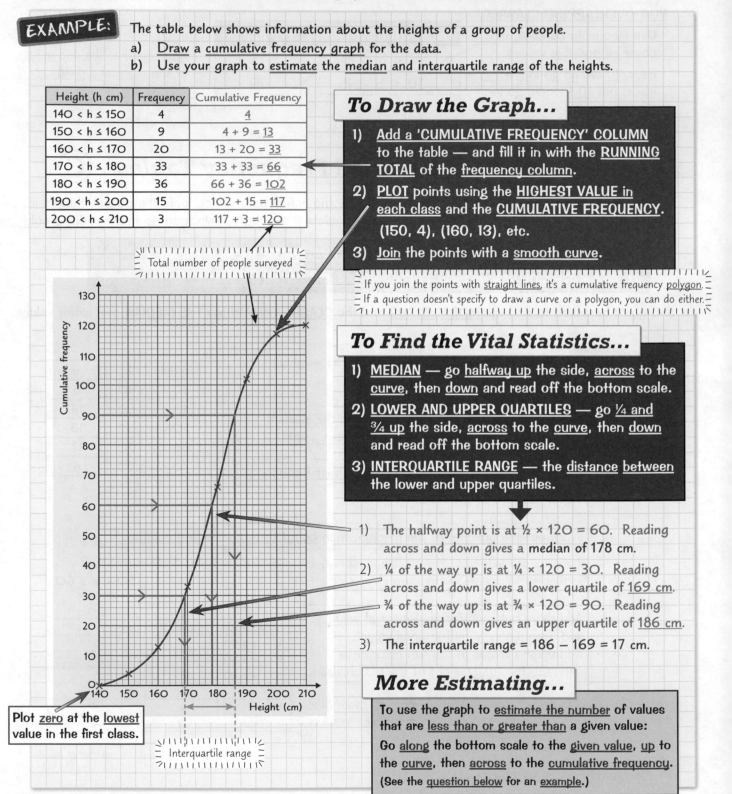

Plot zero at the lowest value in the first class.

Interquartile range

To Draw the Graph...

1) Add a 'CUMULATIVE FREQUENCY' COLUMN to the table — and fill it in with the RUNNING TOTAL of the frequency column.
2) PLOT points using the HIGHEST VALUE in each class and the CUMULATIVE FREQUENCY. (150, 4), (160, 13), etc.
3) Join the points with a smooth curve.

If you join the points with straight lines, it's a cumulative frequency polygon. If a question doesn't specify to draw a curve or a polygon, you can do either.

To Find the Vital Statistics...

1) MEDIAN — go halfway up the side, across to the curve, then down and read off the bottom scale.
2) LOWER AND UPPER QUARTILES — go ¼ and ¾ up the side, across to the curve, then down and read off the bottom scale.
3) INTERQUARTILE RANGE — the distance between the lower and upper quartiles.

1) The halfway point is at ½ × 120 = 60. Reading across and down gives a median of 178 cm.
2) ¼ of the way up is at ¼ × 120 = 30. Reading across and down gives a lower quartile of 169 cm.
¾ of the way up is at ¾ × 120 = 90. Reading across and down gives an upper quartile of 186 cm.
3) The interquartile range = 186 – 169 = 17 cm.

More Estimating...

To use the graph to estimate the number of values that are less than or greater than a given value:
Go along the bottom scale to the given value, up to the curve, then across to the cumulative frequency.
(See the question below for an example.)

How do you make a total run...

Time to try another lovely Exam Practice Question.

Q1 a) Draw a cumulative frequency diagram for this data. [3 marks]
b) Use your diagram to estimate the number of fish with a length of more than 50 mm. [2 marks]

Length of fish (l mm)	Frequency
0 < l ≤ 20	4
20 < l ≤ 40	11
40 < l ≤ 60	20
60 < l ≤ 80	15
80 < l ≤ 100	6

Histograms and Frequency Density

A histogram is just a bar chart where the bars can be of different widths. This changes them from nice, easy-to-understand diagrams into seemingly incomprehensible monsters (and an examiner's favourite).

Histograms Show Frequency Density (A)

1) The vertical axis on a histogram is always called frequency density. You work it out using this formula:

Frequency Density = Frequency ÷ Class Width

Remember... 'frequency' is just another way of saying 'how much' or 'how many'.

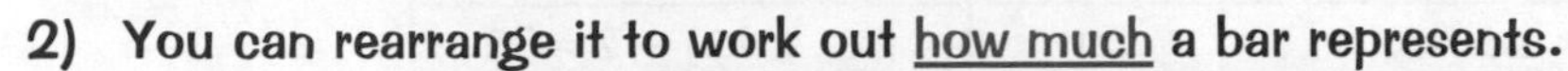

2) You can rearrange it to work out how much a bar represents.

Frequency = Frequency Density × Class Width = AREA of bar

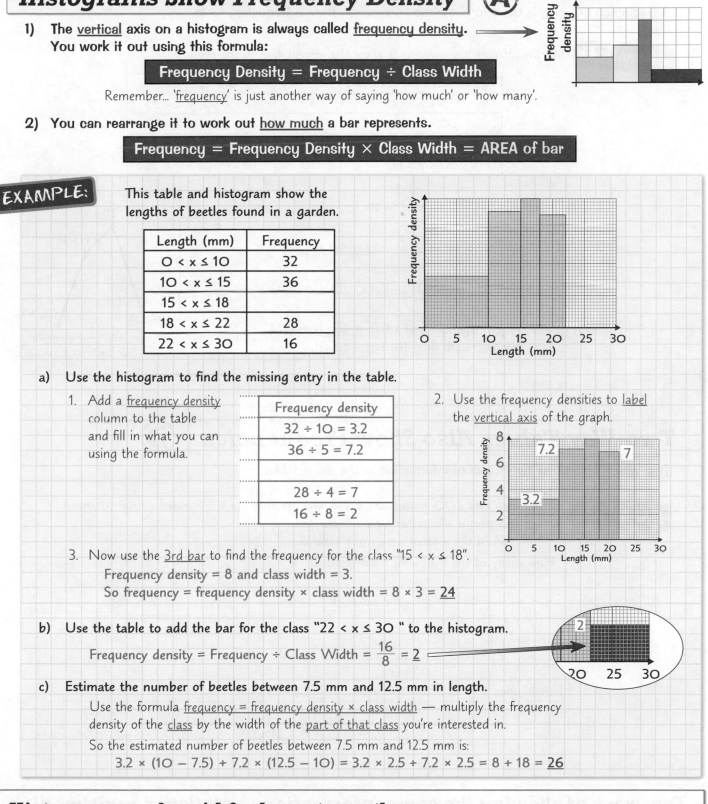

EXAMPLE: This table and histogram show the lengths of beetles found in a garden.

Length (mm)	Frequency
$0 < x \leq 10$	32
$10 < x \leq 15$	36
$15 < x \leq 18$	
$18 < x \leq 22$	28
$22 < x \leq 30$	16

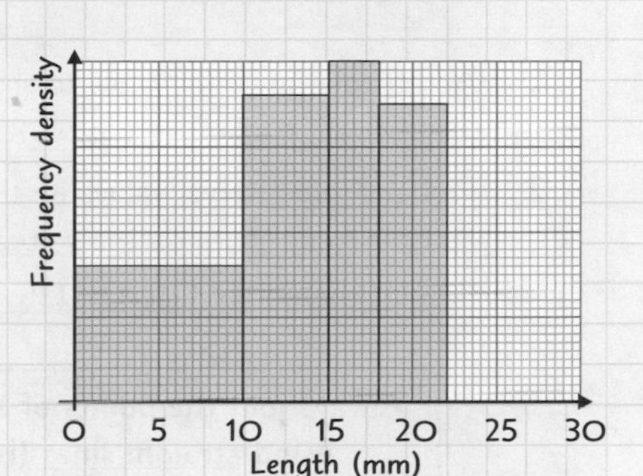

a) Use the histogram to find the missing entry in the table.

1. Add a frequency density column to the table and fill in what you can using the formula.

Frequency density
32 ÷ 10 = 3.2
36 ÷ 5 = 7.2
28 ÷ 4 = 7
16 ÷ 8 = 2

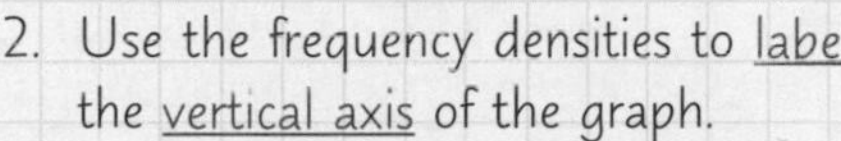

2. Use the frequency densities to label the vertical axis of the graph.

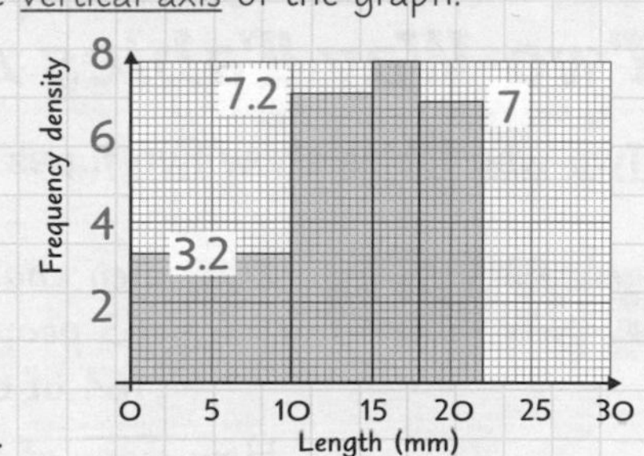

3. Now use the 3rd bar to find the frequency for the class "$15 < x \leq 18$".
   Frequency density = 8 and class width = 3.
   So frequency = frequency density × class width = 8 × 3 = 24

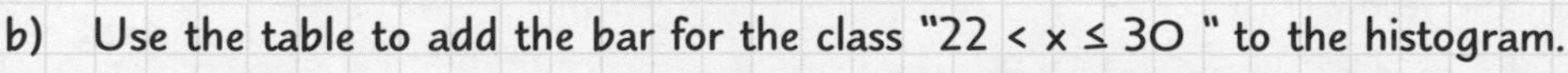

b) Use the table to add the bar for the class "$22 < x \leq 30$" to the histogram.

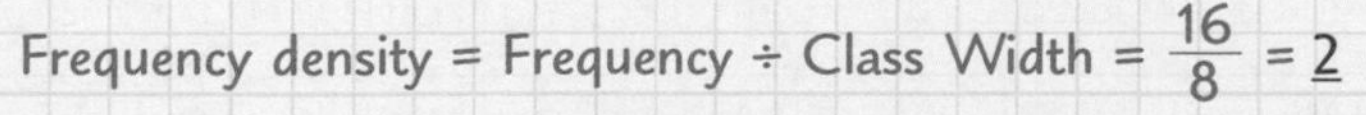

Frequency density = Frequency ÷ Class Width = $\frac{16}{8}$ = 2

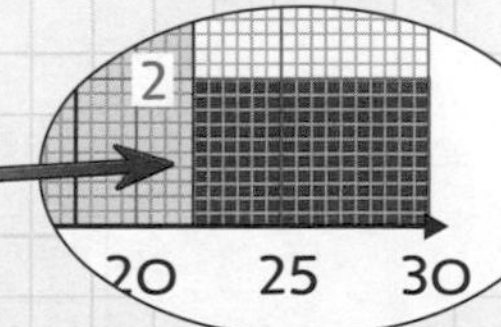

c) Estimate the number of beetles between 7.5 mm and 12.5 mm in length.

Use the formula frequency = frequency density × class width — multiply the frequency density of the class by the width of the part of that class you're interested in.

So the estimated number of beetles between 7.5 mm and 12.5 mm is:

3.2 × (10 – 7.5) + 7.2 × (12.5 – 10) = 3.2 × 2.5 + 7.2 × 2.5 = 8 + 18 = 26

Histograms — horrid foul creatures they are...

Although they look very like harmless bar charts, histograms are actually pretty unpleasant. Make sure you get your head around the method above...

Q1 This table shows information about the lengths of slugs in a garden. Draw a histogram to represent the information. [4 marks] (A)

Length (mm)	Frequency
$0 < x \leq 40$	20
$40 < x \leq 60$	45
$60 < x \leq 65$	15
$65 < x \leq 100$	70

Other Graphs and Charts

You're nearly at the end of graphs and charts now. Nearly... but not quite.

Frequency Polygons Show Frequencies

A frequency polygon is used to show the information from a frequency table.

EXAMPLE: Draw a frequency polygon to show the information in the table.

Age (a) of people at concert	Frequency	mid-interval value
20 < a ≤ 30	12	(20 + 30) ÷ 2 = 25
30 < a ≤ 40	21	(30 + 40) ÷ 2 = 35
40 < a ≤ 50	18	(40 + 50) ÷ 2 = 45
50 < a ≤ 60	10	(50 + 60) ÷ 2 = 55

1. Add a column to the table to show the mid-interval values.
2. Plot the mid-interval values on the horizontal axis and the frequencies on the vertical axis.
   So plot the points (25, 12), (35, 21), (45, 18) and (55, 10).

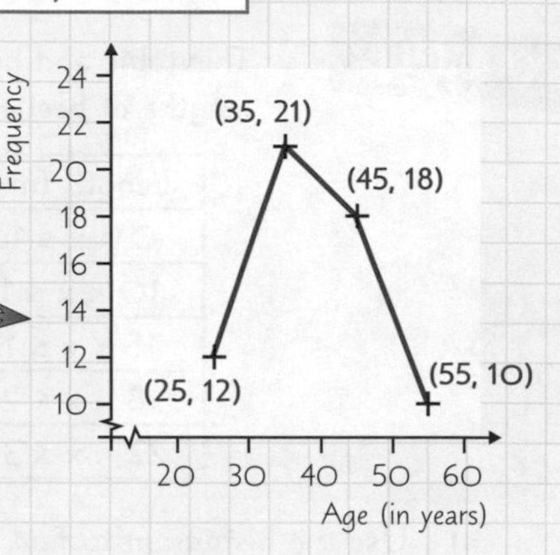

Always join the points of a frequency polygon using straight lines (i.e. not a curve).

Two-Way Tables Also Show Frequencies

Two-way tables show two types of information in the same table.

200 men and 200 women are asked whether they are left-handed or right-handed.

- 63 people altogether were left-handed.
- 164 of the women were right-handed.

How many of the men were right-handed?

1. Draw a 2 × 2 table to show the info from the question — this is in the yellow cells.
2. Then fill in the gaps by adding and subtracting.

When there's only one thing in a row or column that you don't know, you can always work it out.

	Women	Men	Total
Left-handed	200 – 164 = 36	63 – 36 = 27	63
Right-handed	164	200 – 27 = 173	164 + 173 = 337
Total	200	200	200 + 200 = 400

Two ways of showing the same information — statistics multimedia...

This is as close to 'art' as you'll get in maths... so if you're at all creative (or if you're not), have a go at this question...

Q1 Complete the two-way table on the right showing how a group of students get to school. [3 marks]

	Walk	Car	Bus	Total
Male	15	21		
Female			22	51
Total	33			100

Scatter Graphs

The last kind of graph in this section is scatter graphs — they're nice, so make the most of them.

Scatter Graphs — Correlation and Line of Best Fit

1) A scatter graph tells you how closely two things are related — the fancy word for this is CORRELATION.
2) If you can draw a line of best fit pretty close to most of your scatter of points, then the two things are correlated.

A line of best fit doesn't have to go through any of the points exactly, but it should go fairly close to most of them.

Strong correlation is when your points make a fairly straight line. This means the two things are closely related to each other.

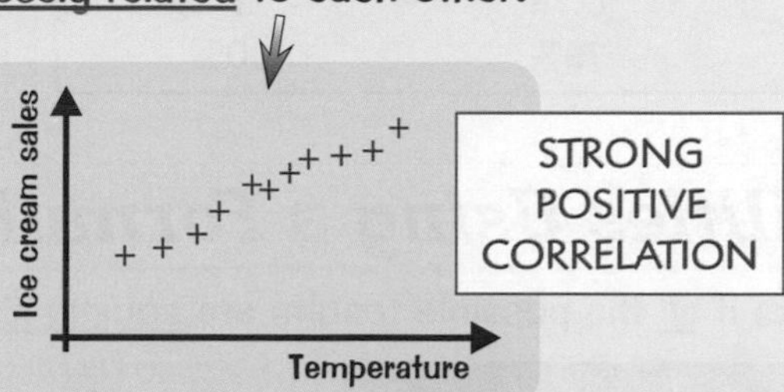

If the points form a line sloping uphill from left to right, then there is positive correlation — this means that both things increase or decrease together.

Weak correlation means your points don't line up quite so nicely (but you still need to be able to see where you'd draw a line of best fit).

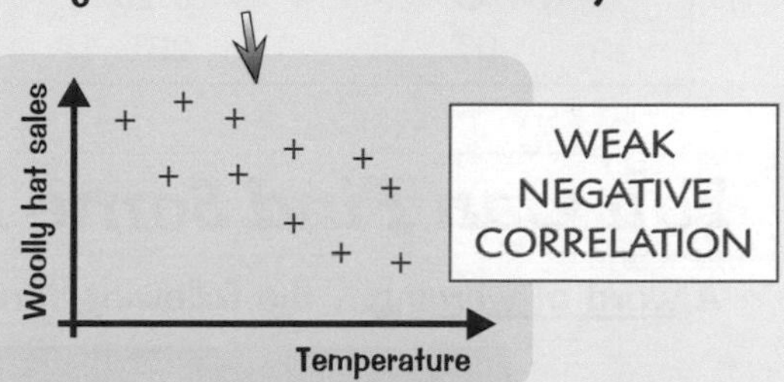

If the points form a line sloping downhill from left to right, then there is negative correlation — this just means that as one thing increases the other decreases.

3) If the points are randomly scattered, and it's impossible to draw a sensible line of best fit, then there's no correlation. Here, newspaper sales and temperature aren't correlated.

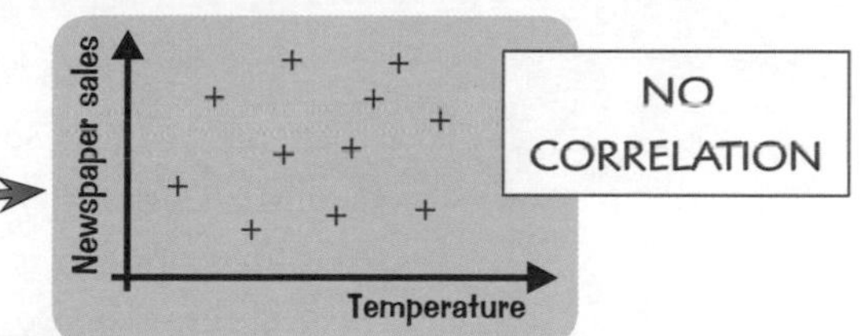

EXAMPLE: The graph below shows the number of zoo visitors plotted against the outside temperature for several Sundays.

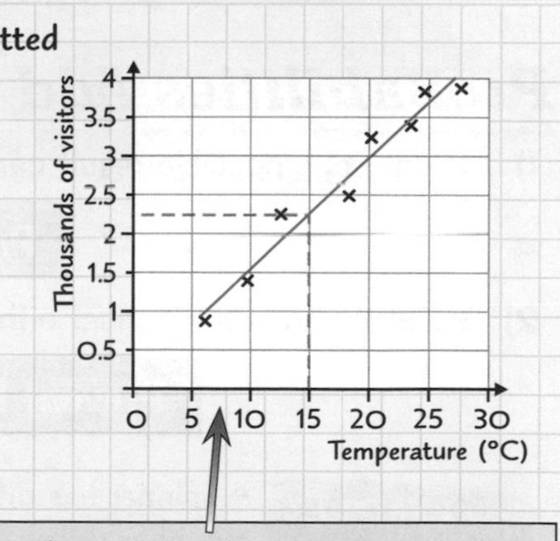

a) Describe the relationship between temperature and the number of visitors to the zoo.

Temperature and the number of visitors have strong positive correlation.

The points make up a fairly straight, uphill line.

b) Estimate how many visitors the zoo would get on a Sunday when the outside temperature is 15 °C.

15 °C corresponds to roughly 2250 visitors.

Draw a line of best fit (shown in blue). Then draw a line up from 15 °C to your line, and then across to the other axis.

Relax and take a trip down Correlation Street...

You need to feel at home with scatter graphs. See how you feel with this question.

Q1 This graph shows Sam's average speed on runs of different lengths.

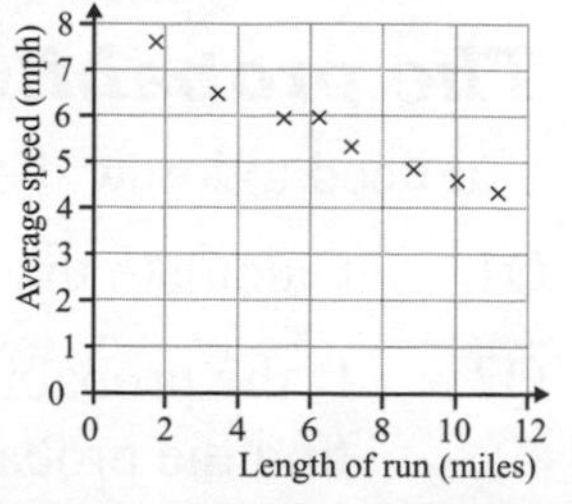

a) Describe the relationship between the length of Sam's run and her average speed. [1 mark]

b) Estimate Sam's average speed for an 8-mile run. [1 mark]

Probability Basics

A lot of people reckon probability is pretty tough. But learn the basics well, and it'll all make sense.

All Probabilities are Between 0 and 1

Probabilities are always between 0 and 1. The higher the probability of something, the more likely it is.

- A probability of ZERO means it will NEVER HAPPEN.
- A probability of ONE means it DEFINITELY WILL.

You can't have a probability bigger than 1.

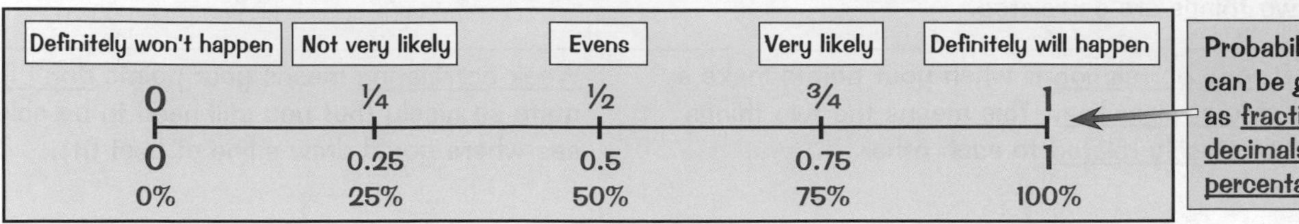

Probabilities can be given as fractions, decimals or percentages.

You Can Find Some Probabilities Using a Formula

A word of warning... the following formula only works if all the possible results are equally likely.

$$\text{Probability} = \frac{\text{Number of ways for something to happen}}{\text{Total number of possible results}}$$

Words like 'fair' and 'at random' show possible results are all equally likely. 'Biased' and 'unfair' mean the opposite.

EXAMPLE: Work out the probability of randomly picking a letter 'P' from the tiles below.

A P P L E P I E

1. There are 3 P's — so there are 3 different ways to 'pick a letter P'.
2. And there are 8 tiles altogether — each of these is a possible result.

$$\text{Probability} = \frac{\text{number of ways to pick a P}}{\text{total number of possible results}}$$

$$= \frac{3}{8} \text{ (or 0.375)}$$

Probabilities Add Up To 1

1) If only one possible result can happen at a time, then the probabilities of all the results add up to 1.

Probabilities always ADD UP to 1.

2) So since something must either happen or not happen (i.e. only one of these can happen at a time):

P(event happens) + P(event doesn't happen) = 1

EXAMPLE: A spinner has different numbers of red, blue, yellow and green sections. What is the probability of spinning green?

Colour	red	blue	yellow	green
Probability	0.1	0.4	0.3	

Only one of the results can happen at a time, so all the probabilities must add up to 1.

P(green) = 1 – (0.1 + 0.4 + 0.3) = 0.2

The probability of this getting you marks in the exam = 1...

You need to know the facts in the boxes above. You also need to know how to use them.

Q1 Calculate the probability of the fair spinner on the right landing on 4. [2 marks]

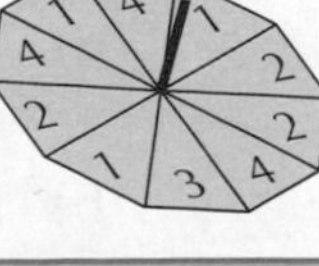

Q2 If the probability of spinning red on a spinner is 0.8, find the probability of spinning any colour except red. [1 mark]

Listing Outcomes and Expected Frequency

With a lot of probability questions, a good place to start is with a list of all the things that could happen (also known as outcomes). Once you've got a list of outcomes, the rest of the question is easy.

Listing All Outcomes: 2 Coins, Dice, Spinners

A sample space diagram is a good way to show all the possible outcomes if there are two activities going on (e.g. two coins being tossed, or a dice being thrown and a spinner being spun, etc.).

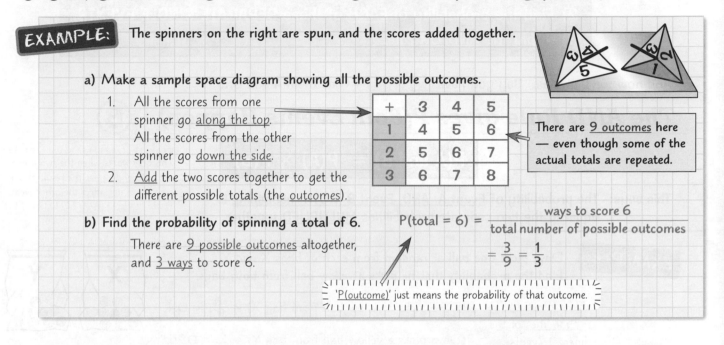

EXAMPLE: The spinners on the right are spun, and the scores added together.

a) Make a sample space diagram showing all the possible outcomes.

1. All the scores from one spinner go along the top. All the scores from the other spinner go down the side.
2. Add the two scores together to get the different possible totals (the outcomes).

+	3	4	5
1	4	5	6
2	5	6	7
3	6	7	8

There are 9 outcomes here — even though some of the actual totals are repeated.

b) Find the probability of spinning a total of 6.

There are 9 possible outcomes altogether, and 3 ways to score 6.

$$P(\text{total} = 6) = \frac{\text{ways to score 6}}{\text{total number of possible outcomes}} = \frac{3}{9} = \frac{1}{3}$$

'P(outcome)' just means the probability of that outcome.

Use Probability to Find an "Expected Frequency"

You can estimate how often you'd expect something to happen if you carry out an experiment n times.

Expected times outcome will happen = probability × number of trials

EXAMPLE: A game involves throwing a fair six-sided dice. The player wins if they score either a 5 or a 6. If one person plays the game 180 times, estimate the number of times they will win.

1. First calculate the probability that they win each game.

$$\text{Probability of winning} = \frac{\text{number of ways to win}}{\text{total number of possible results}} = \frac{2}{6} = \frac{1}{3}$$

2. Then estimate the number of times they'll win in 180 separate attempts.

$$\text{Expected number of wins} = \text{probability of winning} \times \text{number of trials} = \frac{1}{3} \times 180 = \underline{60}$$

When in doubt, make a list...

Don't be fooled by complicated statistics terms — for example, a 'sample space diagram' is basically just a list. Right then... if you're ready to put your skills to the test, try this...

Q1 Two fair 6-sided dice are thrown, and their scores added together.
a) Find the probability of throwing a total of 7. [2 marks]
b) If the pair of dice are thrown 300 times, how many times would you expect a total of 7? [1 mark]

The AND / OR Rules

This page is also about when you have more than one thing happening at a time.

Combined Probability — Two or More Events

1) Always break down a complicated-looking probability question into A SEQUENCE of SEPARATE SINGLE EVENTS.
2) Find the probability of EACH of these SEPARATE SINGLE EVENTS.
3) Apply the AND/OR rule.

And now for the rules. Say you have two events — call them A and B...

The AND Rule gives P(Both Events Happen)

$$P(A \text{ and } B) = P(A) \times P(B)$$

This only works when the two events are independent, i.e. the result of one event does not affect the other event.

This says: The probability of Event A AND Event B BOTH happening is equal to the two separate probabilities MULTIPLIED together.

Dave picks one ball at random from each of bags X and Y. Find the probability that he picks a yellow ball from both bags.

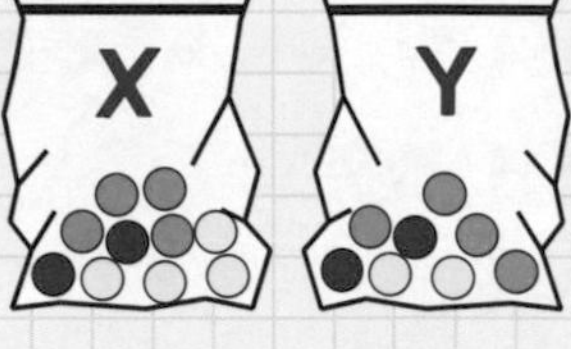

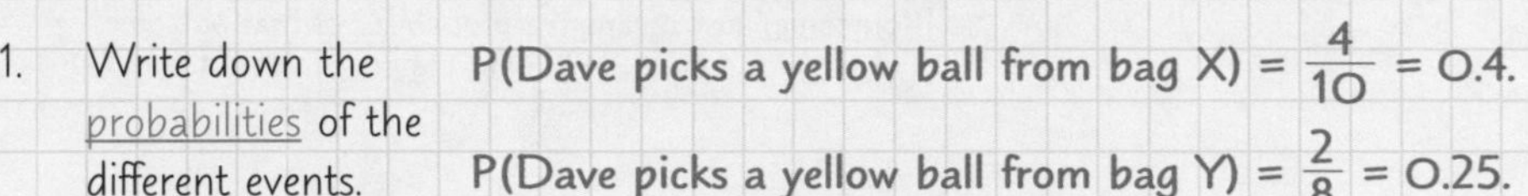

1. Write down the probabilities of the different events.
   P(Dave picks a yellow ball from bag X) = $\frac{4}{10}$ = 0.4.
   P(Dave picks a yellow ball from bag Y) = $\frac{2}{8}$ = 0.25.
2. Use the formula.
   So P(Dave picks a yellow ball from both bags) = 0.4 × 0.25 = 0.1

The OR Rule gives P(At Least One Event Happens)

$$P(A \text{ or } B) = P(A) + P(B)$$

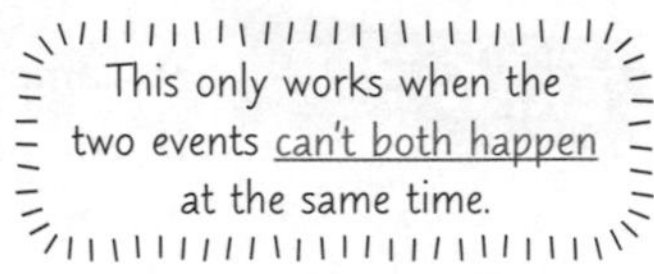

This says: The probability of EITHER Event A OR Event B happening is equal to the two separate probabilities ADDED together.

EXAMPLE:

A spinner with red, blue, green and yellow sections was spun — the probability of it landing on each colour is shown in the table. Find the probability of spinning either red or green.

Colour	red	blue	yellow	green
Probability	0.25	0.3	0.35	0.1

1. Write down the probabilities of the different events. P(lands on red) = 0.25 and P(lands on green) = 0.1.
2. Use the formula. So P(Lands on either red or green) = 0.25 + 0.1 = 0.35

Learn AND remember this — OR you're in trouble...

The way to remember this is that it's the wrong way round — you'd want AND to go with '+' but it doesn't. It's 'AND with ×' and 'OR with +'. Once you've got your head round that, try this Exam Practice Question.

Q1 Two fair six-sided dice are thrown.

a) Find the probability that the first dice lands on either 3 or an even number. [2 marks]

b) Find the probability that both dice land on 6. [2 marks]

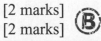

Tree Diagrams

Learn these basic details (which apply to ALL tree diagrams). Then you'll be ready for the one in the exam.

Remember These Four Key Tree Diagram Facts

1) On any set of branches which meet at a point, the probabilities must add up to 1.

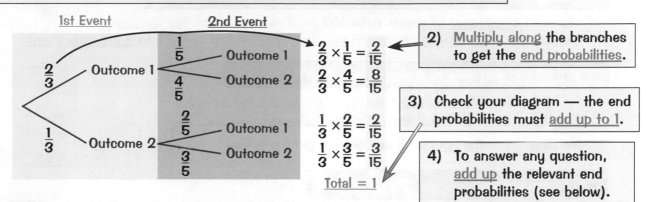

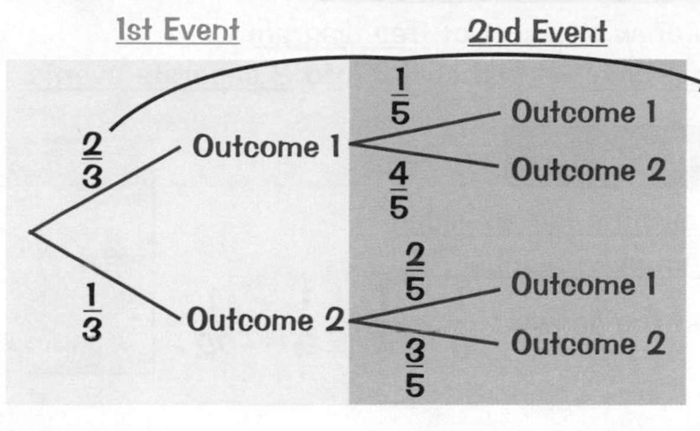

$\frac{2}{3} \times \frac{1}{5} = \frac{2}{15}$

$\frac{2}{3} \times \frac{4}{5} = \frac{8}{15}$

$\frac{1}{3} \times \frac{2}{5} = \frac{2}{15}$

$\frac{1}{3} \times \frac{3}{5} = \frac{3}{15}$

Total = 1

2) Multiply along the branches to get the end probabilities.

3) Check your diagram — the end probabilities must add up to 1.

4) To answer any question, add up the relevant end probabilities (see below).

EXAMPLES:

1. A box contains 5 red discs and 3 green discs. One disc is taken at random and its colour noted before being replaced. A second disc is then taken. Find the probability that both discs are the same colour.

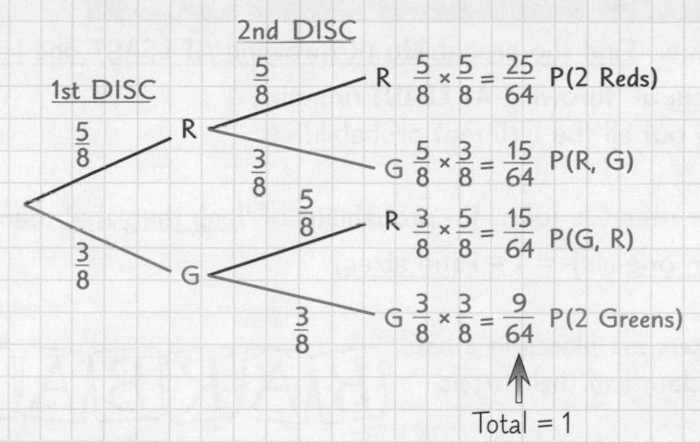

R $\frac{5}{8} \times \frac{5}{8} = \frac{25}{64}$ P(2 Reds)

G $\frac{5}{8} \times \frac{3}{8} = \frac{15}{64}$ P(R, G)

R $\frac{3}{8} \times \frac{5}{8} = \frac{15}{64}$ P(G, R)

G $\frac{3}{8} \times \frac{3}{8} = \frac{9}{64}$ P(2 Greens)

Total = 1

The probabilities for the 1st and 2nd discs are the same. This is because the 1st disc is replaced.

1. P(both discs are red) = $\frac{25}{64}$

P(both discs are green) = $\frac{9}{64}$

2. So P(both discs are same colour)

$= \frac{25}{64} + \frac{9}{64} = \frac{34}{64} = \frac{17}{32}$

2. A box contains 5 red discs and 3 green discs. Two discs are taken at random without replacement. Find the probability that both discs are the same colour.

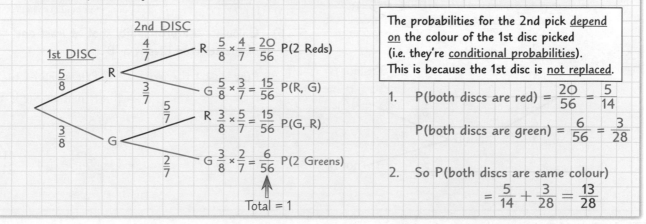

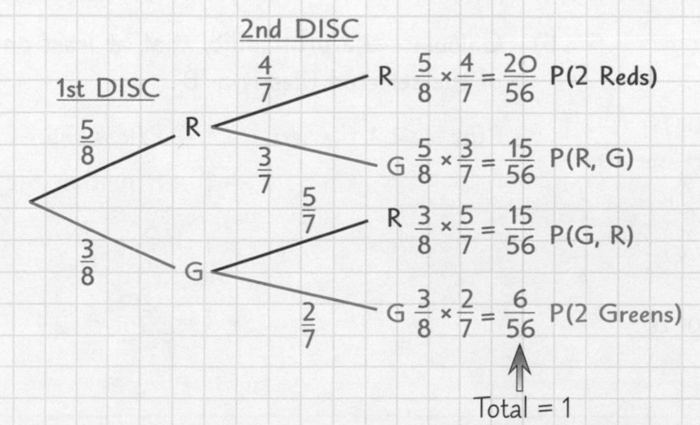

R $\frac{5}{8} \times \frac{4}{7} = \frac{20}{56}$ P(2 Reds)

G $\frac{5}{8} \times \frac{3}{7} = \frac{15}{56}$ P(R, G)

R $\frac{3}{8} \times \frac{5}{7} = \frac{15}{56}$ P(G, R)

G $\frac{3}{8} \times \frac{2}{7} = \frac{6}{56}$ P(2 Greens)

Total = 1

The probabilities for the 2nd pick depend on the colour of the 1st disc picked (i.e. they're conditional probabilities). This is because the 1st disc is not replaced.

1. P(both discs are red) = $\frac{20}{56} = \frac{5}{14}$

P(both discs are green) = $\frac{6}{56} = \frac{3}{28}$

2. So P(both discs are same colour)

$= \frac{5}{14} + \frac{3}{28} = \frac{13}{28}$

Please don't make a bad tree-based joke. Oak-ay, just this once...

How convenient — answers growing on trees. Learn the routine, and then have a go at this...

Q1 A bag contains 6 red balls and 4 black ones. If two girls each pluck out a ball at random (without replacement), find the probability that they get different coloured ones. [3 marks]

Tree Diagrams

Here's another page all about tree diagrams. This is excellent news, as tree diagrams are really useful.

Four Extra Details for the Tree Diagram Method:

1) Always break up the question into a sequence of separate events.

You need a sequence of events to be able to draw any sort of tree diagram.
For example... '3 coins are tossed at the same time' — just split it into 3 separate events.

2) Don't feel you have to draw complete tree diagrams.

For example... 'What is the probability of throwing a fair six-sided dice 3 times and getting 2 sixes followed by an even number?'

The diagram on the right is all you need to get the answer: $\frac{1}{6} \times \frac{1}{6} \times \frac{1}{2} = \frac{1}{72}$

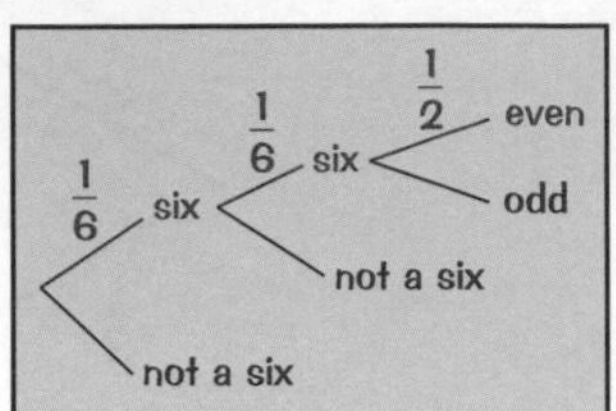

3) Watch out for conditional probabilities.

This is where the probabilities on a set of branches change, depending on the result of the previous event. For example... if you're picking things at random (e.g. cards from a pack, or balls out of a bag) without replacing your earlier picks.

See the last example on p115.

4) With 'AT LEAST' questions, it's always (1 – probability of 'LESS THAN that many'):

For example... 'I throw 3 fair six-sided dice. Find the probability of throwing AT LEAST one six.'
There are in fact quite a few different ways of 'throwing AT LEAST one six', and you could spend a long time working out all the different probabilities.
The clever trick you should know is this:
The probability of 'AT LEAST something or other' is just: 1 – probability of 'less than that many'.
So... P(at least one six) = 1 – P(less than one six) = 1 – P(no sixes).

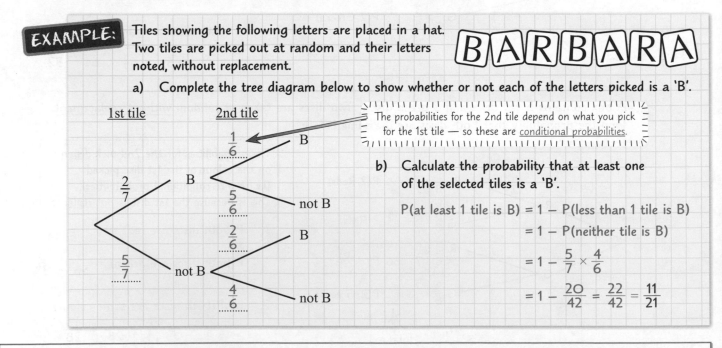

Tiles showing the following letters are placed in a hat. Two tiles are picked out at random and their letters noted, without replacement.

a) Complete the tree diagram below to show whether or not each of the letters picked is a 'B'.

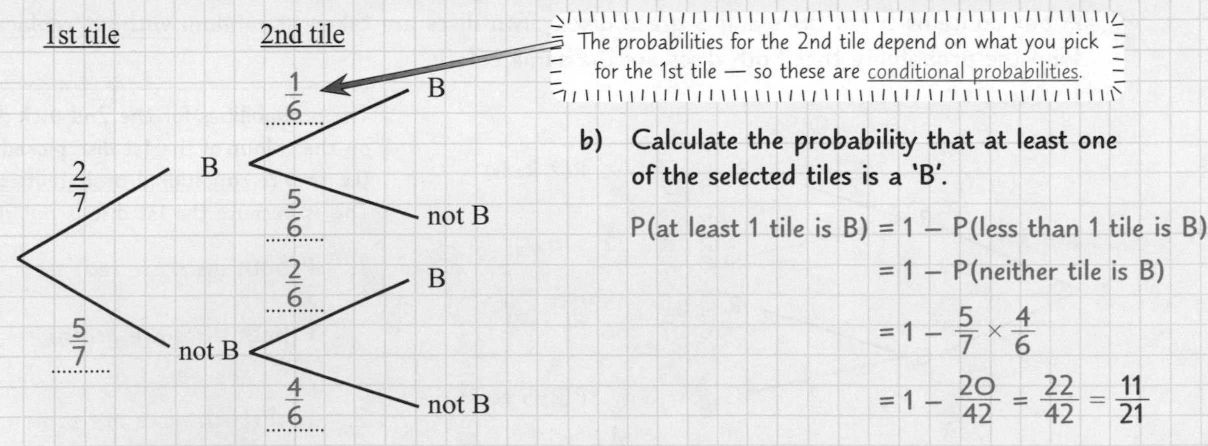

b) Calculate the probability that at least one of the selected tiles is a 'B'.

P(at least 1 tile is B) = 1 – P(less than 1 tile is B)
= 1 – P(neither tile is B)
$= 1 - \frac{5}{7} \times \frac{4}{6}$
$= 1 - \frac{20}{42} = \frac{22}{42} = \frac{11}{21}$

Things that grow on trees summary: answers = yes, money = no...

With probability questions that seem quite hard, drawing a tree diagram is usually a good place to start. Try it with the (quite hard) Exam Practice Question below...

Q1 The spinner on the right is spun twice. Use a tree diagram to find the probability that the score on the second spin is greater than the score on the first spin. [3 marks]

Relative Frequency

This isn't how often your granny visits. It's a way of working out probabilities. Since you asked, my granny visits twice a year. She'd like to visit more, but sleeping on the blow-up bed plays havoc with her back.

Fair or Biased?

The probability of rolling a three on a normal dice is $\frac{1}{6}$ — you know that each of the 6 numbers on the dice is equally likely to be rolled, and there's only 1 three.

BUT this only works if it's a fair dice. If the dice is a bit wonky (the technical term is 'biased') then each number won't have an equal chance of being rolled. This is where relative frequency comes in — you can use it to estimate probabilities when things might be wonky.

Do the Experiment Again and Again and Again...

You need to do an experiment over and over again and count how often an outcome happens (its frequency). Then you can do a quick calculation to find the relative frequency of something.

$$\text{Relative frequency} = \frac{\text{Frequency}}{\text{Number of times you tried the experiment}}$$

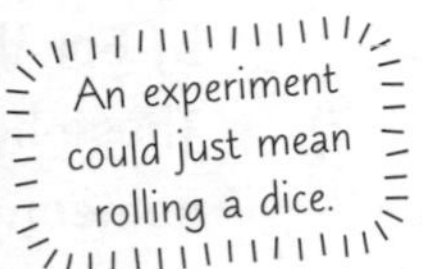

An experiment could just mean rolling a dice.

You can use the relative frequency of an outcome to estimate its probability.

EXAMPLE: The spinner on the right was spun 100 times. Use the results in the table below to estimate the probability of getting each of the scores.

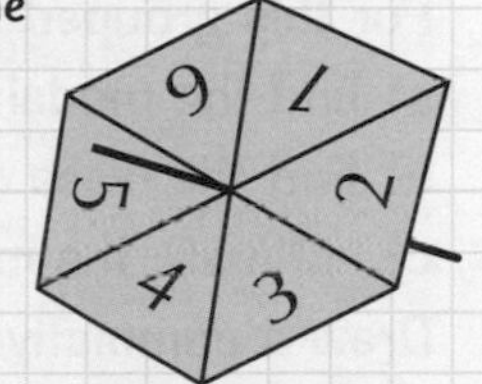

Score	1	2	3	4	5	6
Frequency	10	14	36	20	11	9

Divide each of the frequencies by 100 to find the relative frequencies.

Score	1	2	3	4	5	6
Relative Frequency	$\frac{10}{100} = 0.1$	$\frac{14}{100} = 0.14$	$\frac{36}{100} = 0.36$	$\frac{20}{100} = 0.2$	$\frac{11}{100} = 0.11$	$\frac{9}{100} = 0.09$

The MORE TIMES you do the experiment, the MORE ACCURATE your estimate of the probability will be. If you spun the above spinner 1000 times, you'd get a better estimate of the probability of each score.

If your answers are far away from what you'd expect, then you can say that the dice is probably biased.

Do the above results suggest that the spinner is biased?

Yes, because the relative frequency of 3 is much higher than you'd expect, while the relative frequencies of 1, 5 and 6 are much lower.

For a fair 6-sided spinner, you'd expect all the relative frequencies to be about 1 ÷ 6 = 0.17(ish).

This is a tough topic — make sure you revise it relatively frequently...

If a coin/dice/spinner is fair, then you can tell the probability of each result basically 'just by looking at it'. But if it's biased, then you have no option but to use relative frequencies to estimate probabilities.

Q1 Sandro threw a dice 1000 times and got the results shown in the table below.

Score	1	2	3	4	5	6
Frequency	140	137	138	259	161	165

a) Find the relative frequencies for each of the scores 1-6. [2 marks]

b) Do these results suggest that the dice is biased? Give a reason for your answer. [1 mark]

Revision Questions for Section Six

Here's the inevitable list of straight-down-the-middle questions to test how much you know.

- Have a go at each question... but only tick it off when you can get it right without cheating.
- And when you think you could handle pretty much any statistics question, tick off the whole topic.

Sampling and Collecting Data (p100-102)

1) What is a sample and why does it need to be representative?
2) Say what is meant by a random sample.
3) This table shows information about some students. If a stratified sample of 50 students is taken, how many boys should be in the sample?

Boys	Girls
80	120

4) List the four key things you should bear in mind when writing questionnaire questions.

Finding Average and Spread (p103-105)

5) Write down the definitions for the mode, median, mean and range.
6) a) Find the mode, median, mean and range of this data: 2, 8, 11, 15, 22, 24, 27, 30, 31, 31, 41
   b) For the above data, find the lower and upper quartiles and the interquartile range.
7) These box plots show information about how long it took someone to get to work in summer and winter one year. Compare the travel times in the two seasons.

Winter
Summer
0 10 20 30 40 50 mins

Frequency Tables and Cumulative Frequency (p106-108)

8) For this grouped frequency table showing the lengths of some pet alligators:
   a) find the modal class,
   b) find the class containing the median,
   c) estimate the mean.

Length (y, in m)	Frequency
$1.4 \le y < 1.5$	4
$1.5 \le y < 1.6$	8
$1.6 \le y < 1.7$	5
$1.7 \le y < 1.8$	2

9) Draw a cumulative frequency graph for the data in the above grouped frequency table.

More Graphs and Charts (p109-111)

10) How do you work out what frequency a bar on a histogram represents?
11) Draw a frequency polygon to show the data in the table on the right.

Time (t, in secs) taken to run 100 m	Frequency
$10 \le t < 12$	2
$12 \le t < 14$	8
$14 \le t < 16$	7
$16 \le t < 18$	3

12) 125 boys and 125 girls were asked if they prefer Maths or Science. 74 of the boys said they prefer Maths, while 138 students altogether said they prefer Science. How many girls said they prefer Science?
13) Sketch graphs to show:
    a) weak positive correlation, b) strong negative correlation, c) no correlation

Easy Probability (p112-113)

14) I pick a random number between 1 and 50. Find the probability that my number is a multiple of 6.
15) What do the probabilities of all possible outcomes of an experiment add up to (if none of them can happen together)?
16) Write down the formula for estimating how many times you'd expect something to happen in n trials.

Harder Probability (p114-117)

17) I throw a fair six-sided dice twice. Find P(I throw a 6 and then an even number).
18) I throw a fair six-sided dice. Find P(I throw either a 5 or a multiple of 3).
19) I pick a card at random from a normal pack of cards. I make a note of it, but don't replace it before I then pick a second card. Use a tree diagram to find the probability of me getting two kings.
20) When might you need to use relative frequency to find a probability?

Answers

Get the full versions of these answers online
Step-by-step worked solutions to these questions, with a full mark scheme, are included as a printable PDF with your free Online Edition — you'll find more info about how to get hold of this at the front of this book.

Section One

Page 2 — Calculating Tips

Q1 –4

Page 3 — Calculating Tips

Q1 0.02530405844

Page 5 — Multiples, Factors and Prime Factors

Q1 $990 = 2 \times 3 \times 3 \times 5 \times 11$

Q2 $160 = 2 \times 2 \times 2 \times 2 \times 2 \times 5$

Page 6 — LCM and HCF

Q1 36 **Q2** 12

Page 8 — Fractions

Q1 a) $\frac{17}{32}$ **b)** $\frac{2}{3}$

c) $\frac{167}{27} = 6\frac{5}{27}$ **d)** $-\frac{43}{12} = -3\frac{7}{12}$

Q2 180

Page 9 — Fractions, Decimals and Percentages

Q1 a) $\frac{4}{10} = \frac{2}{5}$ **b)** $\frac{2}{100} = \frac{1}{50}$

c) $\frac{77}{100}$ **d)** $\frac{555}{1000} = \frac{111}{200}$

e) $\frac{56}{10} = \frac{28}{5}$

Q2 a) 57% **b)** $\frac{6}{25}$ **c)** 90%

Page 11 — Fractions and Recurring Decimals

Q1 $\frac{14}{111}$

Q2 Let $r = 0.\dot{0}\dot{7}$.
Then $100r - r = 7.\dot{0}\dot{7} - 0.\dot{0}\dot{7}$
$\Rightarrow 99r = 7 \Rightarrow r = \frac{7}{99}$

Q3 $\frac{5}{111} = \frac{45}{999} = 0.\dot{0}4\dot{5}$

Page 13 — Percentages

Q1 549 ml **Q2** 27%

Q3 £20 500 **Q4** £209

Page 14 — Compound Growth and Decay

Q1 £486.20 **Q2** £99 396.26

Page 16 — Ratios

Q1 a) 5:7 **b)** 2:3 **c)** 3:10

Q2 21 bowls of porridge

Q3 £3500, £2100, £2800

Page 17 — Rounding Numbers

Q1 a) 3.57 **b)** 0.05
c) 12.910 **d)** 3546.1

Page 18 — Rounding Numbers

Q1 a) 568 **b)** 23400
c) 0.0456 **d)** 0.909

Q2 Answer should be either 5 (if rounded to 1 s.f.) or 6 (if rounded to nearest integer).

Page 19 — Bounds

Q1 a) x — l.b. = 2.315 m, u.b. = 2.325 m
y — l.b. = 0.445 m, u.b. = 0.455 m
b) max = 4.572 (to 3 d.p.), min = 4.513 (to 3 d.p.)

Page 21 — Standard Form

Q1 8.54×10^5; 1.8×10^{-4}

Q2 0.00456; 270 000

Q3 a) 2×10^{11} **b)** 6.47×10^{11}

Revision Questions — Section One

Q1 a) Whole numbers — either positive or negative, or zero
b) Numbers that can be written as fractions
c) Numbers which will only divide by themselves or 1

Q2 a) 169 **b)** 7 **c)** 3 **d)** 125

Q3 a) $210 = 2 \times 3 \times 5 \times 7$
b) $1050 = 2 \times 3 \times 5 \times 5 \times 7$

Q4 a) 14 **b)** 40

Q5 Divide top and bottom by the same number till they won't go any further.

Q6 a) $8\frac{2}{9}$ **b)** $\frac{33}{7}$

Q7 Multiplying: Multiply top and bottom numbers separately.
Dividing: Turn the second fraction upside down, then multiply.
Adding/subtracting: Put fractions over a common denominator, then add/subtract the numerators.

Q8 a) $\frac{14}{99}$ **b)** $3\frac{1}{7}$ **c)** $\frac{11}{24}$ **d)** $7\frac{11}{20}$

Q9 a) Divide the top by the bottom.
b) Put the digits after the decimal point on the top, and a power of 10 with the same number of zeros as there were decimal places on the bottom.

Q10 a) (i) $\frac{4}{100} = \frac{1}{25}$ (ii) 4%
b) (i) $\frac{65}{100} = \frac{13}{20}$ (ii) 0.65

Q11 Fractions where the denominator has prime factors of only 2 or 5 give terminating decimals. All others give recurring decimals.

Q12 Let $r = 0.\dot{5}\dot{1}$.
Then $100r - r = 51.\dot{5}\dot{1} - 0.\dot{5}\dot{1}$
$\Rightarrow 99r = 51 \Rightarrow r = \frac{51}{99} = \frac{17}{33}$

Q13 To find x as a percentage of y, make sure both amounts are in the same units, then divide x by y and multiply by 100.

Q14 percentage change = (change ÷ original) × 100

Q15 17.6 m

Q16 6% simple interest pays £59.62 more (to the nearest penny)

Q17 240

Q18 1. Add up the parts
2. Divide to find one part
3. Multiply to find the amounts

Q19 600, 960, 1440

Q20 a) 427.96 **b)** 428.0
c) 430 **d)** 428.0

Q21 Estimates should be around 20-24.

Q22 The upper and lower bounds are half a unit either side of the rounded value.

Q23 132.2425 m²

Q24 1. The front number must always be between 1 and 10.
2. The power of 10, n, is how far the decimal point moves.
3. n is positive for big numbers, and negative for small numbers.

Q25 a) 9.7×10^5 **b)** 3.56×10^9
c) 2.75×10^{-6}

Q26 a) 1.5875×10^3 **b)** 2.739×10^{12}

Section Two

Page 23 — Sequences

Q1 a) 7n – 5
b) 51
c) No, as the solution to 7n – 5 = 63 does not give an integer value of n.

Answers

Page 24 — Powers and Roots

Q1 a) e^{11} **b)** f^4
c) g^3 **d)** $6h^7j^2$

Q2 a) 125 **b)** $\frac{1}{5}$ **c)** 2

Page 25 — Algebra Basics

Q1 $3x + 8y$

Page 26 — Multiplying Out Brackets

Q1 a) $-15x + 10y$ **b)** $2x^2 - 10x$

Q2 a) $y^2 - y - 20$ **b)** $4p^2 - 12p + 9$

Page 27 — Factorising

Q1 a) $7(x - 2)$ **b)** $3y(2x + 5y)$

Q2 $2(2x + y)(2x - y)$

Q3 $\frac{6}{x + 7}$

Page 28 — Manipulating Surds

Q1 a) $2 + 3\sqrt{2}$ **b)** $9 - 4\sqrt{5}$

Q2 $\frac{2\sqrt{3}}{21}$

Page 29 — Solving Equations

Q1 $x = 2$

Q2 $y = 4$

Q3 $x = 6$

Page 30 — Solving Equations

Q1 $x = \pm 6$

Q2 $x = 8$

Page 31 — Rearranging Formulas

Q1 $q = 7(p - 2r)$

Q2 $z = \frac{3x - y}{2}$

Page 32 — Rearranging Formulas

Q1 a) $y = \pm 2\sqrt{x}$ **b)** $y = \frac{xz}{x - 1}$

Page 33 — Factorising Quadratics

Q1 $(x + 5)(x - 3)$

Q2 $x = 4$ or $x = 5$

Page 34 — Factorising Quadratics

Q1 $(2x + 3)(x - 4)$

Q2 $x = \frac{2}{3}$ or $x = -4$

Q3 $(3x + 2)(x + 10)$

Q4 $x = -\frac{2}{5}$ or $x = 3$

Page 35 — The Quadratic Formula

Q1 $x = 0.39$ or $x = -10.39$

Q2 $x = 1.46$ or $x = -0.46$

Page 36 — Completing the Square

Q1 $(x - 6)^2 - 18$

Q2 $(x + 5)^2 - 18 = 0$, so $x = -5 \pm 3\sqrt{2}$

Page 37 — Quadratic Equations — Tricky Ones

Q1 $x = \frac{1 \pm \sqrt{3}}{2}$

Page 38 — Algebraic Fractions

Q1 $\frac{3b^2}{a}$

Q2 $\frac{x(x + 3)}{4}$

Q3 $\frac{x - 11}{(x + 4)(x - 1)}$

Page 39 — Inequalities

Q1 a) $x < 4$ **b)** $x \geq 3$

Q2 $x \leq 4$

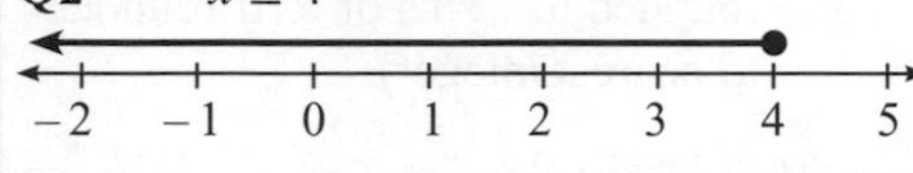

Page 40 — Graphical Inequalities

Q1

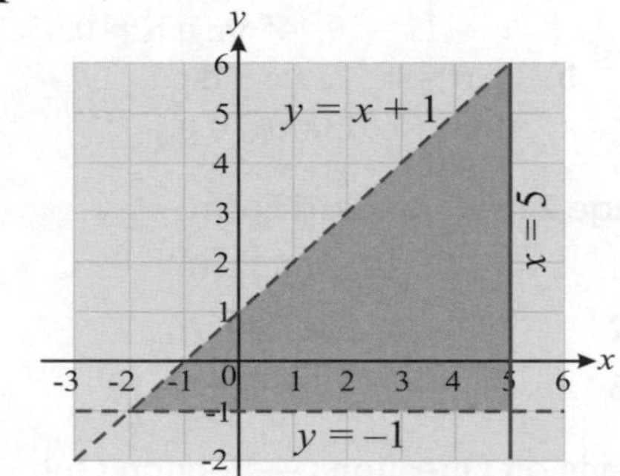

Page 41 — Trial and Improvement

Q1 $x = 3.6$ **Q2** $x = 5.5$

Page 42 — Simultaneous Equations and Graphs

Q1 $x = 2, y = 4$

Q2 $y = 3$

Page 43 — Simultaneous Equations

Q1 $x = 5, y = 3$

Q2 $x = 3, y = -1$

Page 44 — Simultaneous Equations

Q1 $x = 0, y = 4$ and $x = 6, y = 40$

Q2 $x = 1, y = -1$ and $x = -4, y = 14$

Page 45 — Direct and Inverse Proportion

Q1 $s = 4$

Page 46 — Proof

Q1 Take two consecutive even numbers, 2n and 2n + 2. Then 2n + (2n + 2) = 4n + 2 = 2(2n + 1), which is even.

Q2 $(a + b)(a - b) \equiv a^2 - ab + ab - b^2 \equiv a^2 - b^2$ (this is the proof of the difference of two squares)

Revision Questions — Section Two

Q1 a) $2n + 5$ **b)** $-3n + 14$

Q2 Yes, it's the 5th term.

Q3 a) x^9 **b)** y^2 **c)** z^{12}

Q4 $5x - 4y - 5$

Q5 a) $6x + 3$ **b)** $x^2 - x - 6$

Q6 a) $(x + 4y)(x - 4y)$
b) $(7 + 9pq)(7 - 9pq)$
c) $12(x + 2y)(x - 2y)$

Q7 a) $3\sqrt{3}$ **b)** 5

Q8 a) $x = 2$ **b)** $x = \pm 3$

Q9 a) $p = -\frac{4y}{3}$ **b)** $p = \frac{qr}{q + r}$

Q10 a) $x = -3$ and $x = -6$
b) $x = 4$ and $x = -\frac{3}{5}$

Q11 $x = \frac{-b \pm \sqrt{b^2 - 4ac}}{2a}$

Q12 a) $x = 1.56$ and $x = -2.56$
b) $x = 0.27$ and $x = -1.47$
c) $x = 0.44$ and $x = -3.44$

Q13 a) $x = -6 \pm \sqrt{21}$ **b)** $x = 3 \pm \sqrt{11}$

Q14 $\frac{3x + 1}{(x + 3)(x - 1)}$

Q15 $x \geq -2$

Q16

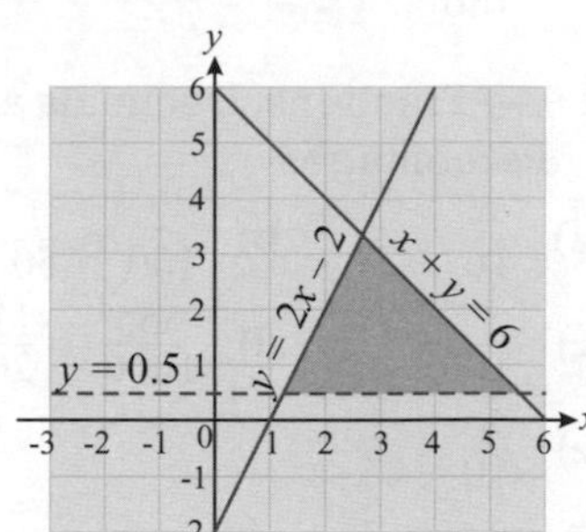

Q17 $x = 4.1$

Q18 Sketch the graphs and find the coordinates of the points where the graphs cross.

Q19 $x = -2, y = -2$ and $x = -4, y = -8$

Q20 $y = kx^2$

Q21 $p = 72$

Q22 Take an even number, $2p$, and an odd number, $2q + 1$. Their product is $2p \times (2q + 1) = 4pq + 2p = 2(2pq + p)$, which is even.

Section Three

Page 48 — X, Y and Z Coordinates

Q1 (0.5, −1)

Q2 (0, 4, 2)

Page 49 — Straight-Line Graphs

Q1 $y = -2$

Answers

Page 50 — Plotting Straight-Line Graphs

Q1

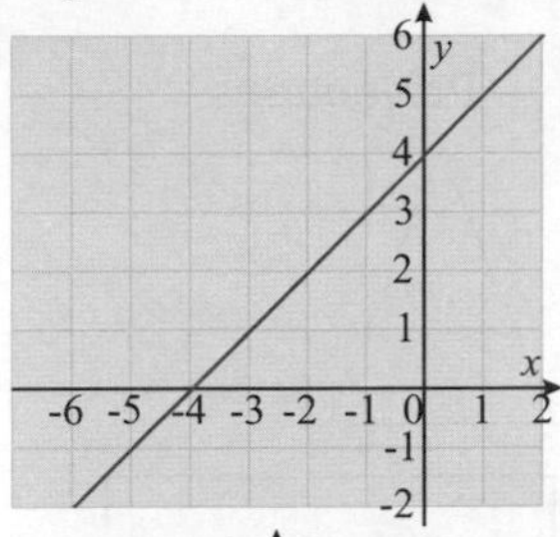

Q2

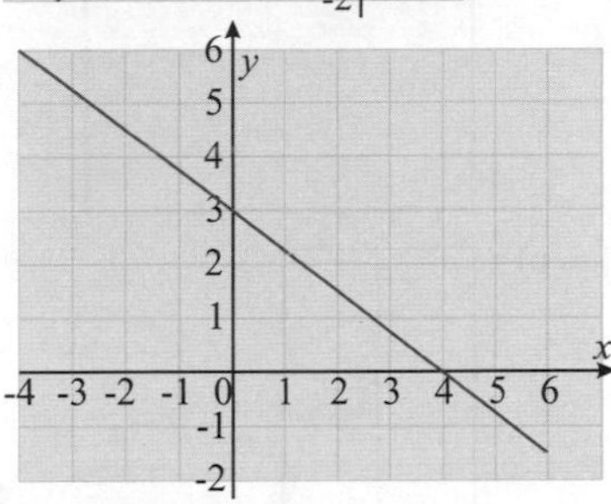

Page 51 — Finding the Gradient

Q1 gradient = –5

Page 52 — "y = mx + c"

Q1

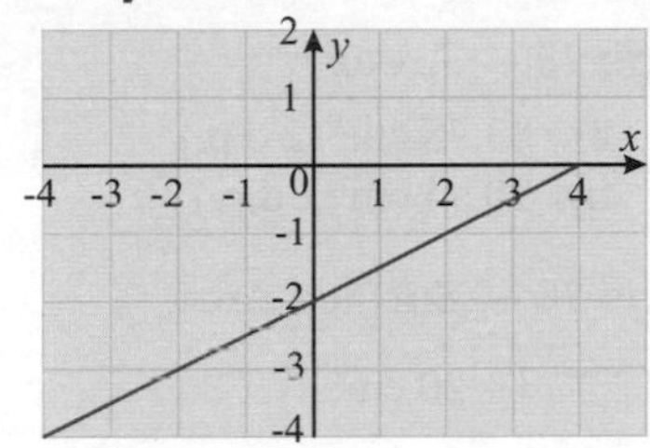

Q2 $y = \frac{1}{2}x + 5$

Page 53 — Parallel and Perpendicular Lines

Q1 $y = -x + 5$

Q2 Write both lines in the form $y = mx + c$: $y = -5x + 2$ and $y = \frac{1}{5}x + \frac{3}{5}$. So the gradients are –5 and $\frac{1}{5}$. Multiplying these together gives $-5 \times \frac{1}{5} = -1$, so the lines are perpendicular.

Page 54 — Quadratic Graphs

Q1 a)

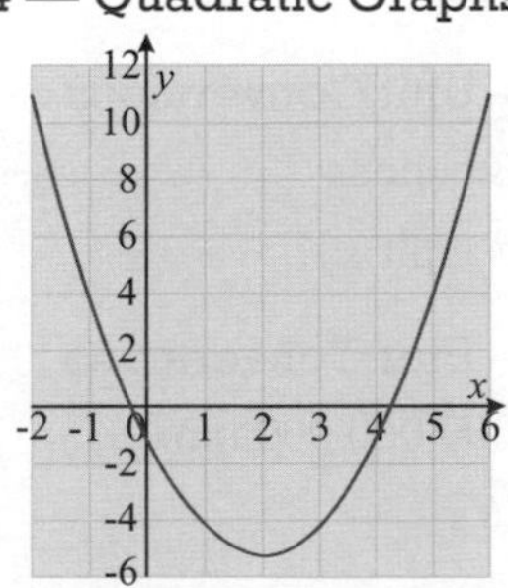

b) $x = -1.2$ (accept –1 to –1.4), $x = 5.2$ (accept 5 to 5.4).

Page 55 — Harder Graphs

Q1 a)

x	–2	–1	0	1	2
y	–3	2	1	0	5

b)

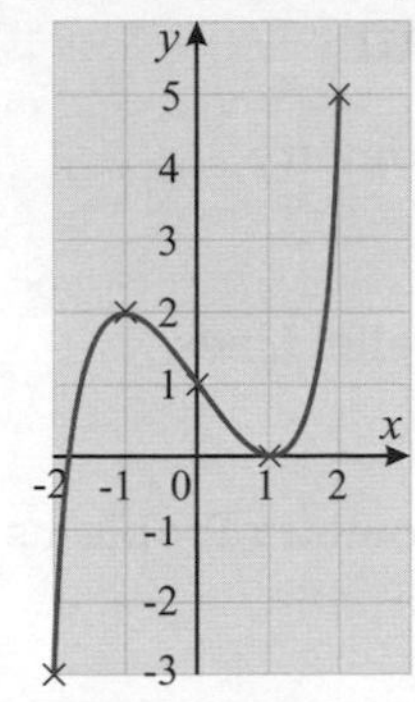

Page 56 — Harder Graphs

Q1 a)

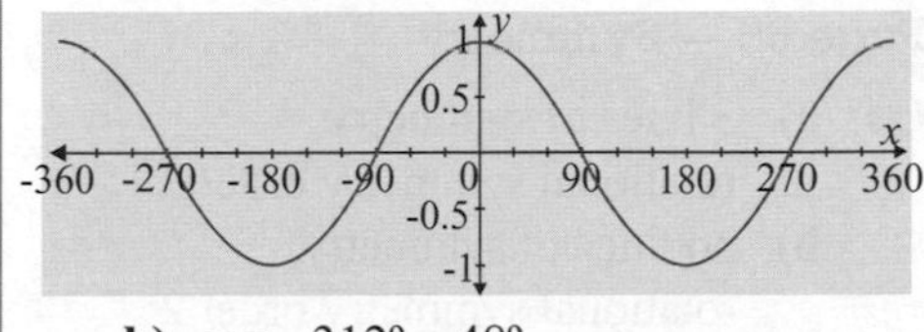

b) $x = -312°, -48°$

Page 58 — Graph Transformations

Q1 a) (8, 3) **b)** (4, –1) **c)** (4, 1.5)

Page 59 — Real-Life Graphs

Q1

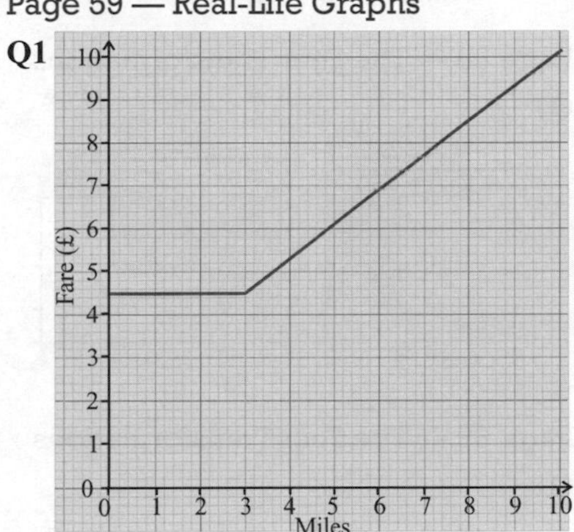

Revision Questions — Section Three

Q1 (2.5, –0.5)

Q2 A(4, 5, 0), B(4, 5, 5)

Q3

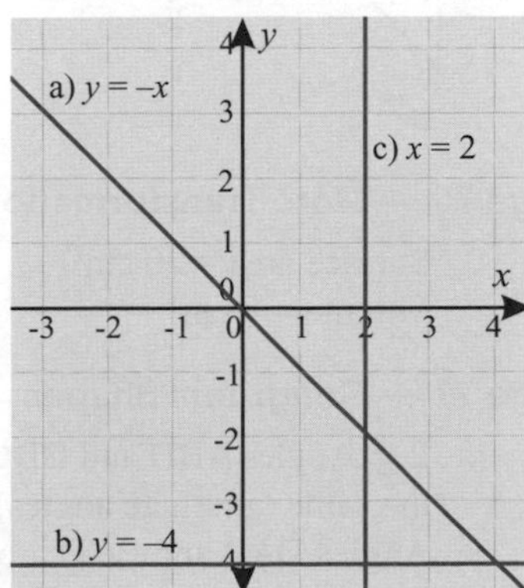

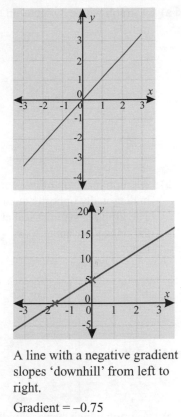

Q4 Straight-line equations just contain something x, something y and a number. They don't contain any powers of x or y, xy, $1/x$ or $1/y$.

Q5

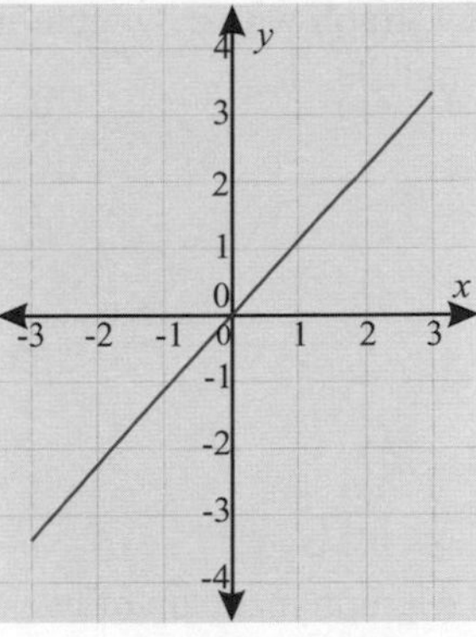

Q6

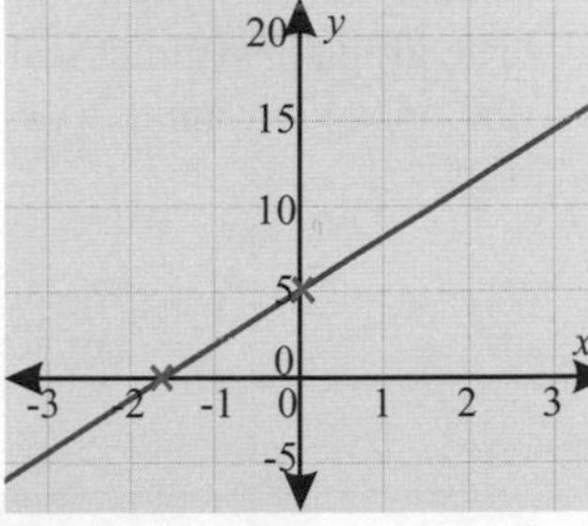

Q7 A line with a negative gradient slopes 'downhill' from left to right.

Q8 Gradient = –0.75

Q9 'm' is the gradient and 'c' is the y-intercept.

Q10

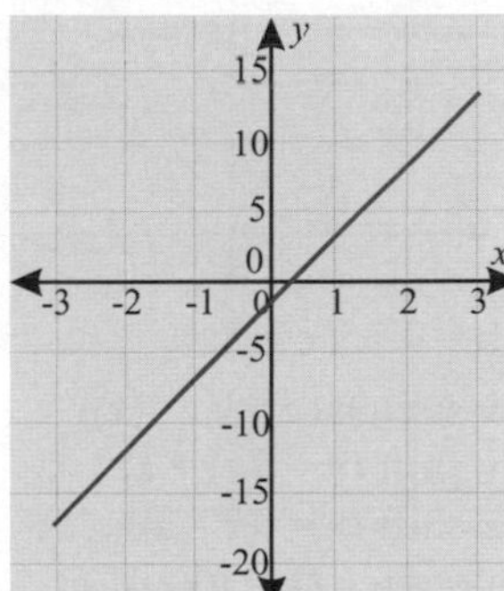

Q11 $y = 2x + 10$

Q12 The gradients of two perpendicular lines multiply together to give –1. The gradients of parallel lines are the same as each other.

Q13 $y = -\frac{1}{2}x + 4$

Q14 They are both symmetrical "bucket shaped" graphs. $y = x^2 + 2x - 8$ is like a "u" whereas $y = -x^2 + 2x - 8$ is like an "n" (or an upturned bucket).

Q15

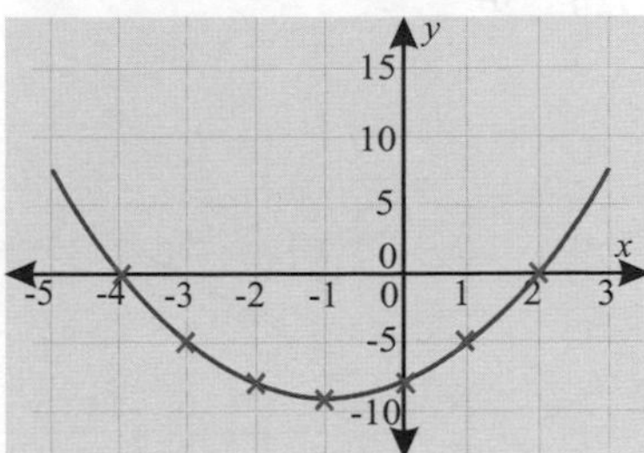

$x = -3.6$ or 1.6 (both ±0.2)

Answers

Q16 a) A graph with a "wiggle" in the middle. E.g.

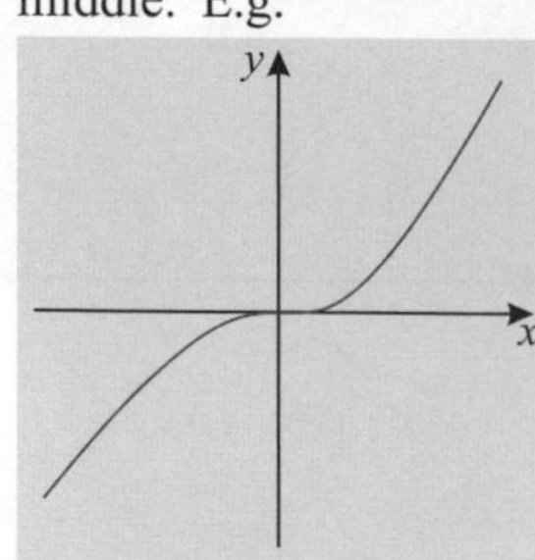

b) A graph made up of two curves in diagonally opposite quadrants. The graph is symmetrical about the lines $y = x$ and $y = -x$. E.g.

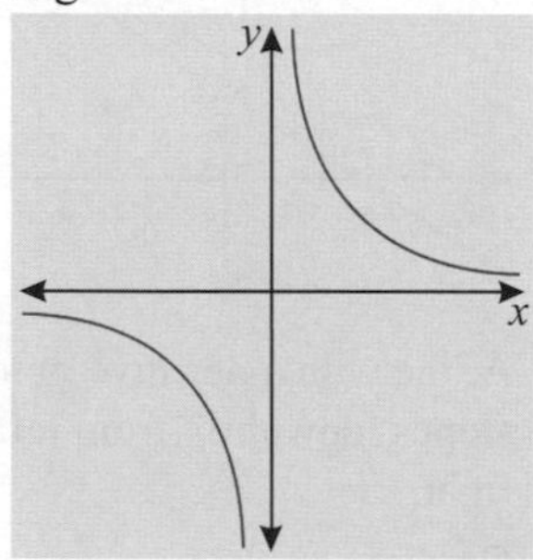

c) A graph which curves rapidly upwards. E.g.

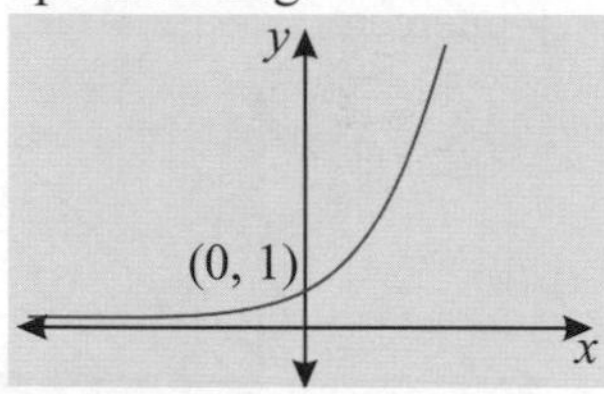

Q17 b = 0.25, c = 8

Q18 y-stretch ($y = \text{k} \times \text{f}(x)$)
y-shift ($y = \text{f}(x) + \text{a}$)
x-shift ($y = \text{f}(x - \text{a})$)
x-stretch ($y = \text{f}(\text{k}x)$)
reflection ($y = -\text{f}(x)$ or $y = \text{f}(-x)$)

Q19 a) $y = (-x)^3 + 1$ is the original graph reflected in y-axis.

b) $y = (x + 2)^3 + 1$ is the original graph shifted by 2 units in the negative x-direction.

c) $y = (3x)^3 + 1$ is the original graph scrunched up horizontally by a factor of 3.

d) $y = x^3 - 1$ is the original graph shifted downwards by 2 units.

Q20

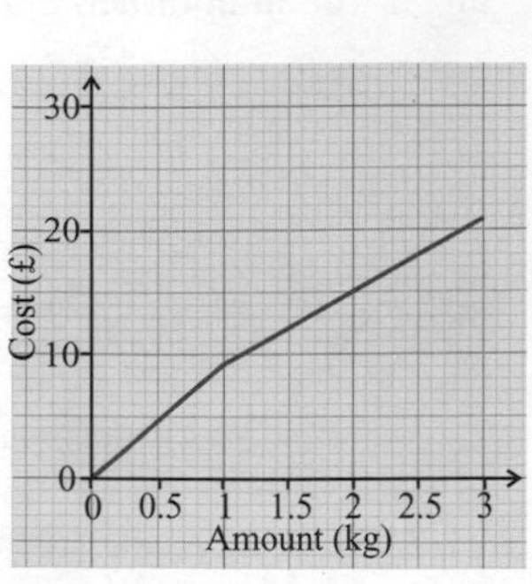

Section Four

Page 61 — Geometry

Q1 $x = 108°$

Page 62 — Parallel Lines

Q1 $x = 116°$

Page 63 — Geometry Problems

Q1 $x = 123°$

Page 64 — Polygons

Q1 144°

Page 65 — Symmetry

Q1 a) 3 lines of symmetry, rotational symmetry order 3
b) no lines of symmetry, rotational symmetry order 2
c) 8 lines of symmetry, rotational symmetry order 8

Page 67 — Circle Geometry

Q1 angle ABD = 63°
angle ACD = 63°

Page 68 — The Four Transformations

Q1

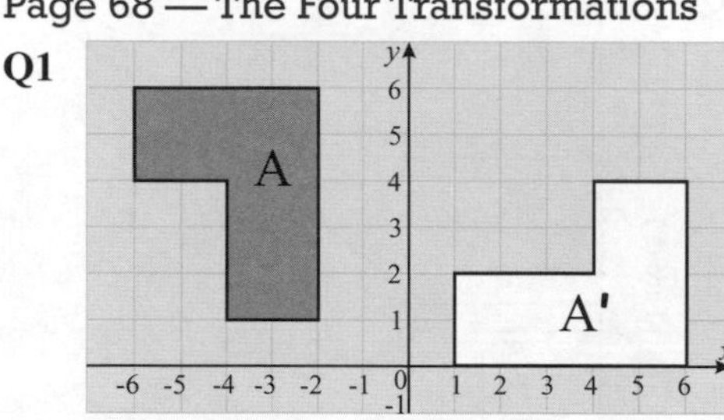

Page 69 — The Four Transformations

Q1

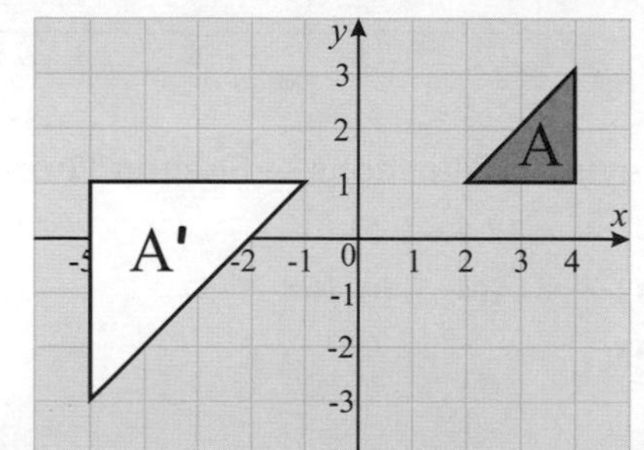

Page 70 — More Transformation Stuff

Q1 Surface area = 20 cm²
Volume = 8 cm³

Page 71 — Congruent Shapes

Q1 E.g. Angles ABD and BDC are the same (alternate angles). Angles DAB and BCD are the same (opposite angles in a parallelogram (ABCD) are equal). Side BD is the same in each shape. So triangles ABD and BCD are congruent as the condition AAS holds.

Page 72 — Similar Shapes

Q1 AD = 50 cm

Page 73 — Projections

Q1 a)

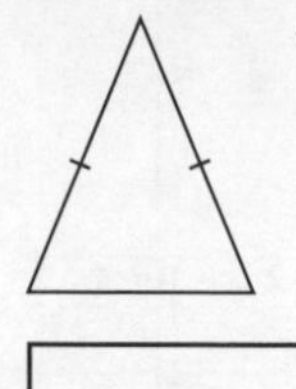

b)

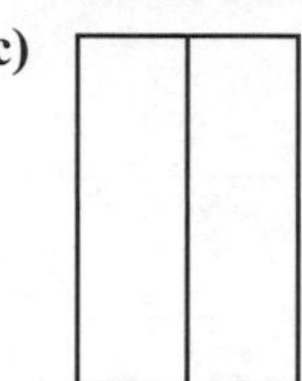

c)

Page 74 — Areas

Q1 $x = 10$

Page 75 — Areas

Q1 a) 83.78 cm² (2 d.p.)
b) 20.94 cm (2 d.p.)

Page 76 — Surface Area

Q1 $l = 20$ cm

Page 77 — Volume

Q1 a) 360 cm³ **b)** 648 g

Page 78 — Volume

Q1 $h = 36$ cm

Page 79 — Density and Speed

Q1 285 kg (3 s.f.)
Q2 101.25 km

Page 80 — Distance-Time Graphs

Q1 a) 15 minutes
b) 6 km

Page 81 — Unit Conversions

Q1 5.8 gallons
Q2 1.8 mph

Page 82 — Unit Conversions

Q1 3 000 000 000 mm³
Q2 $267

Answers

Page 83 — Triangle Construction

Q1

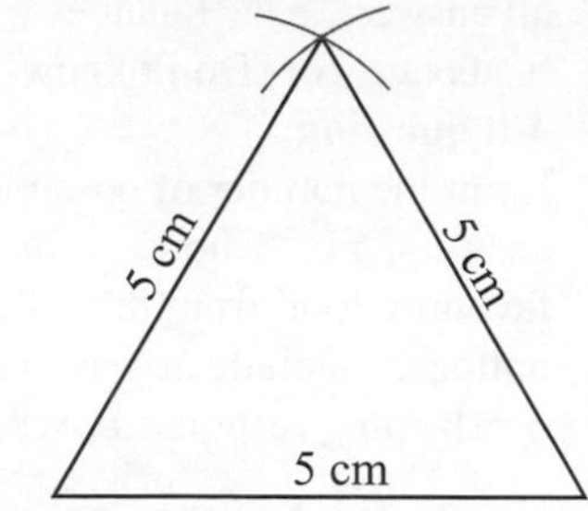

Q2

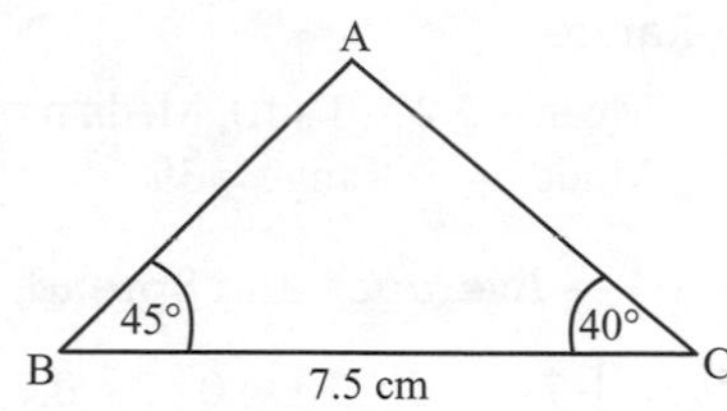

Page 85 — Loci and Constructions

Q1

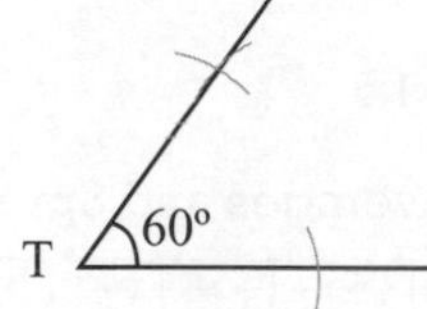

Page 86 — Loci and Construction — Worked Examples

Q1 Shaded area = where public can go

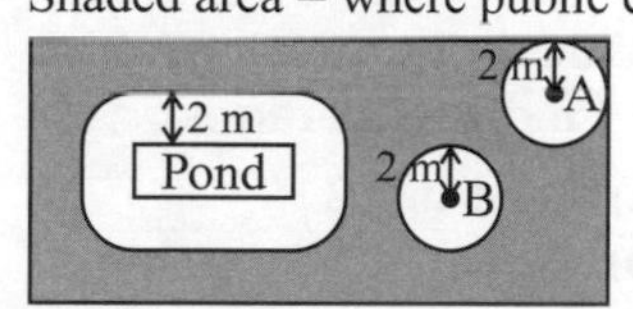

Page 87 — Bearings

Q1 297°

Q2 29.2 km

Revision Questions — Section Four

Q1 360°

Q2 a) $x = 154°$ **b)** $y = 112°$
c) $z = 58°$

Q3 60°

Q4 lines of symmetry = 3
order of rotational symmetry = 3

Q5 90°

Q6 a) $x = 53°$
b) $y = 69°$
c) $z = 33°$

Q7 a) Translation by vector $\begin{pmatrix} -2 \\ -4 \end{pmatrix}$
b) Reflection in the y-axis

Q8

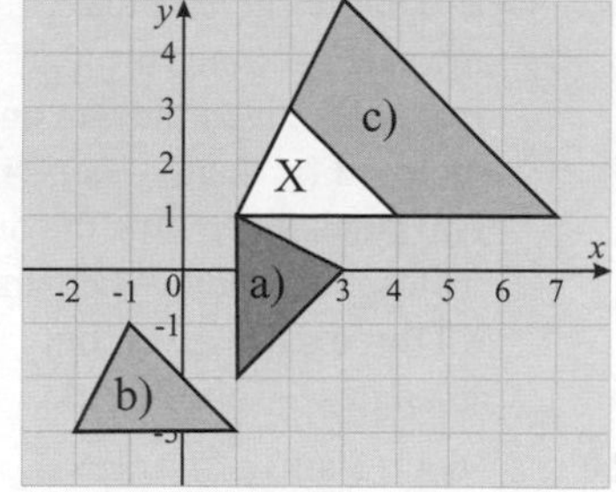

Q9 80 cm^2

Q10 SSS, AAS, SAS, RHS

Q11 E.g. angles ACB and ACD are right angles (as it's a perpendicular bisector of a chord)
AB = AD (they're both radii)
CB = CD (as the chord is bisected)
So the condition RHS holds and the triangles are congruent.

Q12 $x = 2.5$ cm

Q13 The view from directly above an object.

Q14

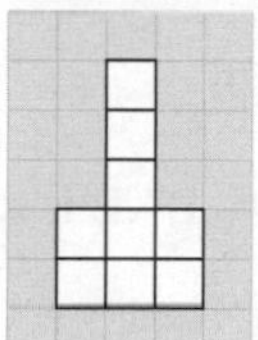

Q15 $A = \frac{1}{2}(a + b) \times h_V$

Q16 220 cm^2

Q17 Circumference = 16π cm, area = 64π cm^2

Q18 39.27 cm^2

Q19 Surface area = $4\pi r^2$

Q20 75π cm^2

Q21 396 cm^3

Q22 129.85 cm^3 (2 d.p.)

Q23 12 500 cm^3

Q24 42 mph

Q25 The object has stopped.

Q26 36 mph

Q27 a) 5600 cm^3 **b)** 3.6 kg
c) 10.8 km/h **d)** 5 690 000 cm^2

Q28 7.5 miles

Q29

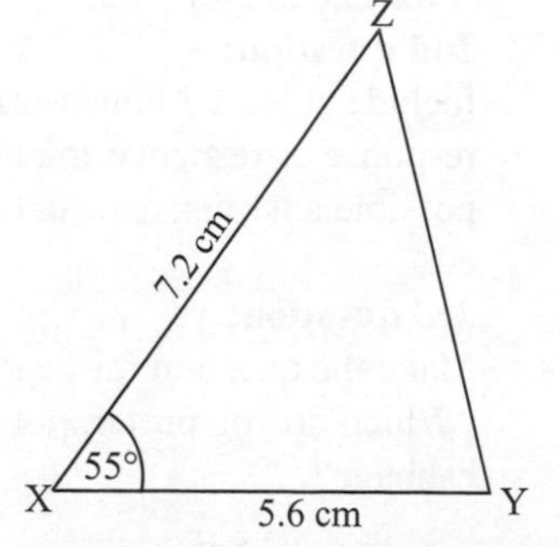

Q30

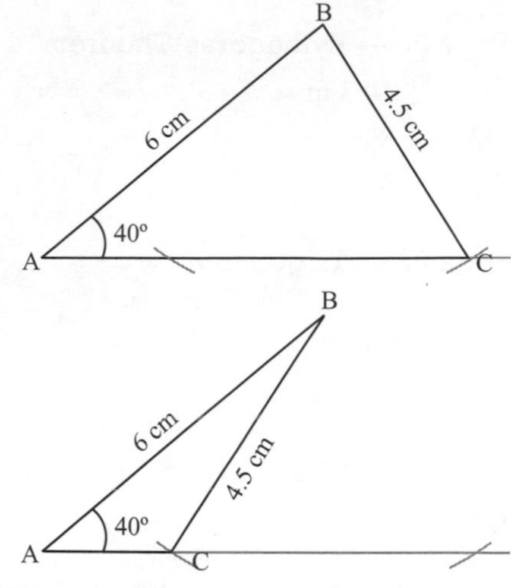

Not full size

Q31 A circle

Q32

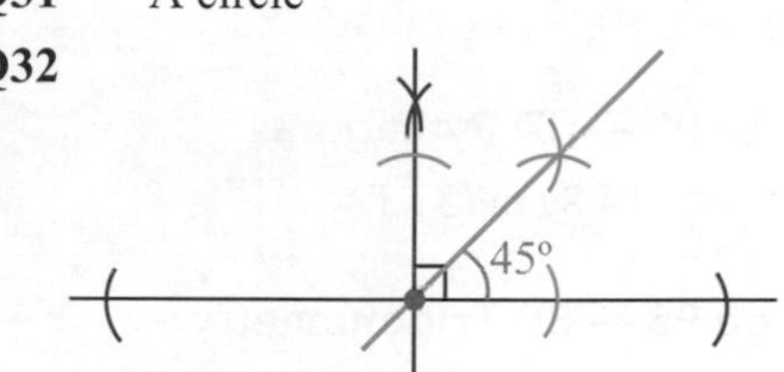

Q33

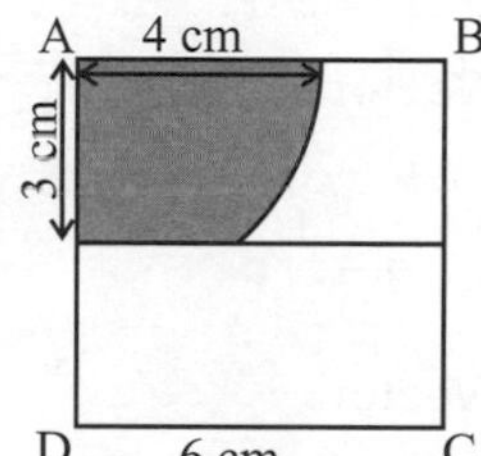

Q34 Put your pencil on the diagram at the point you're going FROM. Draw a northline at this point. Draw in the angle clockwise from the northline — this is the bearing you want.

Q35

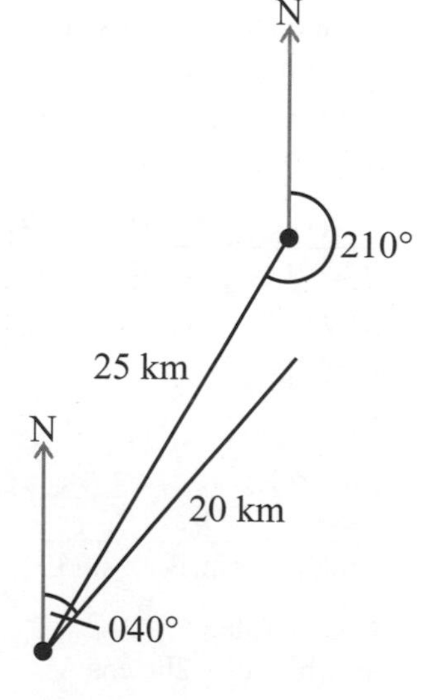

Answers

Section Five

Page 90 — Pythagoras' Theorem

Q1 10.3 m

Q2 8 m

Q3 5

Page 92 — Trigonometry — Sin, Cos, Tan

Q1 27.1°

Q2 2.97 m

Page 93 — The Sine and Cosine Rules

Q1 32.5 cm²

Page 94 — The Sine and Cosine Rules

Q1 20.5 cm (3 s.f.)

Q2 59.5° (3 s.f.)

Page 95 — 3D Pythagoras

Q1 14.8 cm (3 s.f.)

Page 96 — 3D Trigonometry

Q1 17.1° (3 s.f.)

Page 97 — Vectors

Q1 $\overrightarrow{AB} = \mathbf{p} - 2\mathbf{q}$

$\overrightarrow{NA} = \mathbf{q} - \frac{1}{2}\mathbf{p}$

Page 98 — Vectors

Q1 $\overrightarrow{AB} = \mathbf{a} - \mathbf{b}$

$\overrightarrow{DC} = \frac{3}{2}\mathbf{a} - \frac{3}{2}\mathbf{b} = \frac{3}{2}(\mathbf{a} - \mathbf{b})$

ABCD is a trapezium.

Revision Questions — Section Five

Q1 $a^2 + b^2 = c^2$

You use Pythagoras' theorem to find the missing side of a right-angled triangle.

Q2 4.72 m

Q3 7.8

Q4

O
T × A

Q5 33.4°

Q6 5.77 cm

Q7 21.5°

Q8 Sine rule:

$$\frac{a}{\sin A} = \frac{b}{\sin B} = \frac{c}{\sin C}$$

Cosine rule:

$a^2 = b^2 + c^2 - 2bc \cos A$

Area $= \frac{1}{2}ab \sin C$

Q9 Two angles given plus any side — sine rule.
Two sides given plus an angle not enclosed by them — sine rule.
Two sides given plus the angle enclosed by them — cosine rule.
All three sides given but no angles — cosine rule.

Q10 56.4° (3 s.f.)

Q11 6.84 cm (3 s.f.)

Q12 48.1 cm² (3 s.f.)

Q13 $a^2 + b^2 + c^2 = d^2$

Q14 11.9 m (3 s.f.)

Q15 15.2° (3 s.f.)

Q16 54°

Q17 Multiplying by a scalar changes the size of a vector but not its direction.

Q18 a) $\overrightarrow{AX} = \frac{1}{3}\mathbf{a}$

b) $\overrightarrow{DX} = \frac{4}{3}\mathbf{a} - \mathbf{b}$

$\overrightarrow{XB} = \frac{8}{3}\mathbf{a} - 2\mathbf{b}$

c) $\overrightarrow{XB} = 2\,\overrightarrow{DX}$, so DXB is a straight line.

Section Six

Page 100 — Sampling and Bias

Q1 Two from, e.g: sample too small, motorways not representative of average motorist, only done at one time of day and in one place.

Page 101 — Sampling Methods

Q1 2 middle managers

Page 102 — Collecting Data

Q1 **1st question:**
Question should include a time frame, e.g. "How much do you spend on food each week?"
Include at least 3 non-overlapping response boxes, covering all possible answers, e.g. '£0 to £20', 'over £20 to £40', etc.
2nd question:
Include at least 3 non-overlapping response boxes, covering all possible answers, e.g. '0-1', '2-3', etc.
3rd question:
Make the question fair, e.g. "Which do you prefer, potatoes or cabbage?"
Response boxes should cover all answers, e.g. 'Potatoes', 'Cabbage', or 'Don't know'.
4th question:
Limit the number of possible answers, e.g. "Choose your favourite food from the following options." Include at least 3 non-overlapping response boxes.

Page 103 — Mean, Median, Mode and Range

Q1 Mean = 5.27 (3 s.f.), Median = 6
Mode = –5, Range = 39

Page 104 — Averages and Spread

Q1 a)

0	7
1	1 4 6
2	2 6
3	0

Key: 0 | 7 = 0.7

b) IQR = 1.5

Page 105 — Averages and Spread

Q1

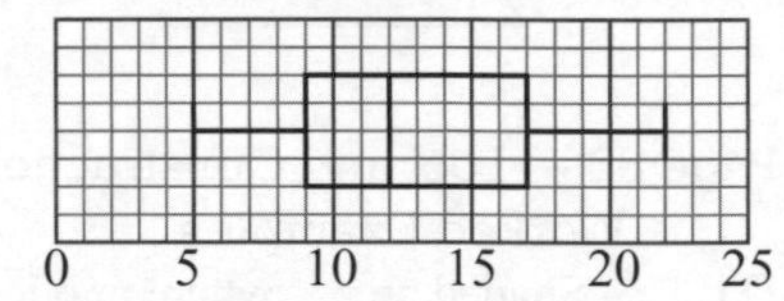

Page 106 — Frequency Tables — Finding Averages

Q1 a) Median = 2

b) Mean = 1.66

Page 107 — Grouped Frequency Tables

Q1 17.4 cm (3 s.f.)

Page 108 — Cumulative Frequency

Q1 a)

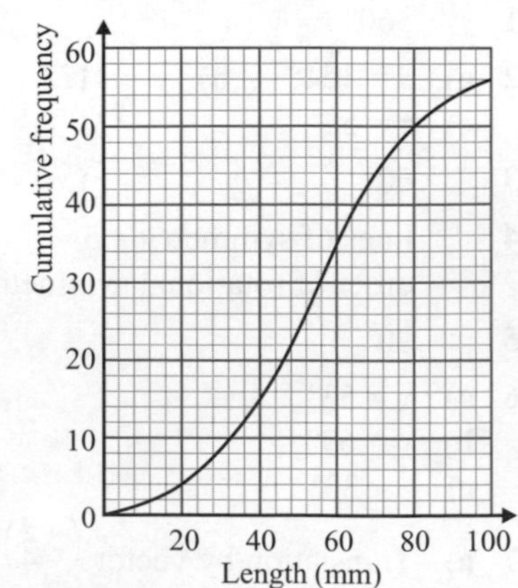

b) Answer in the range 30–34.

Answers

Page 109 — Histograms and Frequency Density

Q1

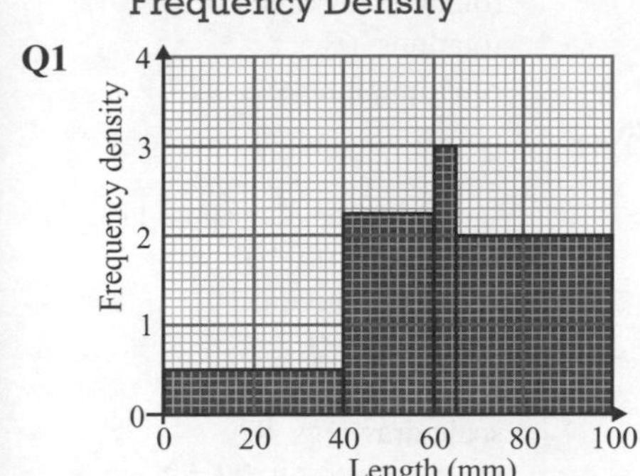

Page 110 — Other Graphs and Charts

Q1

	Walk	Car	Bus	Total
Male	15	21	13	49
Female	18	11	22	51
Total	33	32	35	100

Page 111 — Scatter Graphs

Q1 a) Strong negative correlation

b) Approximately 5 mph

Page 112 — Probability Basics

Q1 $\frac{3}{10} = 0.3$

Q2 0.2

Page 113 — Listing Outcomes and Expected Frequency

Q1 a) $\frac{1}{6}$

b) Approximately 50 times

Page 114 — The AND / OR Rules

Q1 a) $\frac{2}{3}$

b) $\frac{1}{36}$

Page 115 — Tree Diagrams

Q1 $\frac{8}{15}$

Page 116 — Tree Diagrams

Q1

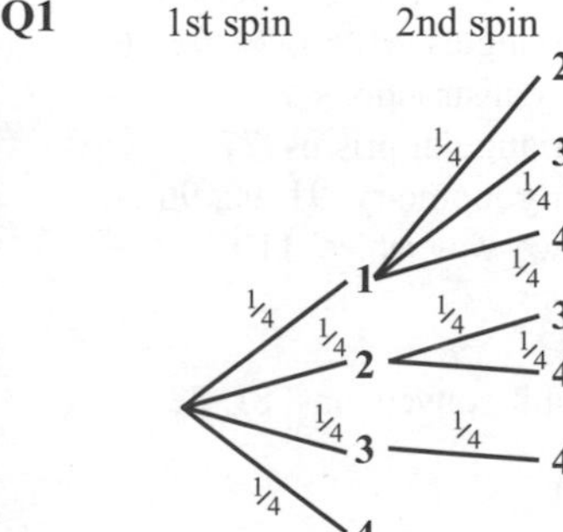

P(2nd spin greater than 1st spin) = $\frac{3}{8}$

Page 117 — Relative Frequency

Q1 a)

Score	Relative frequency
1	0.14
2	0.137
3	0.138
4	0.259
5	0.161
6	0.165

b) Yes, because the relative frequency for 4 is much higher than you'd expect from a fair dice (which is 1 ÷ 6 = 0.166...).

Revision Questions — Section Six

Q1 A sample is part of a population. Samples need to be representative so that conclusions drawn from sample data can be applied to the whole population.

Q2 A random sample is one where every member of the population has an equal chance of being in it.

Q3 20

Q4 Questions need to be:
(i) clear and easy to understand
(ii) easy to answer
(iii) fair (i.e. not leading or biased)
(iv) easy to analyse afterwards

Q5 The mode is the most common value.
The median is the middle value when the data has been arranged in order of size.
The mean is the total of the data values divided by the number of data values.
The range is the difference between the highest and lowest data values.

Q6 a) Mode = 31, Median = 24
Mean = 22, Range = 39

b) Lower quartile = 11
Upper quartile = 31
Interquartile range = 20

Q7 The median time in winter is lower than the median time in summer, so it generally took longer to get to work in the summer.
The range and the IQR for the summer are smaller than those for the winter, so there is less variation in journey times in the summer.

Q8 a) Modal class is: $1.5 \leq y < 1.6$.

b) Class containing median is: $1.5 \leq y < 1.6$

c) Estimated mean = 1.58 m (to 2 d.p.)

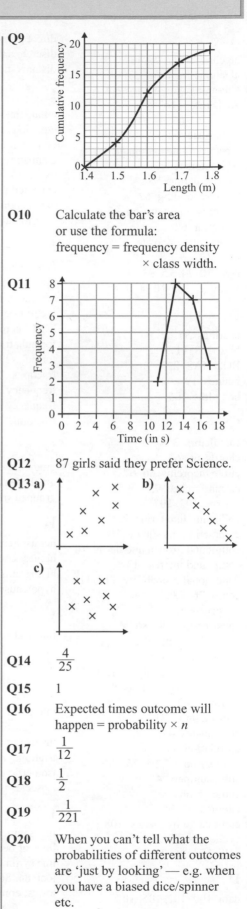

Q9

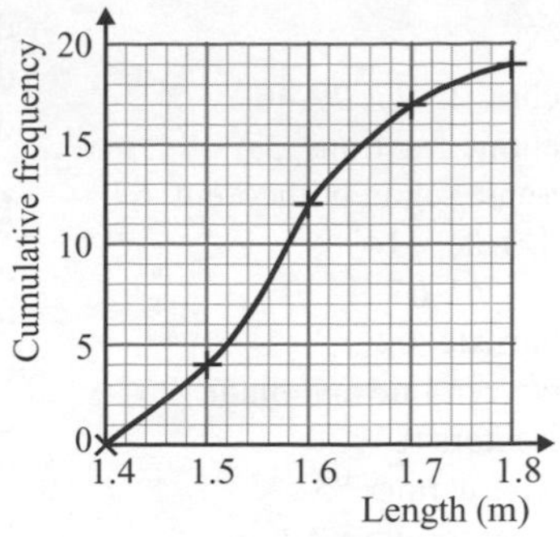

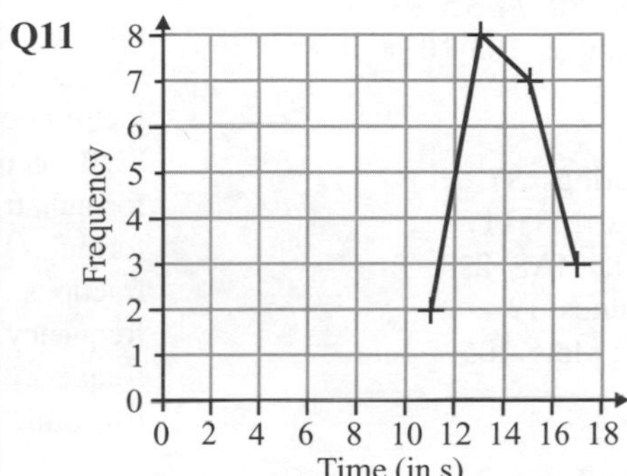

Q10 Calculate the bar's area or use the formula:
frequency = frequency density × class width.

Q11

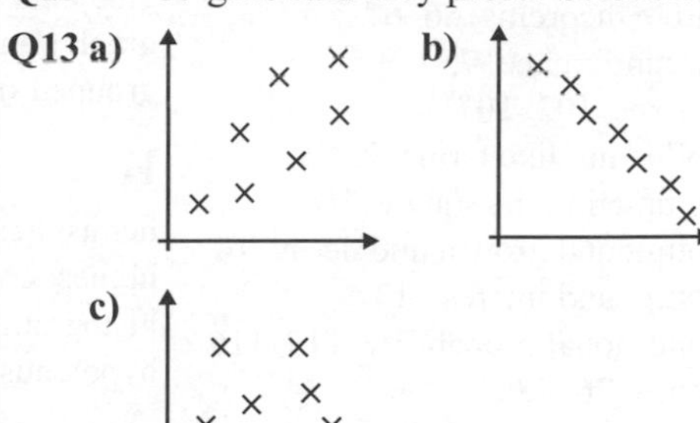

Q12 87 girls said they prefer Science.

Q13 a) **b)** **c)**

Q14 $\frac{4}{25}$

Q15 1

Q16 Expected times outcome will happen = probability × n

Q17 $\frac{1}{12}$

Q18 $\frac{1}{2}$

Q19 $\frac{1}{221}$

Q20 When you can't tell what the probabilities of different outcomes are 'just by looking' — e.g. when you have a biased dice/spinner etc.

Index

A

algebra 25-27, 29-46
algebraic fractions 37, 38
alternate segment theorem 67
AND rule 114
angles 61-63
  alternate 62
  between line and plane 96
  bisectors 84
  constructing 85
  corresponding 62
  depression/elevation 92
  interior 62
  vertically opposite 62
arcs 75
area 70, 74, 75, 93
averages 103-107

B

bearings 87
bias 100, 117
BODMAS 2, 3
bounds 19
box plots 105

C

calculators 3
chords 66, 67
circle theorems 66, 67
circumference 75
classes 102, 107
collecting like terms 25
completing the square 36
compound growth and decay 14
compound interest 13
conditional probability 115, 116
cones 76, 78
congruence 71
constructions 83-86
continuous data 102
conversions 81, 82
  area/volume measurements 82
  decimals to fractions 9
  graphs 82
  speed units 81
coordinates 48
correlation 111
cosine rule 93, 94, 96
cube numbers 4
cubic graphs 55
cuboids 77, 95
cumulative frequency 108
cyclic quadrilaterals 66
cylinders 76, 77

D

data 100, 102, 105, 107
decimal places 17
decimals 9, 10
denominators 7, 8
density 79
difference of two squares (D.O.T.S.) 27
direct proportion 45
discrete data 102
distance-time graphs 80

E

enlargements 69, 70
equations of straight lines 49, 52
equilateral triangles 65, 85
estimating 18
exact answers 28
expected frequency 113
exponential graphs 55
exterior angles 61, 64

F

factorising 27
factorising quadratics 33, 34
factors 5, 6, 15
factor tree 5
FOIL method 26
formula triangles 79, 91, 92
formulas 31
fractions 7-11
frequency 106-108, 110, 117
frequency polygons 110
frustums 78

G

gradients 51-53, 59, 80
graph transformations 57, 58
grouped frequency tables 107

H

hemispheres 76, 78
highest common factor (HCF) 6
histograms 109
hypotenuse 71, 91

I

imperial units 81
improper fractions 7
inequalities 39, 40
  on a graph 40
integers 4
interior angles 61, 62, 64
interquartile range 104, 105, 108
inverse proportion 45
irrational numbers 4
isometric paper 73
isosceles triangles 61, 63, 65

L

line segment 48
lines of best fit 111
line symmetry 65
loci 84, 86
lowest common multiple (LCM) 6, 8

M

mean 103, 105-107
median 103-108
metric units 81
mid-interval values 107, 110
midpoint of a line segment 48
mixed numbers 7
mode 103, 105-107
multiples 5, 6
multiplying out brackets 26

N

negative numbers 25
nets 76
northline 87
numerators 7

O

OR rule 114
outcomes 113

P

parallel lines 53, 62, 63, 85
parallelograms 98
percentages 9, 12, 13
perpendicular bisectors 84
perpendicular lines 53
pi (π) 4
polygons 64, 65
population 100
powers 4, 24
prime factors 5, 10
prime numbers 4
prisms 77
probability 112-117
projections 73
proof 46
proportion 9, 45
proportional division 16
pyramids 78, 95, 96
Pythagoras' theorem 90, 96

Q

quadratics 33-37, 54
  completing the square 36
  graphs 54
  quadratic formula 35
quadrilaterals 61, 74
qualitative data 102
quantitative data 102
quartiles 104, 105, 108
questionnaires 102

R

random sampling 100, 101
range 103-106
rationalising the denominator 28
rational numbers 4
ratios 15, 16, 98
real-life graphs 59
rearranging formulas 31, 32
reciprocal graphs 56
recurring decimals 4, 9-11
reflections 58, 68
regular polygons 64, 65
relative frequency 117
right-angled triangles 65, 90-92, 96
roots 4, 24
rotational symmetry 65
rotations 68
rounding numbers 17-19

S

sample space diagrams 113
sampling 100, 101
  simple random 101
  stratified 101
scalars 97
scale drawings 87
scale factors 69, 70, 72
scalene triangles 65
scatter graphs 111
sectors 75
segments 75
sequences 23
significant figures 18
similarity 72
simple interest 13
simplifying 25
simultaneous equations 42-44
  graphs 42
sin and cos graphs 56
sin, cos and tan 91, 92
sine rule 93, 94, 96
solving equations 29, 30, 54
speed 79-81
spheres 76, 78
spread 104, 105
square numbers 4
square roots 28, 30
standard form 20, 21
stem and leaf diagrams 104
straight-line graphs 49-53
surds 4, 28
surface area 70, 76
symmetry 65

T

tangents 66, 67
terminating decimals 4, 9, 10
transformations 57, 58, 68-70
translations 68
tree diagrams 115, 116
trial and improvement 41
triangles 61, 65, 74, 90-96
  constructions 83
triangular prisms 77
trigonometry 91, 92, 96
two-way tables 110

U

unit conversions 81, 82

V

vectors 97, 98
volume 70, 77, 78

Y

$y = mx + c$ 52